CONTENTS

FOREWORD BY BRIAN HEAP

'A' Level work, comprising two, three or even four subjects, is a challenging course of study. It follows a period of general education leading to the GCSE in which you have experienced a 'taster' course of up to ten subjects presented to you in a highly structured teaching system. Thereafter, it becomes necessary to make a choice of specialisms for a more concentrated period of two years in which more time will be spent in 'private study' – literally, teaching yourself.

Inevitably, private study is a new experience for most students and the time normally allocated is rarely used to the best advantage. The assimilation of facts, whilst working on your own, can be difficult since it also necessitates identifying the important issues from a range of books and a wealth of information. The framework of your course is naturally vital, not simply in terms of passing your 'A' Level exams but in achieving the right grades you need to enter the university, polytechnic or college degree course of your choice.

My 'A' Level series therefore aims to provide you with this essential framework. This book will give you the support you need to work through your syllabus and to reinforce the knowledge you will need to be sure of success at the end of your school or college career.

Brian Heap

LIST OF SYMBOLS

r	-	typically a discrete variate
x, y	-	typically continuous variates
f	-	frequency
Σ	-	sum of
$\bar{x}$	-	mean of values of x
n	-	number of values
m	-	median
mcv	-	mid-class value
T	-	total of the values of a sample
R	-	range
SIR	-	semi-interquartile range
Q_1	-	lower quartile
Q_3	-	upper quartile
M.D.	-	meon deviation
$\lvert \; \rvert$	-	modulus symbol (making its contents always positive)
S	-	standard deviation of a sample
S^2	-	variance of the sample
μ	-	mean of a population
σ	-	standard deviation of a population
σ^2	-	variance of a population
P	-	probability
P(r)	-	probability of the value r occuring
n!	-	factorial notation short for n(n-1)(n-2)... 2.1.

LIST OF SYMBOLS

$^{n}P_{r}$ - the number of permutations of r objects drawn from n different objects

$^{n}C_{r}$ or $\binom{n}{r}$ - the number of combinations of r objects drawn from n different objects

$P(A|B)$ - conditional probability of event A occuring given that B has already occured

R - typically a discrete random variable

X - typically a continuous random variable

β or p - the parameters of the Binomial Geometric and Proportions distributions

$E[X]$ - expectation (mean) of the random variable X

$V[X]$ - variance of the random variable X

μ - a parameter of Normal and Exponential distributions

σ - the standard deviation of the Normal distribution, and one of its parameters

Z - the standard Normal distribution random variable

N - the number of values in a finite population

χ^2 - (chi squared) a statistical distribution

O, O_r - observed frequency

E, E_r - expected frequency

ν - number of degrees of freedom

λ - the parameter of the Poisson distribution

r - product moment correlation coefficient

r_s - Spearman (Rank) correlation coefficient

d - difference (between two values)

w - weight (as in weighted averages)

1 METHODS OF COLLECTING DATA

1.1 Introduction

All statistical investigations begin by collecting *data* - information in its 'raw' form, sometimes called *raw data.*

First, we must decide the most appropriate *method* of collecting this raw data, bearing in mind financial constraints, nature of the investigation, nature of the data, time constraints, available manpower etc. This chapter considers the various methods in turn, together with their advantages and disadvantages; however, we must first define some special terms that will be used throughout this book.

Before beginning an investigation, we must first decide who or what the investigation will be aimed at, eg an investigation into how intelligent the students of a particular university are. We call the set of all such objects of the investigation the *population* under investigation; each member of this population is called a *unit*. Thus, in the above example, the *population* under investigation is the set of all students attending the university, each particular student being a *unit* of that population.

The raw data will be obtained from some or all of the units of a population (in the above example the raw data may be IQ scores); the data extracted is likely to differ from unit to unit eg the IQ score of one student is likely to be different to the IQ score of another. The 'ways' in which units differ from each other will be called *variates*, eg in the above example, the variate is IQ; we speak of the *values* of the variate, which are likely to differ from unit to unit, eg IQ *scores* are the *values* of the variate IQ, and differ from student to student.

Summarising, then, an investigation begins with the collection of values of variates from some or all of the units of the population of interest; the values so obtained are called the raw data.

1.2 The census

A census involves collecting data from every single unit of the population, eg the national census entails sending out a questionnaire to *every* resident of the UK; it is therefore, the most accurate and complete method of collecting data. If the population is very large, as above, this method of data collection will be time-consuming and costly (this is why a national census occurs only every 10 years, since it takes 10 years to analyse the data!) However, when the population is not too large, the census may be more practicable, eg the pupils of a small school. For most investigations, the population is too large to use the census method (eg the students of a university) and so we must content ourselves with collecting data from a fraction of the population, as described below.

1.3 The survey

The survey involves collecting data from a *representative sample* of the population (a *sample* is a subset of the population), eg a sample of 100 students attending the university; the size of the sample will be determined by factors such as finance available, time, manpower, etc" naturally, the larger the sample, the more representative it will be. The most important factor determining the choice of such a sample is that it is *representative* of the 'underlying' population from which it is drawn: this is so that any findings arising from the *sample* data can then be legitimately generalised to the *population.* For example, in the IQ investigation above, if we chose a sample of 100 students who just all happened to be *female* (ie unrepresentative of the population which is both male and female), any findings about their IQ could not be legitimately generalised to the IQ of *all* students at the university, since sex and IQ may be related *at this particular university* (eg there may be a selective bias towards men). In assessing the following methods of choosing a sample (for a survey), the main criteria is whether the resulting sample will be *representative* of the underlying population.

1.4 Sampling methods

Random sampling

A *random sample* is selected such that every unit of the population has an *equal chance* of being included in it; an example of this *not* being the case would be if the investigator of the IQ study chose 100 students known to himself/herself, which would certainly not give a representative sample. For example, if we wished to choose a random sample of 100 students from a university, we may go to the registrar's office and choose the first 100 names off an alphabetical list of all students. Or, we could put the name of each student on a piece of paper, place them all in a drum, and pick out 100 names. Another method would be to choose 100 enrolment numbers at random using random number tables (or a computer that produced random numbers); the use of random number tables to choose random samples is described later in this chapter.

All of the above methods would produce a *random* sample, but they *may* not produce a *representative* sample. It *may* just so happen that the first 100 names on an alphabetical list of students are of female students. Or, the first 100 names out of the drum *may* all be names of male students. Or the 100 students chosen using their enrolment numbers *may* all be science students. It is unlikely that this will be the case, but it is a possibility.

Systematic sampling

This name is given to any method employed that uses a certain 'system' in selecting a sample, eg selecting every 50th name on the alphabetical list of students of a university with 5,000 students; this would produce a sample of size 100, since 5,000/50 = 100. However, the same criticism applies to this method, since the random nature of the underlying alphabetical list is still being assumed; in fact, it *may* just so happen that every 50th name is that of a male student.

Stratified or proportional sampling

This method incorporates random sampling as described above; in addition, however, the sample is chosen to reflect the underlying population in certain important or relevant ways. Each variate of the population can be thought of as subdividing (stratifying) it into smaller *sub-populations*, eg the variate sex has the two values 'male' and 'female', which divides the population of all students, say, into male students and female students. Now, if the variate under investigation (eg IQ) were related in some way to another variate (eg sex), then we would require that our sample reflect the sub-division of the population by that variate accordingly (ie male/female). Thus, since we know that the population is divided in the ratio 55% females to 45% males, we must ensure that our sample of 100 students comprises 55 females and 45 males (assuming the population of all students at the university in question is similarly subdivided); we could ensure this by firstly choosing the first 55 female names on the alphabetical list, and then, starting at the beginning again, choose the first 45 male names.

If there were thought to be other 'important' variates related to IQ, it would be desirable to reflect the subdivisions they impose in the sample too. For example, if age were thought to be related to IQ, we may wish to ensure that the proportions of students in each of 3 age groups, say, were reflected in the sample. Suppose we divide the students into the 3 age groups 18-21, 22-25, 26 and over; from the registry office, we may find that the proportion of students in these age groups are 40%, 40%, 20%.

Now, of the 55 female students to be selected, we would have to ensure that:

40% of 55 = 22 were aged 18 - 21

40% of 55 = 22 were aged 22 - 25

20% of 55 = 11 were aged 26 and over

We could achieve this by first choosing the first 22 female students aged 18-21 on the alphabetical list, then starting from the beginning again, choosing the first 22 female students aged between 22 and 25, and so on. Similarly, the 45 male students woud have to comprise:

40% of 45 = 18 students aged 18 - 21

40% of 45 = 18 students aged 21 - 25

20% of 45 = 9 students aged 26 and over.

As can be seen, the more 'important' or 'relevant' variates reflected in the sample, the more complicated the selection procedure; furthermore, the larger the sample needed to be selected also, since each extra subdivision of the population further subdivides the sample into smaller parts. Another problem is how to determine which variates are important and which are not; this assumes a prior knowledge, which may not be available in a new kind of investigation, eg it *may* be true that the subject area a student studies in may be related to their IQ, and yet since such a relationship was not suspected or known, this variate was not allowed for, resulting in a (possibly) unrepresentative sample (they may all be science students for example).

Overall, however, this method is the best, even if only the *most* important related variates have been taken into consideration.

Quota sampling

This is similar to stratified sampling, in that important variates have subdivided the sample to be selected into certain categories, eg 'female and aged 18-21'; these categories and the number of units required to be in them, are called a set of *quotas*. The quotas for the above IQ investigation are listed below, together with a *tally* representing the required numbers of units, for each category.

	Female		Male
18 - 21	1111111111111111111111	18 - 21	1111111111111111111111
22 - 25	1111111111111111111111	22 - 25	1111111111111111111111
26 & over	11111111111	26 & over	11111111111

The *difference* between this method and the simple stratified sampling, is that a random method is *not* required to 'fill the quotas', eg the investigator may simply walk around the university asking students to fill in an IQ test sheet until each quota is filled (the obvious drawback to this method is that the investigator could be biased in their selection of students). Thus, although certain important variates may have been allowed for, others, which random sampling would have eliminated, may have not, eg investigator bias - the investigator may only choose his/her friends as subjects (see *interviewer bias*, later).

Cluster sampling

This method could be an adjunct to the above methods and takes into account the natural clustering of the units of a population. For example, an investigation into the relationships between the employees of small businesses would naturally have to involve choosing a sample of small businesses (clusters of employees) and interviewing *all* of their employees, rather than simply selecting a sample of disparate employees.

1.5 The use of random number tables

These will be found on pages 12 and 13 of Cambridge Elementary Statistical Tables, and have been generated by a computer. They comprise lists of single digits, 0 - 9, and were generated in such a way that every digit from 0 - 9 had an equal chance of occupying each space of the page.

Suppose we wish to choose a random sample of 20 digits; we simply start at any point of the page and, reading upwards, downwards, left or right, read off 20 successive digits, eg beginning with the top leftmost digit and reading downwards, we obtain:

2, 7, 9, 2, 3, 4, 4, 1, 0, 3, 0, 6, 6, 8, 0, 2, 4, 4, 2, 4.

We can, however, use the tables to choose random samples of numbers lying between any two given values, and to any specified degree of accuracy, as the examples below illustrate.

Example

Choose a random sample, of size 10, of whole numbers between 0 and 99, inclusive.

Here, we simply read off *pairs* of digits, 00 to 99, starting at any position. Beginning from the leftmost digit of the fourth row, and reading rightwards, we get:

22, 15, 78, 15, 69, 84, 32, 52, 32, 54

Example

Choose five random whole numbers between 25 and 50, inclusive.

Here, we repeat the process in the last example, but exclude any number lying outside the given range. Starting from the same position as above, we get:

32, 32, 37, 38, 37

Example

Choose a random sample of size 8, of numbers (correct to 2 decimal places) lying between 0 and 10.

Since each number written correct to 2 decimal places comprises three digits, eg 2.57, we choose 8 sets of 3 digits, and put a decimal point after the first of each. Beginning at the end of the 10th row and working leftwards, we obtain:

9.43, 7.29, 8.08, 7.71, 3.38, 8.77, 2.21

Example

Choose a random sample, size 5, of numbers (correct to 4 decimal places) lying between 0.0050 and 0.0080; inclusive.

Since each such number comprises three 0s followed by two digits lying between 50 and 80, inclusive, we simply choose 5 of the latter and put 0.00 before them. Beginning at the bottom righthand corner and working upwards, we get:

0.0078, 0.0062, 0.0053, 0.0053, 0.0079

Sampling with or without replacement

We can choose a sample in two ways. Sampling *with replacement* involves choosing a unit from the population, obtaining the necessary data from it, and then *returning* it to the population, eg choosing a ball (unit) from a bag of 10 coloured balls (population), containing 5 red, 4 green and 1 black ball, recording its colour (value of the variate 'colour') and then returning it to the bag *before* making the next selection. Clearly this means that the next unit selected may be the same as the one before, eg in the above example, the first ball selected may have been black, and *also* the second ball selected may be black. Sampling without *replacement* means selecting each unit of the sample, recording the data obtained from it, but *not* returning it to the population, eg in the above example, choosing a ball, recording its colour, but *not* replacing it before the next selection. With this method each unit can only be chosen once, but, naturally, the size of the population is decreased by one, ie it becomes a different population, after each selection eg if the first ball chosen was black, then there is no posibility that the second ball is also black, since there was only one black ball to start with - thus, the population has changed.

If, however, the population is *large*, eg all the students of a university, then there is not much difference between the population before and after a single unit, or, indeed, 100 units, have been selected; to all intents and purposes, sampling with and without replacement is the same when the population is very large. Thus, in the IQ investigation, we would sample without replacement in the knowledge that the population will not be greatly changed.

Exercise 1

1. An investigation into the perspiration rate of humans involves the selection of a sample of 500 people. Given that sex and smoking habits are strongly related to perspiration, how would you stratify the sample (use the data below).

   Male : Female = 45 : 55

   Smokers : Non-smokers = 20 : 30

2. In an experiment into the effects of exercise on heart rate, 600 adults below 60 years of age were selected at random. Of the variates below, which do you think should be used in stratifying the sample? How would you stratify the sample?

   Male : Female = 45 : 55

Age	Percentage of Population
20 - 40	30
40 - 50	20
50 - 60	10
Hair Colour	**Proportion of Population**
Brown	0.6
Fair	0.2
Black	0.2

3. In a large herd of cattle, there are three different breeds, A, B and C, in the ratio 3:2:1. Furthermore, 40% of all cattle are known to be Vitamin B12 deficient. How would you stratify a sample of size 60 to be used for an experiment to determine the effect of a preventive treatment toward this virus?

4. Use random number tables to select the following samples:

   a) size 10, between 5 and 10, correct to 2 decimal places

   b) size 5, between 20 and 50, correct to 1 decimal place

   c) size 9, whole numbers between 151 and 200 inclusive

d) size 15, odd numbers between 20 and 70, inclusive

e) size 4, numbers between 0.57 and 0.88, inclusive correct to 2 decimal places

f) size 8, numbers between 0.384 and 0.987 correct to 3 decimal places

g) size 10, numbers between 0 and 10, correct to 4 decimal places

h) size 5, numbers between 8 and 17, correct to 2 decimal places

i) size 10, numbes between -5 and 5, correct to 1 decimal place

j) size 10, whole numbers between 1234 and 3080, inclusive.

5. Use random number tables to choose a random sample of size 10 from the data below.

   (*Hint*: Number the data below from 00 to 35.)

   Choose a systematic sample of size 6. If the data naturally subdivides into the four categories,

   0.0 - 4.9, 5.0 - 5.9, 6.0 - 7.9, 8.0 - 9.9

   choose a stratified sample of size 12 to reflect this.

DATA					
3.2	4.1	8.1	7.2	5.5	5.7
1.8	2.9	2.4	5.3	6.2	3.8
3.9	4.2	1.3	0.5	5.0	6.5
7.8	9.4	0.7	3.8	5.8	3.1
4.0	8.4	5.3	2.8	6.6	4.5
5.8	2.2	7.5	1.8	5.1	5.0

6. Choose a random sample of size 5, without replacement, from the above data.

2 CLASSIFICATION AND REPRESENTATION OF DATA

2.1 Introduction

Once we have obtained the data, how should we deal with it? Before answering the question we will first consider the *type* of data that may be obtained.

Quantitative vs Qualitative data

If the recorded values of a variate comprising the data are numerical, we call the data *quantitative*, and the variate a *quantitative variate*, eg height, weight, number of teeth, are all quantitative variates. If the values of a variate are not numerical, the variate is called a *qualitative variate*, eg colour of hair, favourite TV movie.

Continuous vs Discrete data

If the recorded values of a variate can be any number in a given range or interval then we call the variate a *continuous variate*, eg the heights of people can take any value between the shortest and tallest people in the world; a certain machine may fill bags of sugar with any weight between 1.99 kg and 2.01 kg. If the values of a variate can only change by *steps*, the variate is called a *discrete variate*, eg the number of eggs laid by hens, shoe sizes, height (measured to the nearest cm). We generally denote *discrete* variates by the letter r, and *continuous* variates by the letter x, though this is not a strict rule.

Now returning to the question posed above, suppose we are presented with a mass of raw data; in its raw form, we can make little sense of it since it is so bulky. Consider the data below, comprising the number of tomatoes yielded by 100 plants

13	14	23	12	13	4	8	12	26	13
16	7	15	15	28	16	23	14	8	12
11	23	21	17	12	17	9	21	28	20
10	25	20	19	15	12	13	27	13	15
11	17	19	28	13	15	19	21	10	25
13	12	15	15	19	20	11	23	14	6
29	20	17	13	15	18	18	22	1	13
7	5	19	23	29	21	19	17	12	11
19	27	16	16	8	23	25	14	3	22
1	18	18	21	24	13	14	9	28	5

A quick glance over the data may indicate that there are not many plants yielding under 10 tomatoes, but beyond such general statements, not much more can be said. One way to make more sense of the data would be to *order* it from largest to smallest, as below.

1	1	3	4	5	5	6	7	7	8
8	8	9	9	10	10	11	11	11	11
12	12	12	12	12	12	12	13	13	13
13	13	13	13	13	13	13	14	14	14
14	14	14	14	15	15	15	15	15	15
15	15	16	16	16	16	17	17	17	17
17	18	18	18	18	19	19	19	19	19
19	19	20	20	20	20	21	21	21	21
22	22	23	23	23	23	23	23	24	25
25	26	27	27	28	28	28	28	29	29

Now we can see more easily the distribution of yields; for instance, there seems to be quite a lot of plants yielding 12, 13 and 23 tomatoes. The data is still rather daunting, however; a further refinement would be to make a tally of the yields as below.

Yield	1	3	4	5	6	7	8	9	10	11	12	13	14	15
Tally	II	I	I	II	I	II	III	II	II	IIII	VII	VIIII	VII	VIII

Yield	16	17	18	19	20	21	22	23	24	25	26	27	28	29
Tally	IIII	V	IIII	VII	IIII	IIII	II	VI	I	II	I	II	IIII	II

Replacing the tally for each yield by a number (the *frequency*, f of the yield), denoting the variate 'yield' by r, we have the *frequency distribution* of yields below.

Yield (r)	1	3	4	5	6	7	8	9	10	11	12	13	14	15
Frequency (f)	2	1	1	2	1	2	3	2	2	4	7	10	7	8

Yield (r)	16	17	18	19	20	21	22	23	24	25	26	27	28	29
Frequency (f)	4	5	4	7	4	4	2	6	1	2	1	2	4	2

We can now see precisely which yields occur most frequently; however, the data is still rather bulky. In order to reduce the size of the data further, we could *group* the yields into six equal *classes*, say, as opposite; this results in a *grouped frequency distribution* of yields.

Yield (r)	Frequency (f)
1 - 5	6
6 - 10	10
11 - 15	36
16 - 20	24
21 - 25	15
26 - 30	9
Total	100

This is the most convenient form in which to present the data; however, there is a price to pay, viz loss of data. From this table alone, we cannot say, for instance, how many plants yielded exactly 15 tomatoes; we can only say how many plants had yields in the range 11 to 15, inclusive, ie specify the *class frequency* of 36. The optimal number of classes to choose has been found to be about 7, though there is no hard and fast rule; too few classes lose too much data, too many classes makes a bulky table.

Now, the class 11 - 15, say, *means* the five yields 11, 12, 13, 14 and 15, and therefore includes the 'end' values 11 and 15; these 'end values' are called *class limits*. Also, notice that the total frequency is 100, the total number of recorded values (yields). Finally, notice that the variate 'yield' is a *discrete, quantitative variate*; hence, we refer to the above distribution as a *discrete (grouped) frequency distribution.*

Let us now consider the *grouped frequency distribution* of a *continuous* variate (x). For example, suppose we obtained the heights (x), in cm, of 200 students, and grouped them into the seven equal classes below.

Height (x cm)	Frequency (f)
140 - 150	8
150 - 160	30
160 - 170	46
170 - 180	52
180 - 190	34
190 - 200	26
200 - 210	4
Total	200

We call this a *continuous (grouped) frequency distribution.* Now this time, since we are dealing with a *continuous* variate, the class 160 - 170, say, contains all of the numbes between 160 and 170, eg 162, 163, 752, etc; *however*, by convention, it *includes* the *lower* class limit, 160 cm, but *excludes* the *upper* class limit, 170 cm. This convention is necessary since, if each class contained *both* its class limits, where would we put the height 170 cm? Thus, we can represent the class 160 - 170 by a straight line, as below:

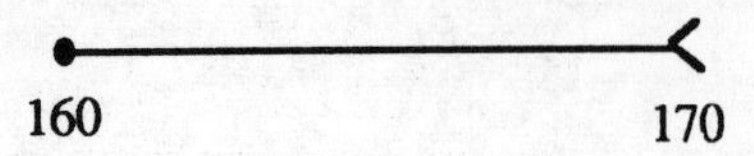

Clearly, this class has a certain size or *width*; indeed, the *class width* here is 170 - 160 = 10 cm.

Furthermore, we can also find the *middle* of the class - called the *mid-class value (MCV)* - by simply averaging the class limits, ie

$$\text{Mid-Class Value} = \frac{160 + 170}{2} = 165 \text{ cm}$$

We will make use of the MCVs in the next chapter.

Notice that the upper class limit of one *continuous* class is the same as the lower class limit of the adjacent class; this is not the case with the *discrete* classes of the first distribution considered, eg the class 11 - 15 has an upper class limit of 15, whereas the adjacent class 16 - 20 has a lower class limit of 16. This fact can always be used to determine the nature of the variate of the distribution and should be looked for before further treatment of the given data.

But why couldn't we do the same with the class 11 - 15 in the first (discrete) distribution of yields considered above? Using the same method of calculation for the class width, this would give a class width of 11 - 15 = 4, which doesn't seem quite right, since the class 11 - 15 contains 5 values 11, 12, 13, 14, 15. In order to calculate the width of this *discrete* class, we must firstly *convert* the *discrete data* to *continuous data.* Notice that between the upper class limit of the class 10 - 14, namely 14, and the lower class limit of the adjacent class 15 - 19, namely 15, there is a gap of 1; if we split this gap in half by *increasing* 14 to 14.5 and *decreasing* 15 to 14.5, we will have eliminated this gap, and so created a continuity from one class to the next.

We can illustrate the conversion of the *discrete* class, 10 - 14, into the *continuous* class 9.5 - 14.5, using the diagram below:

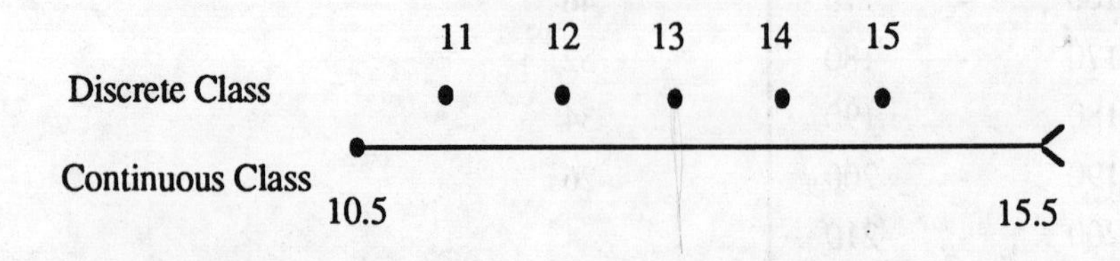

Repeating this process with all of the classes, we obtain the now *continuous* grouped frequency distribution opposite:

Yield (r)			Frequency (f)
0.5	-	5.5	6
5.5	-	10.5	10
10.5	-	15.5	36
15.5	-	20.5	24
20.5	-	25.5	15
25.5	-	30.5	9
Total			100

(*Note*: The extreme two class limits of 1 and 30 have also been altered for the sake of consistency.)

The *discrete* class 11 - 15 has now become the *continuous* class 10.5 - 15.5; thus, the class width is:

$$\text{class - width} = 15.5 - 10.5 = 5$$

which now coincides with the number of values in the class 11 - 15.

The mid-class value of a *discrete* class can be calculated in the same way as for a *continuous* class, without recourse to the above conversion, eg for the class 11 - 15 we have:

$$\text{mid-class value} = \frac{11 + 15}{2} = 13.$$

Using the corresponding continuous class 10.5 - 15.5 gives the same value, however:

$$\text{mid-class value} = \frac{10.5 + 15.5}{2} = 13$$

Histograms

Having tabulated the data in the form of a grouped frequency distribution, we can now set about representing it visually in the form of a *histogram*. Consider the *continuous* grouped frequency distribution of student heights again:

Height (x cm)			Frequency (f)
140	-	150	8
150	-	160	30
160	-	170	46
170	-	180	52
180	-	190	34
190	-	200	26
200	-	210	4
Total			200

A *histogram* of this distribution can be drawn by first marking off the seven classes on an x-axis from 140 to 210; above each class is then constructed a bar of height equal to the frequency of that class, giving the diagram below.

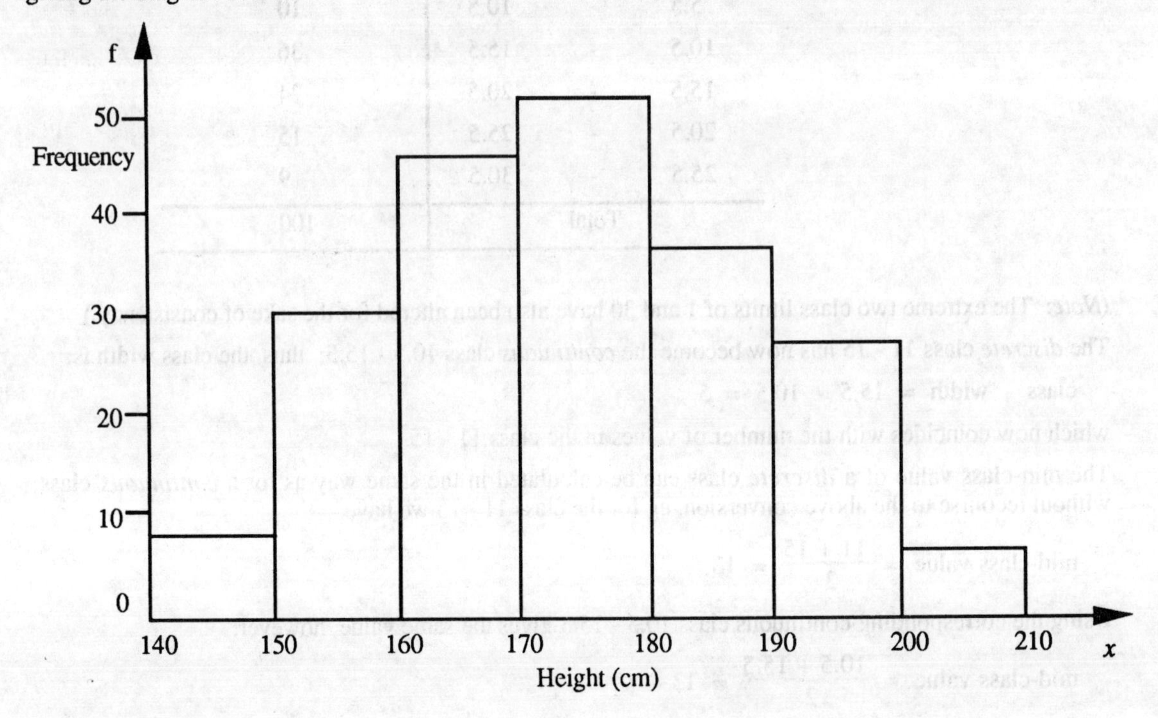

If we now join together the mid-points of the top of the bars, we obtain the *frequency polygon* below.

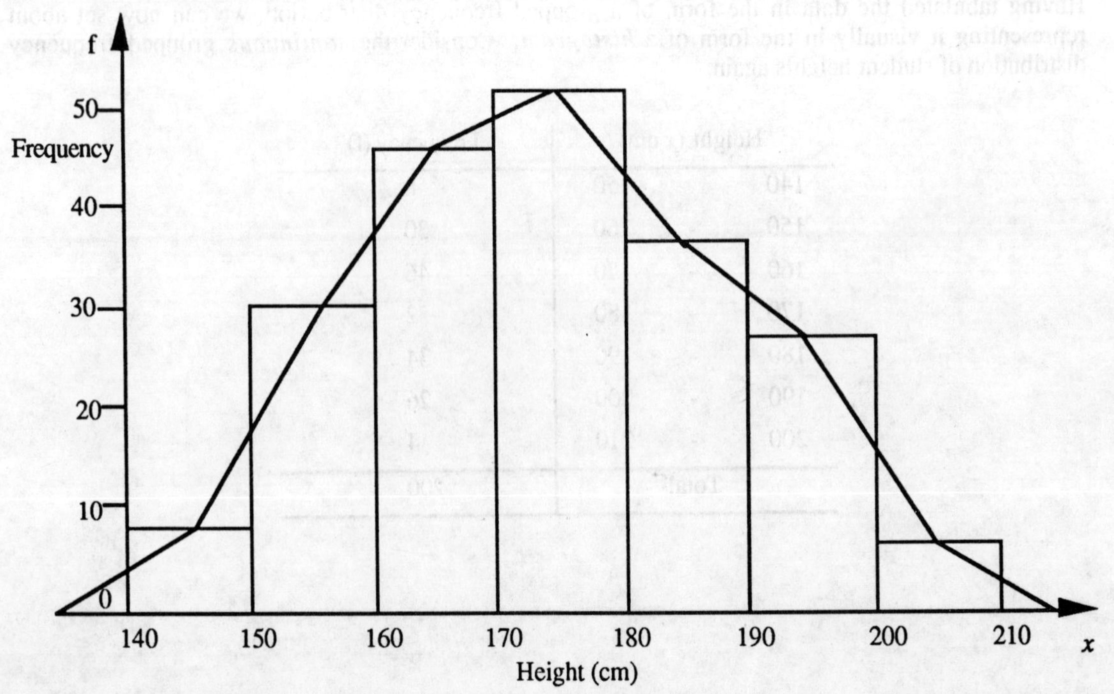

Notice that the polygon meets the x-axis at the mid-points of the two 'imaginary' classes 130-140 and 210-220, representing bars of zero height (since they have no frequency). The frequency polygon indicates the *shape* of the distribution which will be considered later in the chapter.

If we wished to construct a histogram for any given frequency distribution, we must first convert it into a *continuous grouped frequency distribution* as described above; the above construction method can then be applied.

Thus, we can now construct a histogram for our original set of data concerning tomato plant yields; recall the *continuous* distribution below:

Yield (r)	Frequency (f)
0.5 - 5.5	6
5.5 - 10.5	10
10.5 - 15.5	36
15.5 - 20.5	24
20.5 - 25.5	15
25.5 - 30.5	9
Total	100

Using the above construction method, we obtain the histogram below:

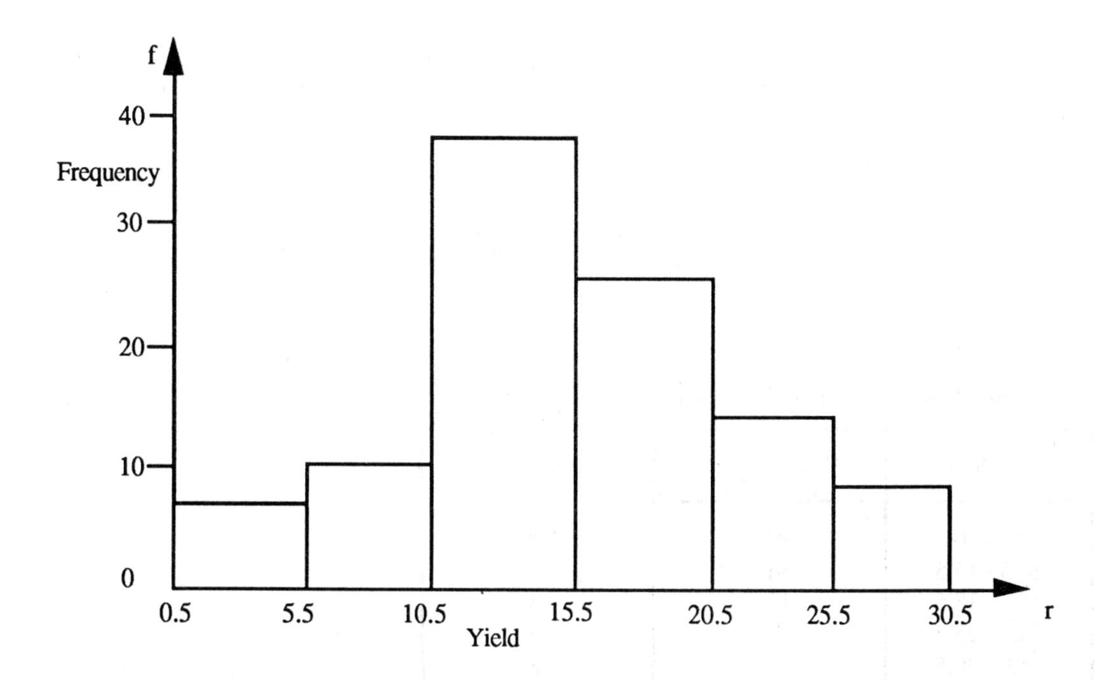

So far, we have only considered grouped frequency distributions with all *equal* classes; suppose, however, we grouped the above tomato yields into the following *unequal* continuous classes:

Yield (r)			Frequency (f)
0.5	-	5.5	6
5.5	-	10.5	10
10.5	-	15.5	36
15.5	-	20.5	24
20.5	-	25.5	15
25.5	-	30.5	9
Total			100

(NB The reader may wish to check the class frequencies by referring to the ungrouped frequency distribution on page 8.)

The construction of a histogram for such distributions with *unequal* classes differs from the above method. We begin, though, as before, by marking off the 5 classes on the *x*-axis, from 0.5 to 30.5. In order to construct the bars above these classes, however, we must use the following rule:

The *area* of the bar above a class must be proportional to the class frequency

Now, the area of a bar is simply its class width multiplied by its height, ie if h is the height of the bar (which we require), then

$$h = \frac{\text{Area of Bar}}{\text{Width of Bar}}$$

A simple way of ensuring that the areas of the bar are proportional to the class frequencies is to actually make them *equal* to the class frequencies; the above formula for calculating the heights of bars then becomes:

$$h = \frac{\text{Class Frequency}}{\text{Class Width}}$$

Consider the first class, 0.5 - 10.5, which has a frequency of 16; then, class width = 10.5 - 0.5 = 10, and so:

$$h = \frac{16}{10} = 1.6$$

Continuing in this way we obtain the heights of all five bars below.

Yield (r)	Frequency (f)	Bar Height (h)
0.5 - 10.5	16	1.6
10.5 - 15.5	36	7.2
15.5 - 18.5	13	4.3
18.5 - 25.5	26	3.7
25.5 - 30.5	9	1.8
Total	100	

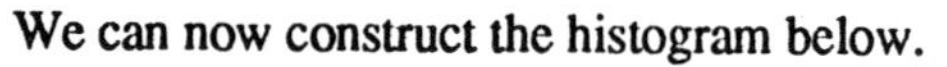

We can now construct the histogram below.

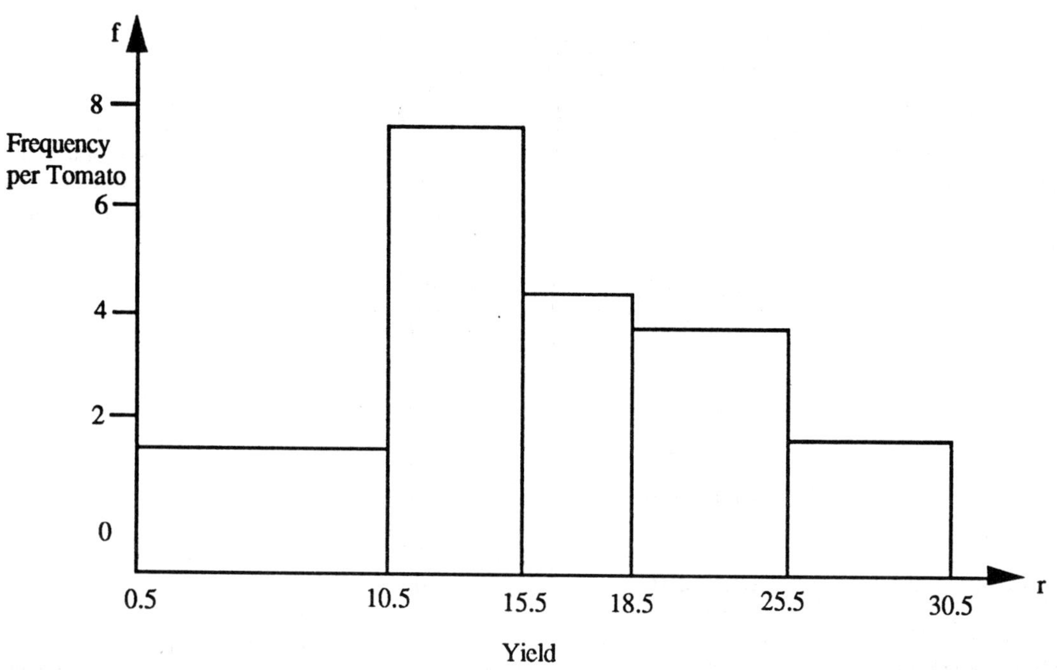

Notice that the vertical axis is no longer simply frequency, but has become frequency per tomato. Furthermore, the total area of all the bars represents the total frequency of 100.

2.2 Cumulative frequency diagrams

An alternative way of tabulating data is in the form of a cumulative frequency distribution. Consider again the continuous grouped frequency distribution of student heights (on page 9) as below.

Height (x cm)	Frequency (f)
140 - 150	8
150 - 160	30
160 - 170	46
170 - 180	52
180 - 190	34
190 - 200	26
200 - 210	4
Total	200

Notice that:

between 140 and 150 cm there are 8 values

between 140 and 160 cm there are 8 + 30 = 38 values

between 140 and 170 cm there are 8 + 30 + 46 = 84 values etc.

Or, in other words:

there are 8 values up to 150 cm

there are 8 + 30 = 38 values up to 160 cm

there are 8 + 30 + 46 = 84 values up to 170 cm etc.

As can be seen, we are adding together (*accumulating*) the frequencies of successive classes to obtain the number of values - the so-called *cumulative frequency* - lying below the upper class limit of each successive class. Summarising these cumulative frequencies, we obtain the *cumulative frequency distribution* below:

Height (x cm)	Cumulative frequency
Up to 150	8
Up to 160	38
Up to 170	84
Up to 180	136
Up to 190	170
Up to 200	196
Up to 210	200

We can now plot these cumulative frequencies against height to obtain the *cumulative frequency diagram* (or *ogive*) of the data. This is done by first marking off the seven classes on the x-axis (as for the histogram) and scaling the cumulative frequencies up the y-axis from 0 to 200. Each cumulative frequency is then plotted above the *upper class limit* of each class (since it indicates the number of values below this *upper* class limit); these points are then joined up by a continuous curve, as in the diagram below.

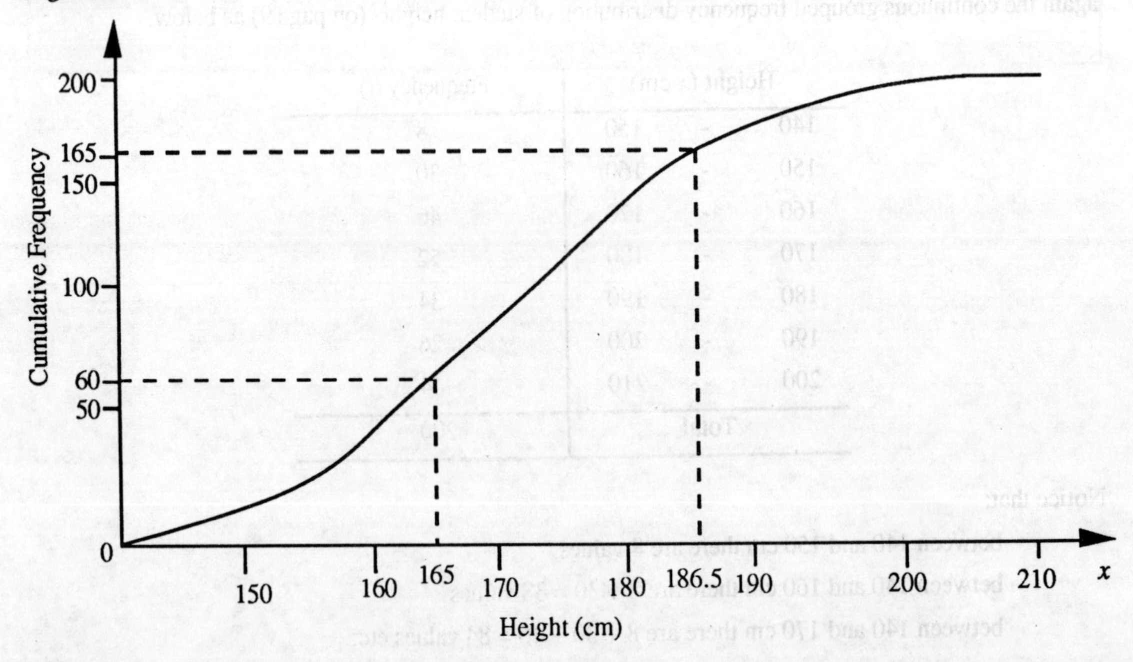

Notice that the curve continues down to meet the x-axis at 140 cm; this is consistent with the fact that there are no values (zero cumulative frequency) below 140 cm.

From this diagram, we can now read off the estimated number of values lying below any given height, eg the number of values below 186.5 cm is approximately 165, ie the number of students shorter than 186.5 cm is approximately 165. By subtracting the cumulative frequency from the total frequency, 200, we obtain the number of value, lying above a given height; thus, the number of students taller than 165 cm is approximately 140 (see Reverse Cumulative Frequency on page 19).

If we divided each cumulative frequency by the total frequency of 200 we would obtain the *relative cumulative frequencies* below. Furthermore, if we expressed each cumulative frequency as a percentage of the total, 200, we would obtain the *percentage cumulative frequencies* below. We can summarise all of the kinds of frequency in the table below.

Up to x cm	Height (x cm)	Frequency	Cumulative Frequency	Relative Cumulative Frequency	Percentage Cumulative Frequency
Up to 150	140 - 150	8	8	8/200 = 0.04	4
Up to 160	150 - 160	30	38	38/200 = 0.19	19
Up to 170	160 - 170	46	84	84/200 = 0.42	42
Up to 180	170 - 180	52	136	136/200 = 0.18	18
Up to 190	180 - 190	34	170	170/200 = 0.85	85
Up to 200	190 - 200	26	196	196/200 = 0.98	98
Up to 210	200 - 210	4	200	200/200 = 1.00	100
	Total	200			

We can now plot either the cumulative frequency, relative cumulative frequency or percentage cumulative frequency against height, depending on the use to which we wish to put the diagram; the following diagram includes all three possible vertical axes and how they coincide with each other.

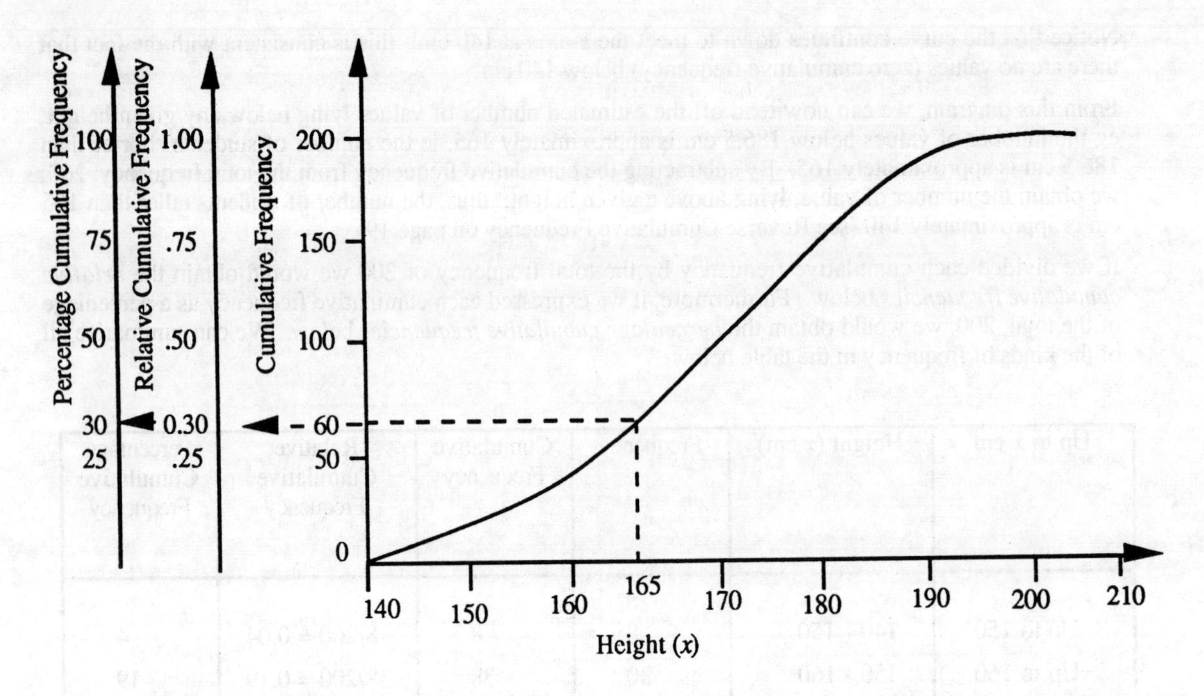

Thus, if we wish to obtain the *percentage* of students shorter than 165 cm, we would read this off the percentage cumulative frequency axis; from the diagram above, we see that this is 30%. However, we could equally as well obtain the *proportion* lying below 165 cm by reading off the relative cumulative frequency axis, which we see is 0.30.

2.3 Reverse cumulative frequency

Consider the following distribution.

Variate (x)	Frequency (f)
0 - 5	3
5 - 10	8
10 - 15	15
15 - 20	12
20 - 30	10
30 and over	
Total	50

It would be impossible to construct an ordinary cumulative frequency diagram for this distribution in view of the 'open ended' nature of the last class, ie there is no upper class limit (and so no cumulative frequency corresponding to it). In such cases we must calculate the *reverse cumulative frequencies* and plot them on a reverse cumulative frequency diagram.

Notice that:

there are all 50 values greater than or equal to 0
there are 47 values greater than or equal to 5
there are 39 values greater than or equal to 10
etc.

We call the number of values greater than or equal to a given value (in this case, the *lower* class limits) the *reverse cumulative frequency* of that value; summarising these, we obtain the following table.

Variate (x)	Reverse Cumulative Frequency
Greater than or equal to 0	50
Greater than or equal to 5	47
Greater than or equal to 10	39
Greater than or equal to 15	24
Greater than or equal to 20	14
Greater than or equal to 30	2

Plotting these *reverse* cumulative frequencies above the *lower* class limits, we obtain the diagram below.

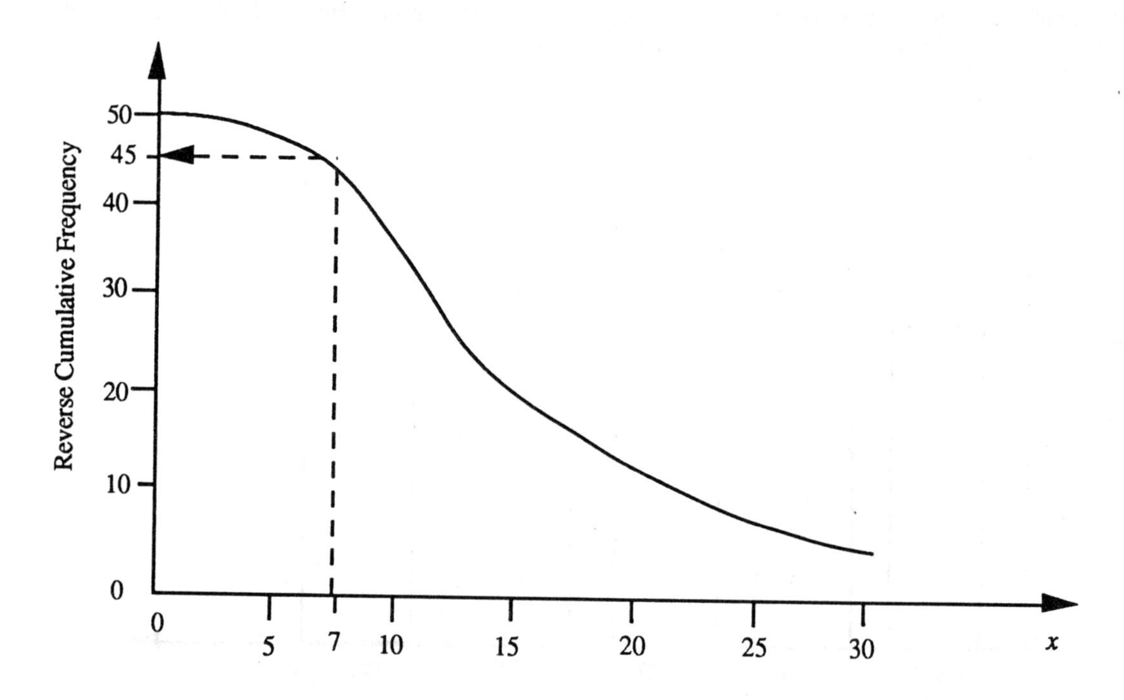

Notice that the curve does not actually meet the x-axis since there is no last upper class limit for it to meet at. From this diagram we can estimate the number of values greater than 7, say, by reading off the vertical axis, which we see is 45.

In a similar way to cumulative frequencies, we can calculate and plot reverse *relative* cumulative frequencies and reverse *percentage* cumulative frequencies.

We will make further use of histograms and cumulative frequency diagrams in the next chapter; however, we now consider some alternative ways of depicting data.

2.4 Bar charts

There are several kinds of bar charts, the choice of which is most suitable depending upon the nature of the data.

Simple bar charts

These can often be used to depict data that varies over time; consider the data below of car sales of a particular model during a five year period.

Year	Sales (100s)
1975	141
1976	173
1977	194
1978	180
1979	220

We simply mark off the 5 years on the x-axis and construct a bar, of uniform width, above each year, with height proportional to the number of cars sold, as shown in the diagram below.

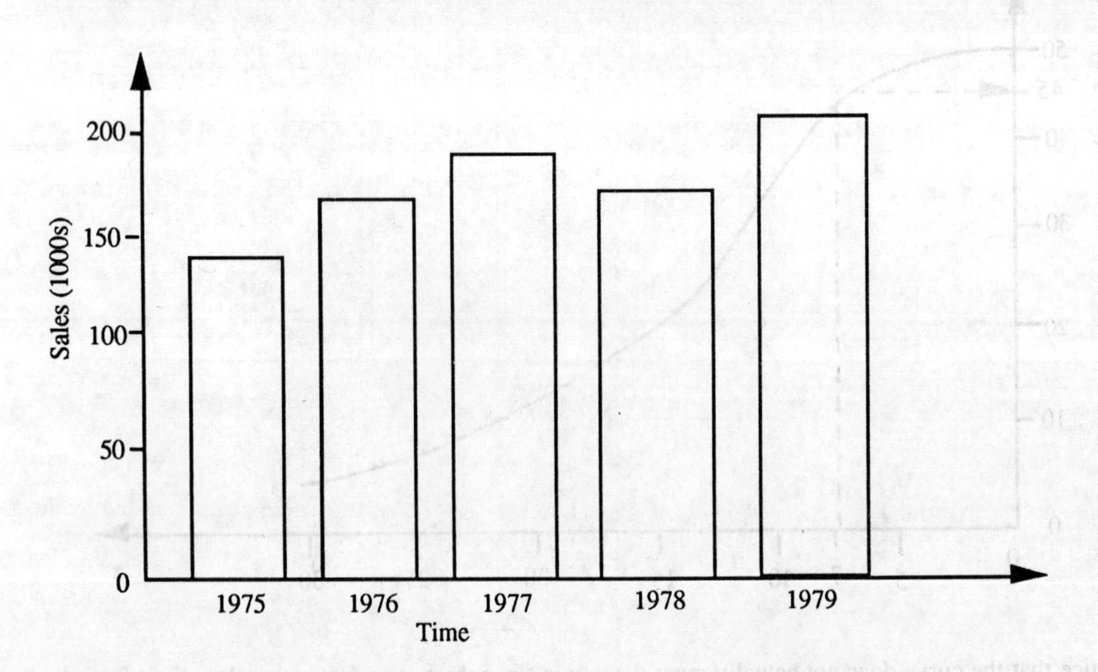

Comparative bar charts

Suppose now, we wish to compare the sales figures for two different makes of car using the same diagram; consider the data below.

Year	Model A Sales (1000s)	Model B Sales (1000s)
1975	141	162
1976	173	170
1977	194	183
1978	180	191
1979	220	200

Here, we construct *two* bars above each year on the x-axis, one for Model A and one for Model B, shading each accordingly, and providing a key to the shading, as in the diagram below.

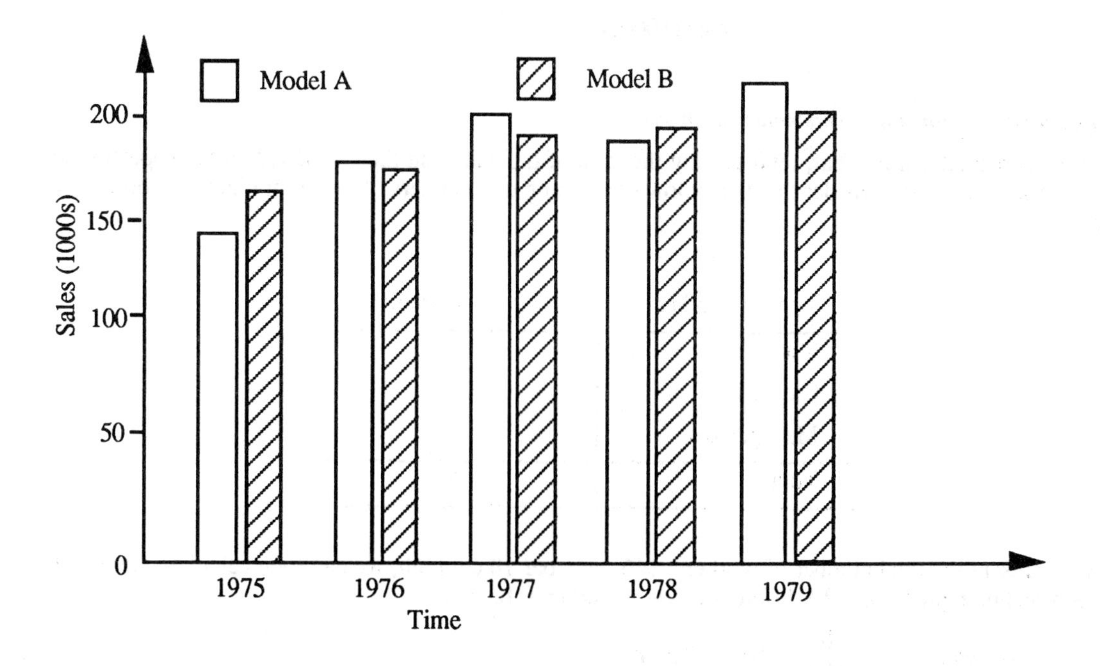

It should be noted that the bars need not be vertical; the sales for Model A could be represented by the horizontal bar chart overleaf, though this is less common since time is usually represented along a horizontal axis.

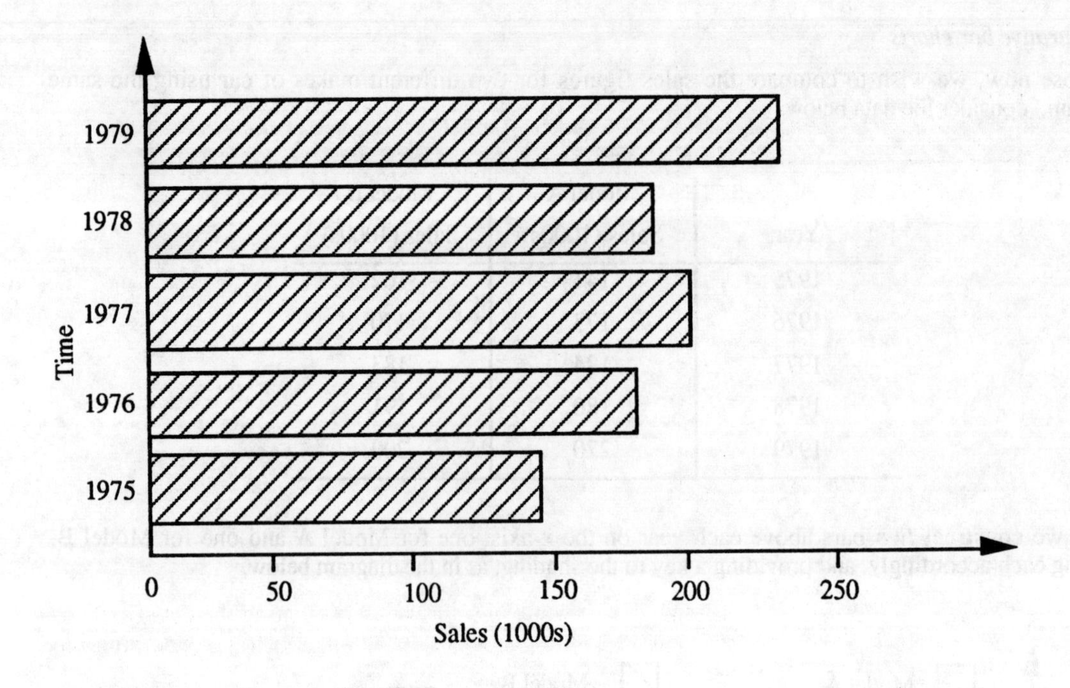

Proportional, sectional or component bar charts

Here we construct a *single* bar, representing the total of the data, and then subdivide it in proportion to the categories of data; consider the data below concerning the numbers of different kinds of worker in a firm.

Type of Worker	No. of Workers
Manual	110
Clerical	29
Management	7
Total	146

If we represent the total number of workers, 146, by a bar 10 cm high, we can then calculate the length of each section representing the different types of worker as below:

Manual: $\frac{110}{146} \times 10 = 7.5$ cm

Clerical: $\frac{29}{146} \times 10 = 2.0$ cm

Management: $\frac{7}{146} \times 10 = 0.5$ cm

Placing the largest section at the bottom, and so on, we obtain in the diagram opposite.

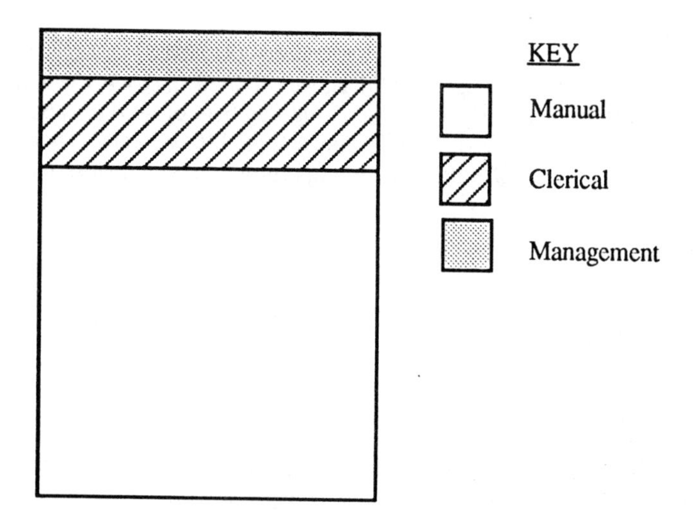

2.5 Line charts

These can be used to represent *discrete* frequency distributions, instead of converting the data firstly to *continuous* (grouped) data and then drawing a histogram. Consider the data below concerning the numbers of eggs laid by hens in a week.

No. of Eggs Laid (x)	0	1	2	3	4	5	6	7	Total
No. of Hens	3	6	5	8	15	0	2	1	50

We simply mark off the values along the x-axis and then construct lines, of height equal to the frequency of each value, above each value, as shown in the diagram below.

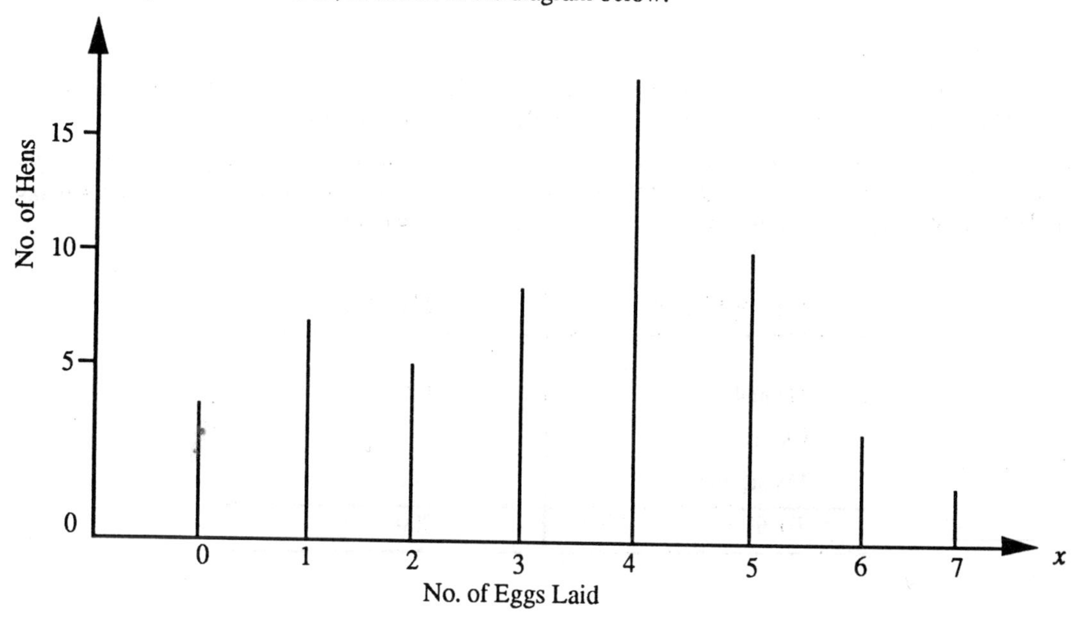

2.6 Pie charts

Instead of subdividing a *bar* in proportion to the categories of data, we subdivide a *circle*, the total *area* of which represents the total data. Consider again the data concerning numbers of workers above. First, construct a circle of radius 3 cm, say, the area of which will represent all 146 workers. Then, the total interior angle of this circle is 360°, which we will subdivide in proportion to the number of each type of worker, as below:

$$\text{Manual:} \quad \frac{110}{146} \times 360 = 271^\circ$$

$$\text{Clerical:} \quad \frac{29}{146} \times 360 = 72^\circ$$

$$\text{Management:} \quad \frac{7}{146} \times 360 = 17^\circ$$

Thus, we can now divide up the circle into 3 sectors, whose subtended angles are 271°, 72° and 17°, representing the Manual, Clerical and Management categories of workers, as below.

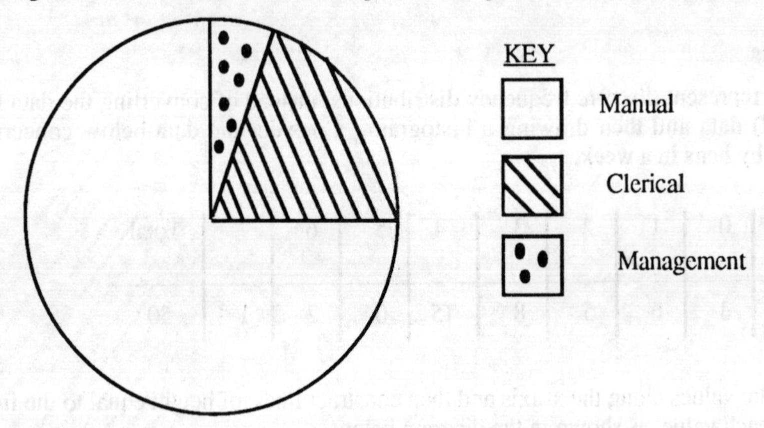

Comparative pie charts

Suppose we now wish to compare the numbers of different types of employees in the firm between the years 1970 and 1980 using *two* pie charts; consider the data below for 1980.

We first draw the above pie chart for 1970. Notice that the total number of employees of the firm has increased to 200; since it is the *area* of the circle which represents the total data, we must draw a bigger circle for 1980. Suppose that the radius of this circle is r, then we have:

Type of Worker	No. of Workers
Manual	120
Clerical	73
Management	7
Total	200

Area of circle for 1970 = $\pi(3)^2 = 9\pi$ represents a total of 140

Area of circle for 1980 = πr^2 represents a total of 200.

Thus, $\frac{\pi r^2}{9\pi} = \frac{200}{146}$, ie $r^2 = \frac{9 \times 200}{146}$

ie $\underline{r = 3.5 \text{ cm}}$

Having constructed a circle of radius 3.5 cm, we can now subdivide it, as before, into 3 sectors, the subtended angles of which are as below:

Manual $\frac{120}{200} \times 360 = 216^o$

Clerical $\frac{73}{200} \times 360 = 131^o$

Management $\frac{7}{200} \times 360 = 13^o$

giving the pie chart for 1980 as in the diagram below

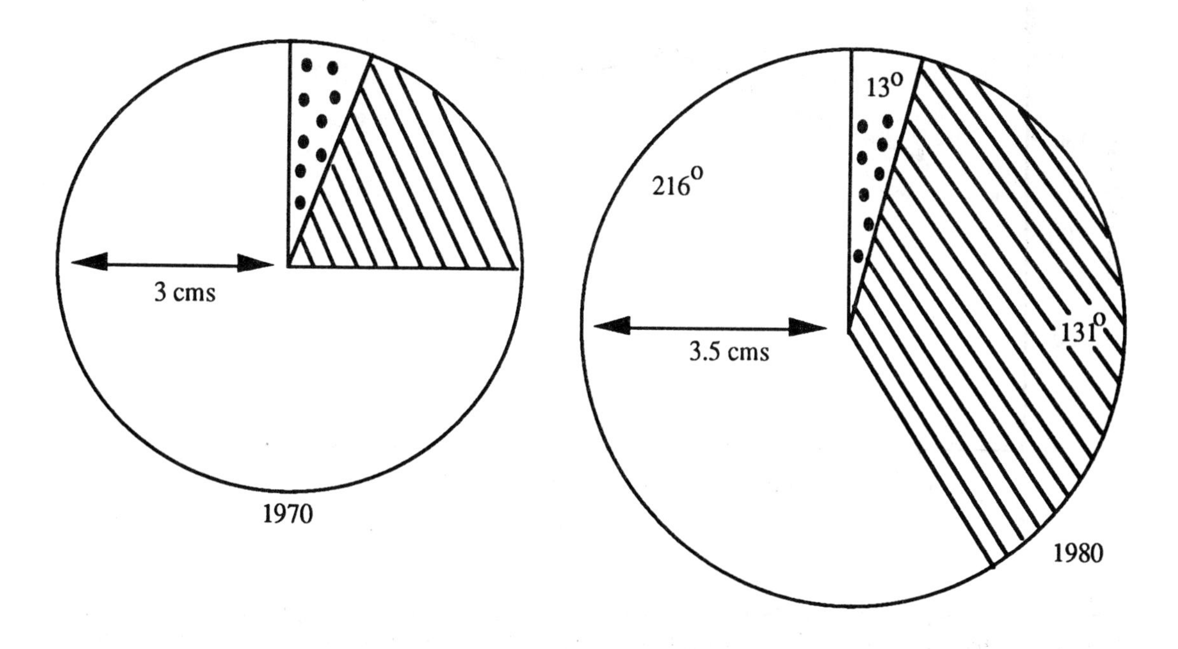

2.7 Graphs

Graphs are most suited to depicting data that changes over time. Consider again the data concerning car sales for Model A above:

Year	Sales (1000s)
1975	141
1976	173
1977	194
1978	180
1979	220

Plotting sales (vertical axis) against time (horizontal axis), we obtain the graph below.

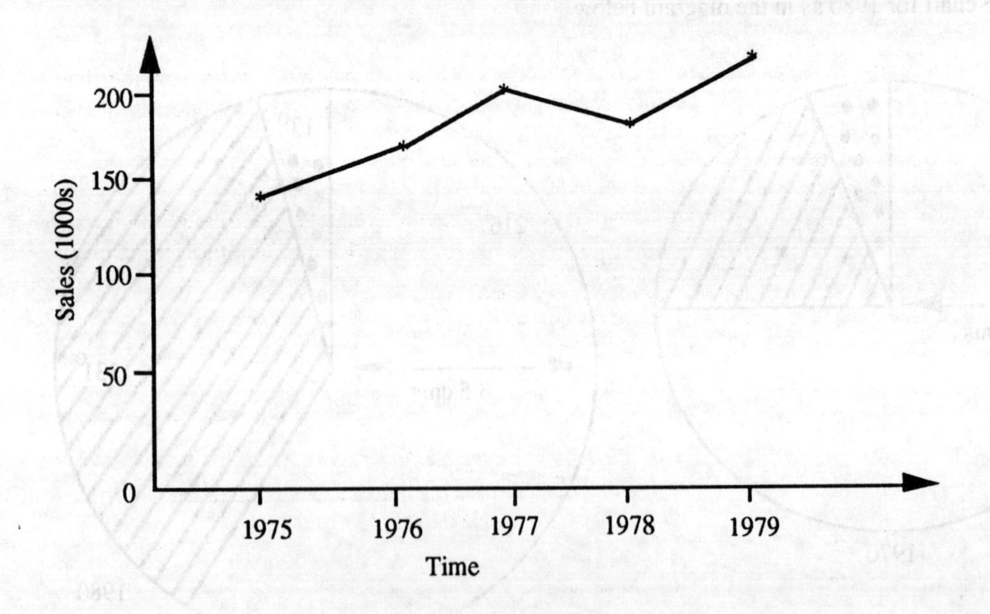

The choice of scale for the vertical axis (sales), however, is important, since it can give rise to a misleading effect. Suppose we begin the vertical scale at 140; the resulting graph is shown opposite.

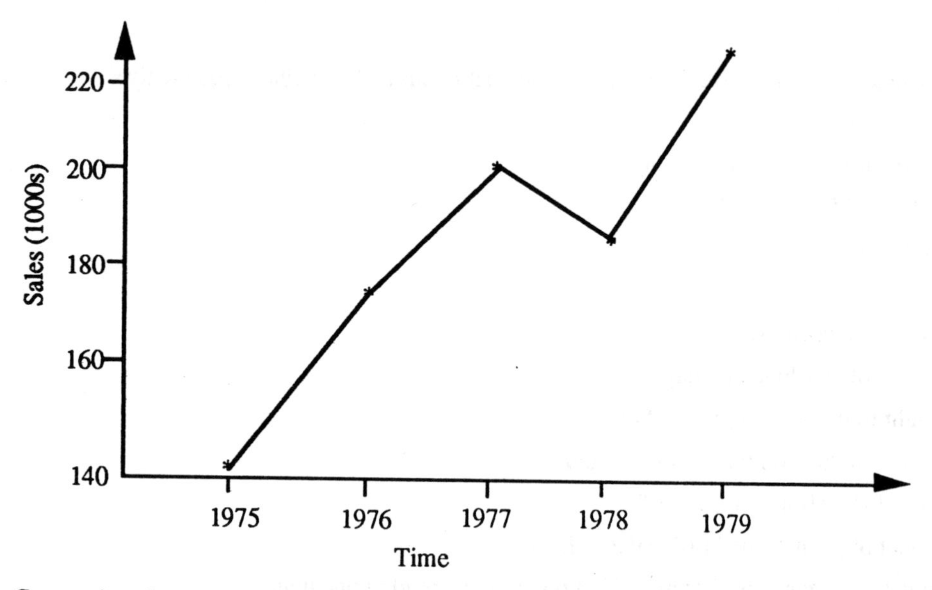

Comparing the graph with the original, we see that the effect of the latter is much more striking than the former, giving the impression of greater increase in sales over the 5 years; this is obviously due to the smaller scale being used.

The shape of a distribution

As mentioned earlier in the chapter, the frequency polygon of a grouped frequency distribution indicates the 'shape' of the distribution; for illustrative purposes, we will depict some commonly occuring shapes using smooth curves in the diagrams below.

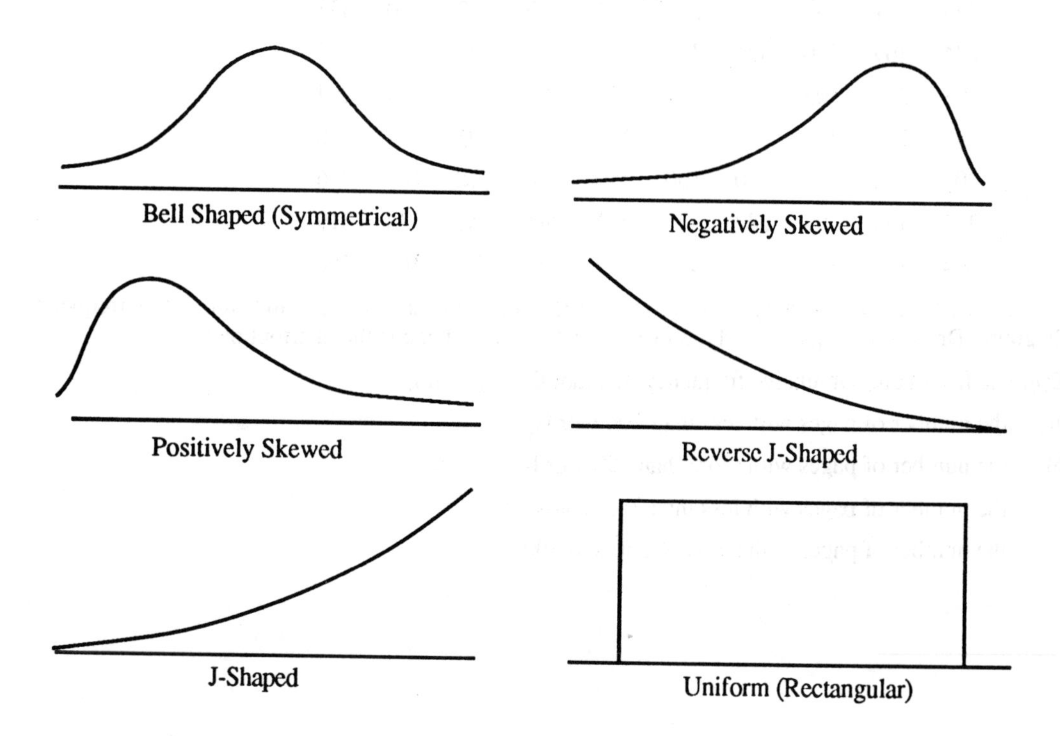

Exercise 2

1. Determine whether the following variates are discrete/continuous and quantitative/qualitative.

   a) height of men;
   b) number of shoes owned by families;
   c) tastes of five different foods;
   d) the emotions;
   e) sex;
   f) left or right-handedness;
   g) the amount of sunshine in a day;
   h) the weight to the nearest gram, of eggs;
   i) the number of hairs on the back of a hand;
   j) car sales of the Mini Metro for 1983;
   k) the amount of paint needed to decorate a house;
   l) the number of green cars passing a fixed point on the road in one hour.

2. The data below consists of the number of words on the 100 pages of a book.

300	250	298	324	61	350	280	29	295	328
200	205	100	12	341	301	284	295	315	319
74	15	204	316	259	180	172	300	309	265
293	241	193	85	260	294	199	272	250	335
284	269	290	321	350	280	32	192	294	2
98	253	305	322	284	288	305	293	104	284
84	291	275	318	299	259	15	105	41	280
307	22	254	300	300	260	190	318	189	246
325	184	294	282	325	284	241	282	324	319
304	100	312	131	341	309	75	179	307	250

   Construct a grouped frequency distribution for the data. Draw a histogram and cumulative frequency diagram. Draw in a frequency polygon for the data. What shape is the distribution?

   Estimate from your cumulative frequency diagram the following:

   a) the number of pages with less than 300 words;
   b) the number of pages with more than 300 words;
   c) the number of pages with less than 250 words;
   d) the number of pages with between 250 and 300 words.

3. For the following distributions state:

   a) whether the variate is continuous or discrete;

   b) the size of the largest class, where appropriate;

   c) where appropriate, the mid-class value of the smallest class.

   Draw a histogram, frequency polygon, and cumulative frequency diagram for each distribution.

   d) State the shape of the distribution.

   i)

x	0	1	2	3	4	5	6
f	2	5	9	7	2	1	0

   ii)

x	0-5	5-10	10-15	15-25	25-30
f	76	41	20	12	4

   iii)

x	3-5	6-8	9-11	12-16	17-19
f	1	2	4	21	13

   iv)

x	f
0.1 - 0.3	5
0.4 - 0.6	16
0.7 - 0.9	29
1.0 - 1.3	28
1.4 - 1.7	15
1.8 - 2.0	7

   v)

x	f
125 - 129	28
130 - 134	51
135 - 139	62
140 - 144	15
145 - 149	9
150 - 154	6
155 & over	14

vi)

x	f
7.12 - 7.50	12
7.50 - 7.92	19
7.92 - 8.42	17
8.42 - 9.00	9

vii)

x	12	13	15	17	18
f	10	22	15	3	1

4. Draw pie charts to compare the crop yields at farms A and B, given the data below.

	Rye	Barley	Corn	Wheat
Farm A	420	590	920	800
Farm B	270	740	900	380

5. The IQs of 200 students were obtained as part of a survey conducted by a research psychologist, and are tabulated below. Construct a histogram, frequency polygon and cumulative frequency diagram for the data. What shape is the distribution? Use your cumulative frequency diagram to estimate the number of students with IQs:

   a) below 118;

   b) above 96;

   c) between 96 and 118.

IQ Score	No. of Students
90 - 94	2
95 - 99	5
100 - 104	7
105 - 109	25
110 - 114	52
115 - 119	48
120 - 124	41
125 & over	20
Total	200

6. The table below shows the quantity, in hundreds of litres, of wheat, barley and oats produced on a certain farm during the years 1971 to 1974.

Crop	Quantity (litres)			
	1971	1972	1973	1974
Wheat	34	43	43	45
Barley	18	14	16	13
Oats	27	24	27	34

a) Construct a sectional bar chart to illustrate these data.

b) For each year express the figure for each crop as a percentage of the annual total and hence construct a percentage bar chart.

c) Comment briefly on the advantages and disadvantages of these methods of illustrating the data.

7. a) State briefly what diagrams you would use to illustrate the following data.

i) The number of marriages in your town for each month of 1977.

ii) The proportion of the harvest yields for 1975 of the cereal crops wheat, barley, oats and rye.

iii) The number of pairs of shoes of different sizes bought by the pupils of your school in 1978.

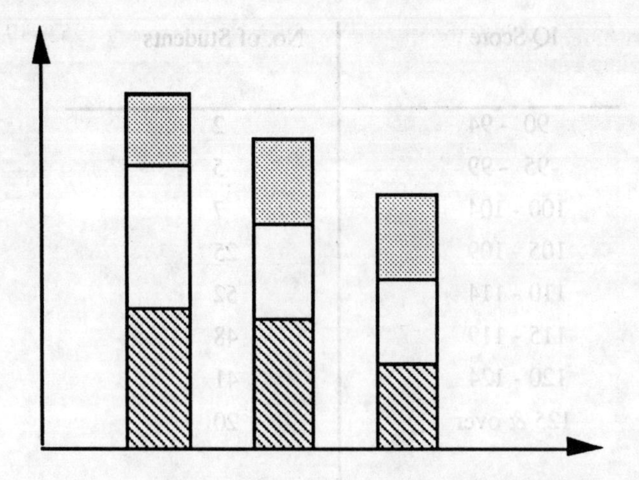

b) The following table gives the percentage composition of meadow hay harvested at different dates.

Date of cutting	Grade Protein	Fat	Soluble Carbohydrates	Fibre	Ash
May 14	17.7	3.2	40.8	23.0	15.2
June 9	11.2	2.7	43.2	34.9	8.0
June 26	8.5	2.7	43.3	38.2	7.3

Illustrate the above data in one diagram.

8. a) State what kind of data are best illustrated by the following diagrams:
   i) pie chart
   ii) bar chart
   iii) histogram

b) The amount of vegetables imported into the United Kingdom in the years 1953, 1957 and 1961 is given in the following table.

Vegetables	Thousand tonnes imported in 1953	1957	1961
Potatoes	121	254	261
Onions	199	220	222
Tomatoes	185	204	158
Others	55	67	88
Total	560	745	729

Use the above table:

i) to represent the proportions of vegetables imported in 1961 in a pie chart of radius 2 units;

ii) to calculate the radii of the circles which would effect a true comparison with the circle already drawn for the fresh vegetables imported in 1953 and 1957 if the areas of the circles are to be proportionate to the total amount of fresh vegetables imported in a year.

9. The table below gives the productions of two basic industries for the European Economic Community (EEC) and the United Kingdom (UK) for the years 1960 and 1970.

	Coal (million tons) 1960	1970	Crude steel (million tons) 1960	970
EEC	19.5	13.7	6	9
UK	17	12.4	2	2.3

Calculate the percentage changes in production between 1960 and 1970. Draw suitable diagrams to compare these percentage changes for the EEC and the UK.

Illustrate by means of pictograms the production of coal and crude steel by the UK for the two given years.

10. The table shows the intelligence quotient (IQ) of 90 pupils at a school.

IQ	65-	75-	85-	95-	105-	115-	125-	135-
No. of pupils	1	7	17	28	22	12	2	1

Each IQ is given correct to one decimal place. Write down the lowest and highest values that would be placed in the class 95-. Compile a cumulative frequency table from the above data and draw a corresponding cumulative frequency curve. (Take 1 cm to represent 5 units on each scale.)

From the graph estimate

a) the median value of the IQ;

b) the semi-interquartile range of the distribution;

c) the percentage of children above average intelligence if an IQ of 100 is taken as average.

11. The table below gives the lifetimes of 430 valves.

Lifetime (to the nearest hour)	Number of valves
300 - 399	20
400 - 499	49
500 - 599	61
600 - 699	83
700 - 799	70
800 - 899	64
900 - 999	51
1,000 - 1,099	23
1,100 - 1,199	7
1,200 - 1,299	2
Total	430

Construct a cumulative frequency distribution from the above data, and from it plot the corresponding cumulative frequency graph. From your graph estimate

a) the median value of the distribution;

b) the semi-interquartile range of the distribution;

c) the percentage of valves expected to burn out in 550 hours;

d) the percentage of the valves sold that the manufacturer will have to replace if he guarantees a valve to last 425 hours?

12. The following table shows the weights, in kilograms, of 250 boys. Each weight was recorded to the nearest 100 grams.

Weight (kg)	Number of boys
44.0 - 47.9	3
48.0 - 51.9	17
52.0 - 55.9	50
56.0 - 57.9	45
58.0 - 59.9	46
60.0 - 63.9	57
64.0 - 67.9	23
68.0 - 71.9	9

Draw the ogive and use it to estimate

a) the semi-interquartile range,

b) the second decile,

c) the eighty-fourth percentile,

d) the percentage of boys weighing over 59 kilograms.

13. The distribution of incomes in a certain establishment was as follows:

Income (£)	2,000-	2,200-	2,400-	2,600-	2,800-	3,000-3,600
No of employees	8	16	36	40	34	26

a) Calculate the mean income.

b) Draw a histogram of the distribution.

c) Using 1 cm to represent £100 and to represent 10 employees, plot an ogive (cumulative frequency curve) of the distribution and hence find

i) the median income

ii) the lower quartile,

iii) the upper quartile,

iv) the semi-interquartile range.

14. The deposit balance of 170 depositors in a rural bank is shown in the table below.

Deposit balance	No. of depositors
less than £25	56
£25 and less than £50	33
£50 and less than £100	28
£100 and less than £150	17
£150 and less than £200	12
£200 and less than £250	11
£250 and less than £500	8
£500 and less than £1,000	5

Calculate

a) the cumulative frequencies,

b) the median of the distribution,

c) the semi-interquartile range of the distribution,

d) the proportion of depositors with a balance in exces of £400.

15. A large store sends out accounts to 80 customers. The table below gives the distribution of the number of days taken to settle accounts.

Number of days		Number of accounts settled
exceeding	not exceeding	
4	8	4
8	12	13
12	16	15
16	20	18
20	24	14
24	28	10
28	32	6

Draw a cumulative frequency polygon for the data. Use your graph to estimate:

a) the semi-interquartile range of the distribution,

b) the percentage of customers who settle their accounts within 21 days.

16. The following table shows the grouped frequency distribution of the hectarage under crops and grass for 200 farms in a certain geographical region.

Hectarage	No. of farms
Under 5	8
5 - under 10	25
10 - under 30	44
30 - under 50	38
50 - under 100	48
100 - under 150	20
150 - under 300	12
300 or more	5

a) Draw, on a sheet of graph paper, the cumulative frequency polygon for the distribution.

b) Determine the median and the quartiles of the distribution.

c) Determine the hectarage under crops and grass which is exceeded by 30 per cent of the farms in the region.

17. The weights of 50 hamsters at 3 months of age are recorded to the nearest gram as below. Represent the data by constructing a histogram. What shape is the distribution?

Weight (x g)	Frequency (f)
20 - 22	1
23 - 25	5
26 - 28	15
29 - 31	18
32 - 34	7
35 - 37	2
38 - 40	2
Total	50

3 MEASURES OF AVERAGE (OR CENTRAL TENDENCY)

Given a set of data, it is useful to know the 'centre' or 'average' of it; a *measure* of average (or *central tendency*) is a single number which in some way indicates the 'centre' of the data. There are three main measures of average, viz mean, median and mode, which we now consider in turn.

3.1 The mean ($\bar{x}$)

consider the following set of values of the variate x:

2 7 4 1 4 5 8

To calculate the *mean* of this set of values, which we denote $\bar{x}$, add them all up and divide by the number of values: there are 7 in this case, ie

$$\bar{x} = \frac{2 + 7 + 4 + 1 + 4 + 5 + 8}{7} = 4\frac{2}{7}$$

We can summarise this calculation in a formula using the Σ notation below. If n is the number of values of the variate x, we have:

$$\boxed{\bar{x} = \frac{\Sigma x}{n}}$$

[Here Σ means 'the sum of' and so Σx means 'the sum of all the values of x'.]

3.2 The mode

The *mode* of a set of values is simply the most frequently occuring value. In the above example, the mode is therefore 4.

There may be more than one mode of a set of values, however. Consider the set of values below:

1 2 2 3 4 4 5 6

Both the values 2 and 4 occur with equally high frequency, and so are the two modes of this set of values. As we shall see later, this can be advantageous.

3.3 The median (M)

The *median*, M, of an *ordered* set of values is the value which has the same number of values lying above it as lying below it. In the above example, we must first order the set of values, from smallest to largest say, as below:

1 2 4 4 5 7 8

We can then see that the median is the 4th value, which is 4, since three values lie to the left of it and three values lie to the right of it, as illustrated below.

1 2 4 4 5 7 8

↑

median

If there is an even number of values in the set, there is no such simple value however. Consider the set of values below:

1 2 3 4 5 6 7 8

We can divide this set of values into two equal groups as below:

1 2 3 4 5 6 7 8

↑

median

Between these two groups lies the median in such cases, which is defined to be the mean of the middle two values, viz 4 and 5. Thus here we have:

$$M = \frac{4+5}{2} = 4.5$$

So we can see that the median of an *ordered* set of values is either the *middle* value or the mean of the *middle two* values, and thus can be loosely thought of as the *middle largest value*.

We now consider how to calculate these three measures of average for data in the form of frequency distributions.

3.4 Ungrouped frequency distributions

Mean ($\bar{x}$)

Consider the discrete frequency distribution of the variate *x* below:

x	1	2	3	4	5	6	7	8	Total
f	6	8	15	20	19	13	4	5	90

Now, the mean ($\bar{x}$) is the sum of all 90 values divided by 90. Thus, since we have 6 values of 1, 8 values of 2, 15 values of 3, etc. then the sum of all 90 values =

6 x 1 + 8 x 2 + 15 x 3 + 20 x 4 + 19 x 5 + 13 x 6 + 4 x 7 + 5 x 8

And so,

$$\bar{x} = \frac{6 \times 1 + 8 \times 2 + 15 \times 3 + 20 \times 4 + 19 \times 5 + 13 \times 6 + 4 \times 7 + 5 \times 8}{90}$$

$$= \frac{388}{90} = 4.31$$

We can summarise this calculation using the formula below

> The mean, $\bar{x}$, of a frequency distribution is given by the formula $\bar{x} = \frac{\Sigma fx}{n}$

where f is the frequency of the value x, and n is the total number of values.

Now, an alternative way of calculating $\bar{x}$ follows from the following argument:

$$\bar{x} = \frac{6 \times 1 + 8 \times 2 + 15 \times 3 + 20 \times 4 + 19 \times 5 + 13 \times 6 + 4 \times 7 + 5 \times 8}{90}$$

$$= \frac{6 \times 1}{90} + \frac{8 \times 2}{90} + \frac{15 \times 3}{90} + \frac{20 \times 4}{90} + \frac{19 \times 5}{90} + \frac{13 \times 6}{90} + \frac{4 \times 7}{90} + \frac{5 \times 8}{90}$$

ie calculate fx for each value, and then divide each such product by n, then add all these up.

This gives the entirely equivalent formla below:

$$\bar{x} = \Sigma \frac{fx}{n}$$

Now since the total number of values, n, is the same as the total frequency, which can be written Σf, an alternative way of writing the first formula is:

$$\bar{x} = \frac{\Sigma fx}{\Sigma f}$$

We thus have the three equivalent formulae for calculating the mean, $\bar{x}$, of a discrete frequency distribution:

$$\bar{x} = \frac{\Sigma fx}{n} = \frac{\Sigma fx}{\Sigma f} = \Sigma \frac{fx}{n}$$

The calculator can be used to calculate $\bar{x}$, as explained in the following note.

Note: Use of the Calculator to Calculate $\overline{x}$

It saves time and gives greater accuracy to calculate $\overline{x}$ using the 'statistics mode' of your calculator. Once in this mode, the data can be entered into the calculator by following the sequence below for each value:

Press value, eg [1]

Press [X]

Press frequency, eg [6]

Press 'data' button (usually with a blue *x* sign)

Repeat for next value.

After the data has been entered, the value of $\overline{x}$ can be obtained by pressing the appropriate button. *Also*, by pressing the button marked n, you can check whether the correct amount of data has been entered, which should equal the total frequency (in the above example, this is 90); this gives greater accuracy overall. If you are using the calculator, be sure to state the formula you are using, including its application to the data, as below, to indicate that you *could* calculate it if long-hand is necessary.

$$\underline{x} = \frac{\Sigma fx}{n} = \frac{6 \times 1 + \ldots.. + 5 \times 8}{90} = 4.31$$

Mode

The mode is, again, the most frequently occuring value and is, thus, the value with the highest frequency. In the above example, this is the value 4 (with a frequency of 20).

Median

In order to calculate the median here, we *could* write out all 90 values in increasing size, as below:

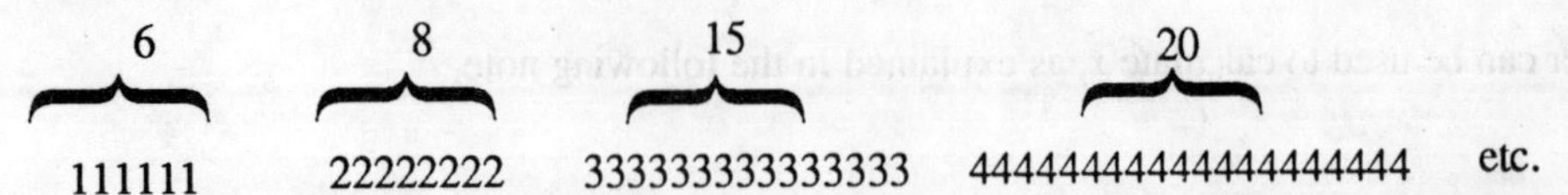

111111 22222222 333333333333333 44444444444444444444 etc.

and then find the middle value(s) of this list. However, since we know that there are 90 values altogether, (an *even* number) then the middle two largest values are the 45th and 46th values of this list, with the median, M being the mean of these. Now, the 45th and 46th values are both 4; this can be seen by accumulating the frequencies of the values until the 45th and 46th values are included, as demonstrated below:

the 1st to the 6th values are 1

the 7th to the 14th values are 2

the 15th to the 29th values are 3

the 30th to the 49th values are 4

[*Note*:

a) If n, the total number of values is *even*, then the middle two values are $\frac{n}{2}$ and $\frac{n}{2} + 1$

(in this case, n = 90 and so the middle two values are $\frac{90}{2} = 45$th and $\frac{90}{2} + 1 = 46$th

b) If n is odd, the middle value is $\frac{n + 1}{2}$ eg if there were 91 values instead, the middle value would be the $\frac{91 + 1}{2} = 46$th value.]

3.5 Grouped frequency distributions

Mean ($\bar{x}$)

Now, as mentioned on page 7, the grouping of data loses some of the data. Consider the distribution of the variate x below.

x	5-10	10-15	15-20	20-30	30-35	Total
f	1	5	9	7	3	25

In the class 10-15, say, there are 5 values; however, we do not know whereabouts they are in this class, ie we do not know their *exact* values. How then can we find the mean of such 'unknown' values? The answer is to assume that they are evenly spread throughout the class, as shown below.

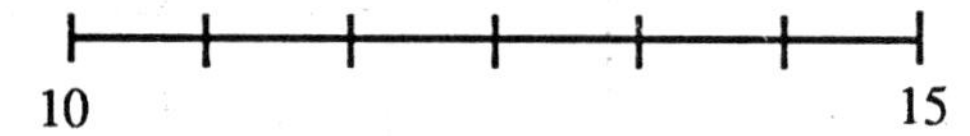

If this were the case, then the mean of the values of each class would be the *mid-class value*, eg in the above class, the mean of the 5 values of x would be 12.5.

Suppose we let those 5 values be x_1, x_2, x_3, x_4, x_5 ;

then, since 12.5 is their mean, we have:

$$12.5 = \frac{x_1 + x_2 + x_3 + x_4 + x_5}{5}$$

ie $x_1 + x_2 + x_3 + x_4 + x_5 = 5 \times 12.5$

In words, this says 'the sum of the 5 values of the class 10-15 is equal to the class frequency 5 times the MCV, 12.5'.

In general, the sum of the values of a class with MCV x_m, and class frequency f, is equal to $f.x_m$.

We now have a method for finding the sum of *all* the values of x. Find the sum of the values of each class and add these together, giving $\Sigma f.x_m$. Finally, to calculate the mean, $\bar{x}$, we must divide this sum by the number of values n, to give the formula below.

The mean, $\bar{x}$, of a grouped frequency distribution with MCVs x_m is given by the formula:

$$\bar{x} = \frac{\Sigma f x_m}{n}$$

Again, replacing n by Σf gives the equivalent formula:

$$\bar{x} = \frac{\Sigma f x_m}{\Sigma f}$$

Also, changing the Σ sign we get another equivalent formula:

$$\bar{x} = \Sigma \frac{f.x_m}{n}$$

Thus, we have the three equivalent formulae:

$$\bar{x} = \frac{\Sigma f x_m}{n} = \frac{\Sigma f x_m}{\Sigma f} = \Sigma \frac{f x_m}{n}$$

In the above example then, we must first calculate the MCV of each class, as below.

x	5-10	10-15	15-20	20-30	30-35	
MCV (x_m)	7.5	12.5	17.5	25	32.5	Total
f	1	5	9	7	3	25

Then, using the first of the above formulae, say, we have:

$$\bar{x} = \frac{\Sigma f x_m}{n} = \frac{1 \times 7.5 + 5 \times 12.5 + 9 \times 17.5 + 7 \times 25 + 3 \times 32.5}{25}$$

$$= 20$$

It must be noted that some distributions contain 'open-ended' classes, as below, for example, for which the MCV *cannot* therefore be calculated. In such cases we *cannot* calculate $\bar{x}$.

x	0.5	5-10	10-15	15 and over
f	2	5	9	3

However, some problems ask you to assume that the open-ended class has a certain width. In the above case, if the last class were assumed to have the same width as the other classes, then its MCV would be 17.5, and we could proceed with the calculation of $\bar{x}$. The alternative to making this assumption would be to calculate the median, or even the mode, as a measure of average.

Mode

Considering the above distribution, we can see that the *class* containing the most values is 15-20; we call this the *modal class*. However, we cannot say which is the most frequent *single* value, since we do not know *any exact* values. It is most likely to be in the modal class (though, of course, not necessarily). As such, there is a *constructional* method of estimating the mode using the histogram of the distribution, as illustrated below. From the diagram, we see that the mode is approximately 17.2.

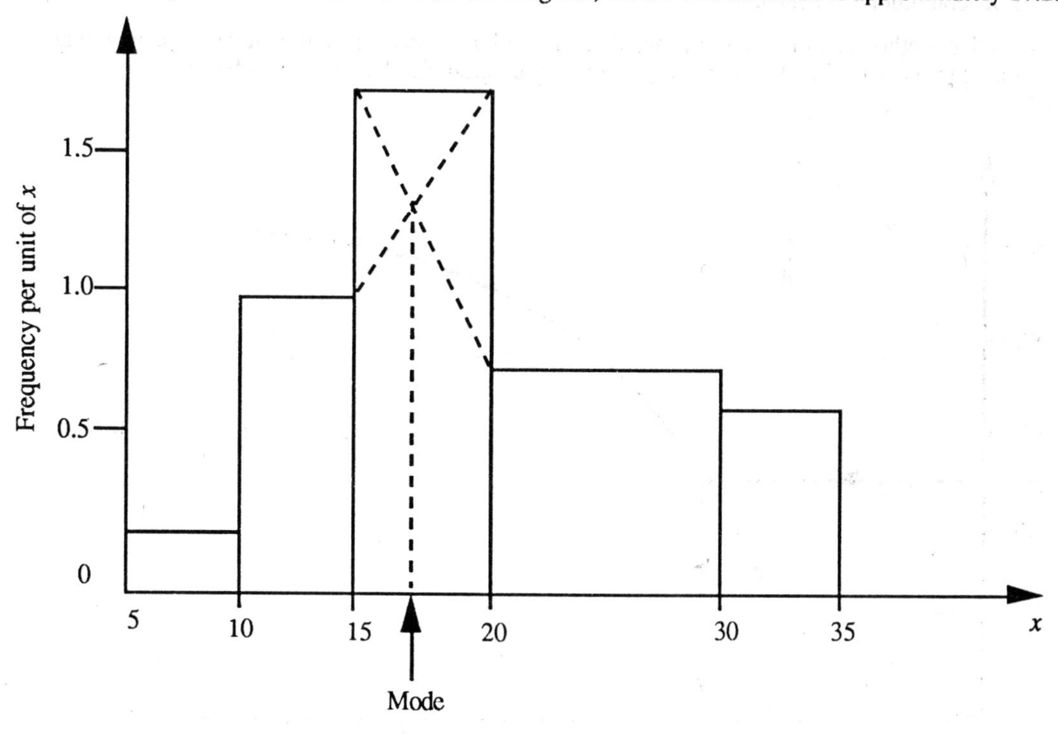

Median

Again, we cannot specify the *exact* value of the median since we do not know *any* exact values; we must estimate it using the assumption that the values are evenly spread out over each class. Considering the above distribution, we know that the median is the middle value which is the 13th value $\left(\text{since } \frac{25+1}{2} = 13\right)$. This lies in the class 15-20, which is illustrated below.

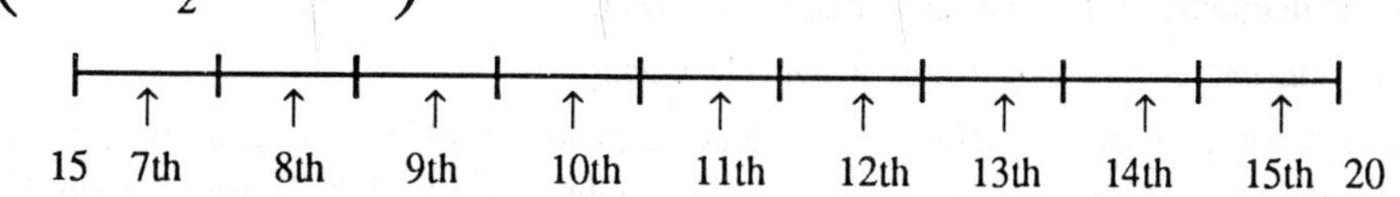

To say that the 9 values are evenly spread out in the interval 15-20 means that each lies at the mid-point of one of 9 equal intervals of the class; each of these small intervals is $\frac{5}{9}$ long (since the class width is 5).

Thus, we can see that the 13th value is $6\frac{1}{2}$ of these intervals into the class, ie

$$\text{Median} = 15 + 6\tfrac{1}{2} \times \tfrac{5}{9} = \underline{18.6}$$

Note that since this is an estimate, we should write:

$$M = 18.6$$

Furthermore, if the distribution is *discrete*, the classes must first be converted to a *continuous* class before applying the above method. This is because of the assumption made above (this does not, of course, apply to the calculation of the *mean*, $\bar{x}$, since the MCV of a discrete class is the *same* as the MCV of its continuous conversion).

An alternative method of estimating the median is through the use of the cumulative frequency diagram for the distribution. Consider the percentage frequency diagram for the above distribution.

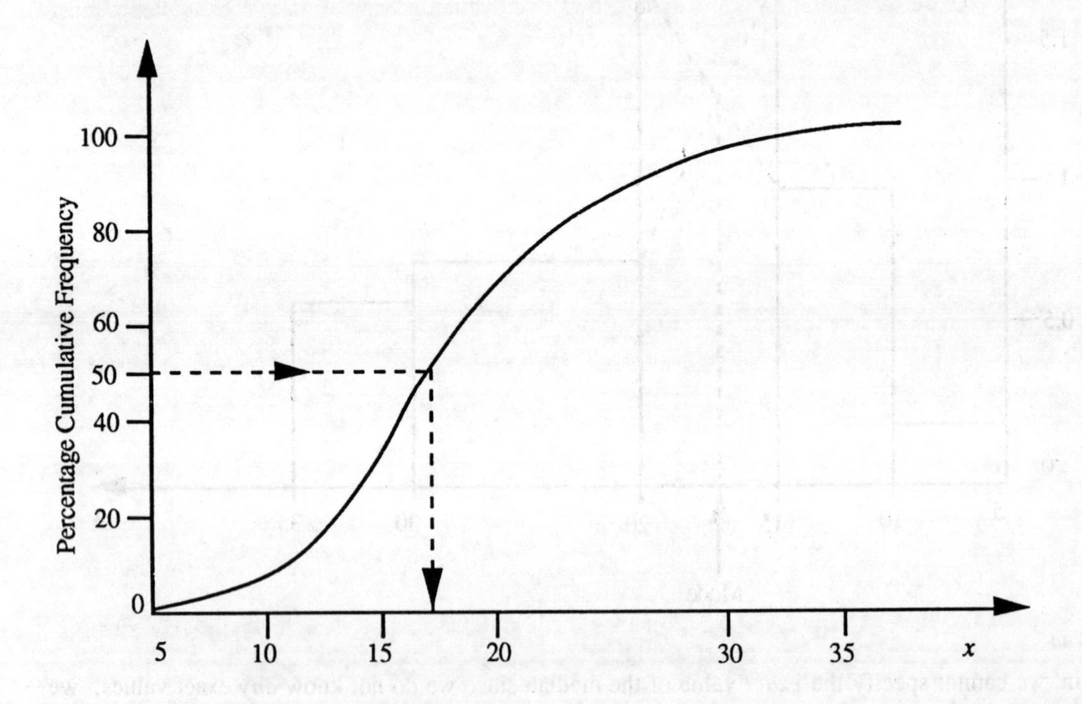

Because the median, M, is that value below which the same number of values lie as above it, it corresponds to a percentage cumulative frequency of 50%, ie half the values lie below it, half the values lie above it. From the diagram, we see that this value is M = 18.3.

The advantages and disadvantages of the measures of central tendency

The choice of which measure to use for a given set of values depends on which most truly reflects the 'centre' of the data; we now consider the advantages and disadvantages of the three measures using this as a criterion.

Mode

Advantages:

i) Quick and easy to calculate

ii) It is the only measure able to reflect data that is naturally divided into two or more groups.

Consider the data below.

x	1	2	3	4	5	6	7	8	9
f	2	5	3	1	2	4	5	2	2

The shape of this distribution is as below, and shows two distinct groups of data, each having a most frequent value.

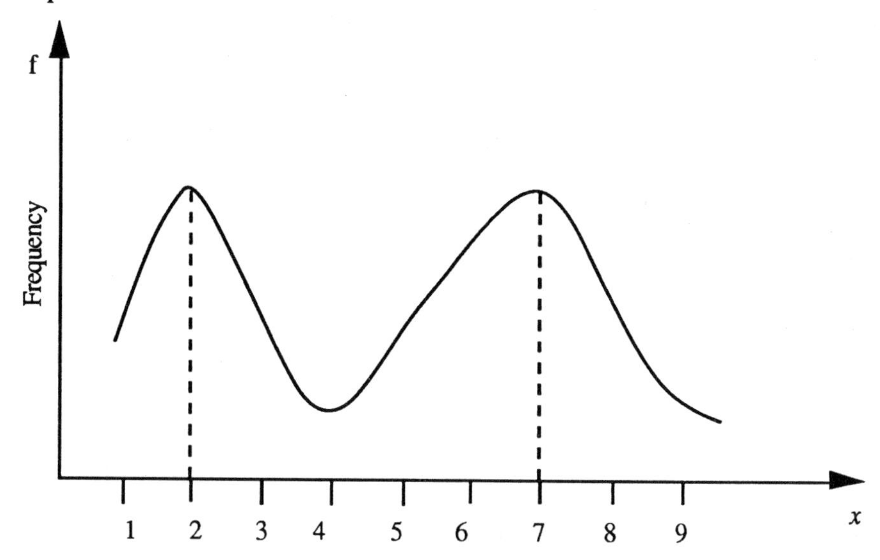

The modes of this data are 2 and 7. Either of the other two measures of average would give a *single* value that would not reflect the divided nature of the data.

Disadvantage

It only uses a few values in its calculation. Consider the data below:

2 2 109 107 101 100 103 108 99 102

The mode of this data is 2 which clearly does not reflect the centre of the data; it only makes use of the two values of 2 in its calculation.

Mean

Advantages:

i) Uses *all* of the data in its calculation, and so 'summarises' *all* of the data.

ii) Is part of a well-developed theory of statistics in view of its mathematical properties.

Disadvantages:

It is *sensitive to extreme values*. Consider the data below.

1 1 1 1 1 1 1 1,000

The single value of 1,000 is called an *extreme value* (extremely large in this case), and since the mean uses *all* of the values in its calculation, this single, extremely large value will 'drag' the mean up beyond a central position, ie

$$\bar{x} = \frac{1 + 1 + 1 + 1 + 1 + 1 + 1 + 1000}{8} = 125.9$$

Median

Advantage:

It is *not* sensitive to extreme values. This is because the value of the median is dependent upon its *position*, ie where the 'middle' of the data is. Suppose we calculated the median for the above set of data, first with the extreme value removed, and then with it included, as below:

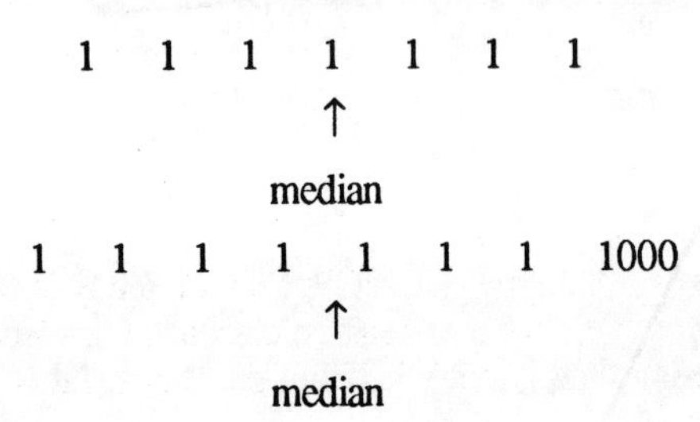

As can be seen, the median simply moved up half a space when the extreme value is included. In both cases the median is 1.

Disadvantage:

It only uses a few values in its calculation (like the mode), and thus could be in error. Consider the data below:

1 1 1 50 1000 1000 1000

The median of this data is 50 (and thus uses only the *value* 50 in its calculation).

Now, if we replaced this single value of 50 by the value 500, the median immediately takes this new view, thus showing immense variability. This is because it only uses one or two actual *values* in its calculation, relying solely on their 'middle' position.

The relationship between the mean, median and mode

If the data is symmetrically distributed then the mean, median and mode will roughly coincide (see diagram (a) below). If, however, the data has a *skewed* distribution then the mean will be dragged upwards (in the case of positive skew) or downwards (in the case of negative skew), relative to the mode, due to the influence of the extreme values (represented by the 'tail' of the distribution), and is illustrated in diagrams (b) and (c) below. The median will always lie between the mode and mean in such cases, since it only moved half a space with the addition of each extreme value (as demonstrated above). This relationship is given in the diagram opposite for each type of skew.

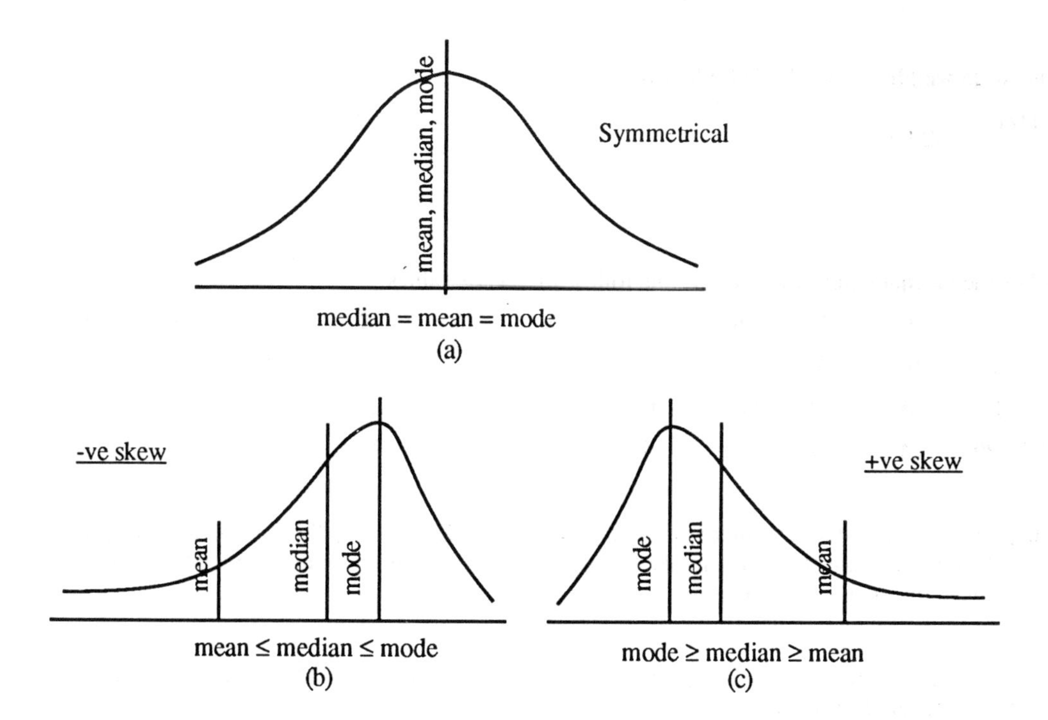

Thus, if the data is *symmetrically* distributed, the most suitable measure of central tendency is the *mean.* However, if the data has a *skewed* distribution, the most appropriate measure of central tendency is the *median.* In the case of a *'multi-modal'*, (ie has more than one mode) distribution, the *mode* is the most appropriate measure.

A special problem involving the mean

The following type of problem involving the mean, $\bar{x}$, sometimes occurs in examinations, together with a similar 'special problem' which is illustrated at the end of the next chapter. Consider the following example.

Example

50 girls and 30 boys were weighed during a routine physical examination at a school. The girls had an average weight of 55 kg, whilst the boys had an average weight of 47 kg. What is the average weight of all 80 children?

In order to calculate the average weight of all 80 children, we must find their total weight and divide this by 80. Now, let T_B and T_G be the total weights of the boys and girls respectively: then we have:

$$55 = \frac{T_G}{5} \quad \text{and} \quad 47 = \frac{T_B}{30}$$

ie $T_G = 50 \times 55$ and $T_B = 47 \times 30$

ie $T_G = 2{,}750$ and $T_B = 1{,}410$

Thus, the total weight of all 80 children is:

$$T_G + T_B = 2{,}750 + 1{,}410 = 4160$$

Hence, the mean weight ($\bar{x}$) of all 80 children is:

$$\bar{x} = \frac{4160}{80} = 52 \text{ kg}$$

Exercise 3

1. Calculate the mean, median and mode of the following sets of values:

a)	1	3	9	4	2	1	5	8	7	5	1
b)	2	9	2	3	9	7	4	2	2	9	3
c)	0.3	1.8	2.5	3.1	4.2	5.1	0.8	1.7			
d)	129	84	265	93	81	82	94	106			

2. Calculate the mean, median and mode of the distributions in Exercise 2, No. 3. Use the histograms for these distributions to estimate the mode where necessary. Use the cumulative frequency diagrams also to estimate the medians where appropriate.

3. a)-f) Use the cumulative frequency diagrams of Exercise 2, Nos. 10, 11, 12, 13, 14 and 16 to estimate the median of the data.

4.

Class	Frequency
0 and up to 2	176
2 and up to 10	222
10 and up to 20	343
20 and up to 30	285
30 and up to 40	312
40 and up to 50	306
50 and up to 60	296
60 and up to 70	182
70 and up to 80	100
80 and up to 110	28

For the above table calculate:

a) the arithmetic mean and

b) the median.

(Give your answers correct to one decimal place.)

5. Estimate the mode of the following distribution both diagrammatically and by calculation.

Length (cm)	0-	5-	10-	15-	20 and over
Frequency	8	12	6	3	1

6.

Mark	2	3	4	5	6	7	8	9	10
No. of pupils	1	6	6	9	14	21	11	10	2

The table gives the distribution of marks in a test. What is the mean mark obtained?

7.

No. of days	0-	50-	100-	150-	200-	250-	300-365	Total
No. of children	29	24	30	20	25	17	11	156

The table gives the ages in days (over 15 years) of children in the fifth form at a particular school. What is the median age of children in the fifth form at that school?

8. The table below shows the number of calls per day received at a fire station over a given period. Calculate the mean number of calls per day.

No. of calls per day	0	1	2	3	4	5
No. of days	75	62	32	16	9	6

9. The number of bracts on specimens of a particular species of plants are tabulated below. Determine the mode.

No. of bracts	4	5	6	7	8	9	10	11	12	13	14	15
No. of specimens	1	1	1	2	12	10	6	6	2	2	1	1

10. The table below shows the number of calls per day received by a fire station over a given year. Determine the median of this distribution.

No. of calls per day	0	1	2	3	4	5	6 and over
No. of days	139	102	57	30	19	12	6

11. The mean age of five boys is 12 years. The mean age of these five boys and a further 10 girls is found to be 14 years. What is the mean age of the 10 girls?

12. The number of goals scored in 20 football matches is recorded below.

No. of goals	0	1	2	3	4	Total
No. of matches	1	5	9	3	2	20

Calculate:

a) mean number of goals scored

b) modal number of goals scored

c) median number of goals scored.

If the results of a further 10 matches averaged 3 goals per match, what is the average number of goals scored for all 30 matches?

13. In three successive periods of 10 days, the average numbers of hours of sunshine were 11, 15 and 9 hours. What is the average number of hours of sunshine per day over the 30 day period?

14. In an attempt to compare the standards of referees in a football league, the referee is given a grade from 0 to 6 by the secretary of each of the home and away teams after each match. The table below shows the distribution of the grades given to a particular referee after 4 matches:

Grade	0	1	2	3	4	5	6
Frequency	1	0	3	2	0	0	2

State the mode of the grades.

After one further match the mean of all the grades for this referee was still the same as it was after 4 matches. If the secretary of the home team gave a grade of 5, what grade did the secretary of the away team give?

15. Which measure of average is best suited to each of the distributions in Exercise 2, No. 3?

4 MEASURES OF DISPERSION (OR SPREAD OR VARIATION)

4.1 Introduction

Whereas the last chapter was concerned with various measures of the centre or average of a set of values, this chapter considers several measures of how 'spread out' or 'dispersed' the values are.

Consider the following two sets of values:

A:	11.10	10.90	10.95	11.15	11.10	10.96	11.00	11.05	10.83
B:	6	3	5	9	12	15	13	17	19

The shapes of these sets of data are illustrated in the diagram below.

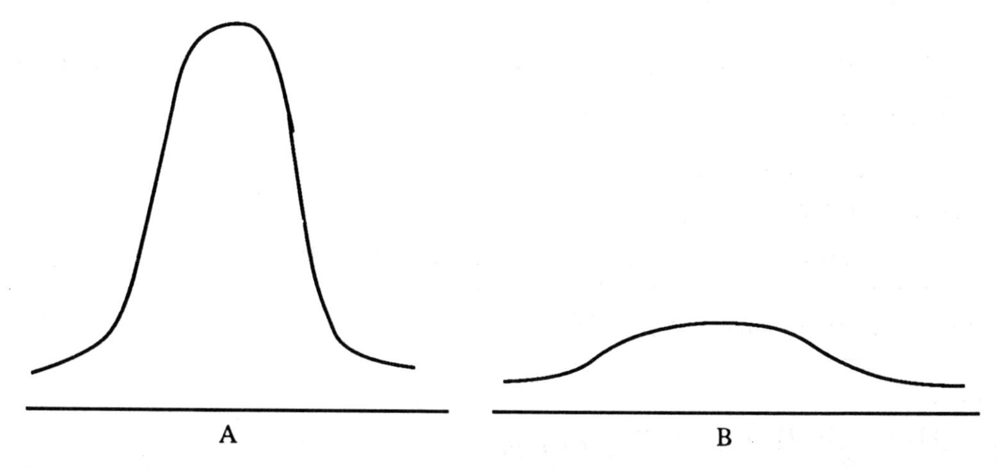

Both sets of data have means of 11. The values of set A are closely bunched together, and cluster around the mean of 11. In contrast, the values of set B are spread out, ie they vary greatly from one another, and are well dispersed. What we wish to do here is *measure* the degree to which the values are spread out or dispersed. This can be done in several ways, each of which has various advantages and disadvantages over the others.

4.2 The range (R)

The range, R, of a set of values of the variate x is simply the smallest value subtracted from the largest value. In the above two sets of data, we have:

Set A: Range = 11.15 - 10.83 = 0.32

Set B: Range = 19 - 3 = 16

Thus, we can see that in these cases the range certainly does reflect the 'variation' or dispersion of the data.

4.3 The semi-interquartile range (SIR)

Before we can define the semi-interquartile range, we must firstly define the upper and lower quartiles, Q_3 and Q_1, respectively. Recall that the median, M, (which is, in fact, the same as Q_2) is the value below which 50% of the values lie. In a similar way, the lower quartile Q_1 is the value below which 25% of the values lie, and the upper quartile Q_3 is the value below which 75% of the values lie. Ordering the values in sets A and B above, we have:

Set A:	10.83	10.9	10.95	10.96	11.00	11.05	11.10	11.10	11.15
Set B:	3	5	6	9	12	13	15	17	19

For the data in Set A, we kow that the median M is 11.00; this value has 4 values lying below it and 4 values lying above it. Now, the lower quartile Q_1 is the middle value of these 4 values below the median, and therefore lies between the values 10.9 and 10.95, thus:

$$Q_1 = \frac{10.9 + 10.95}{2} = \frac{21.85}{2} = \underline{10.925}$$

In a similar manner Q_3 lies in the middle of the 4 values, above the median, and therefore lies between the values 11.10 and 11.10, thus:

$$Q_3 = \frac{11.10 + 11.10}{2} = \underline{11.10}$$

We can now define the <u>interquartile range</u> (IR) as the difference between the upper and lower quartiles, ie

$$\boxed{\text{Interquartile Range (IR)} = Q_3 - Q_1}$$

In this case, the interquartile range is:

$$IR = 11.10 - 10.9125 = \underline{0.1875}$$

We finally define the *semi-interquartile range* (SIR) as half the interquartile range, ie

$$\boxed{\text{Semi-Interquartile Range (SIR)} = \frac{Q_3 - Q_1}{2}}$$

In this case,

$$SIR = \frac{11.10 - 10.9125}{2} = \frac{0.1875}{2} = 0.09375$$

Repeating this process with the data in set B, we find that the SIR = 5.25. Thus we see that the semi-interquartile range does, indeed, reflect the dispersion of these two sets of data.

Alternatively, we could construct a percentage cumulative frequency diagram for the data and read off Q_1 and Q_3 corresponding to percentage cumulative frequencies of 25% and 75%. Consider again the example on page 39. The percentage cumulative frequency diagram is presented below.

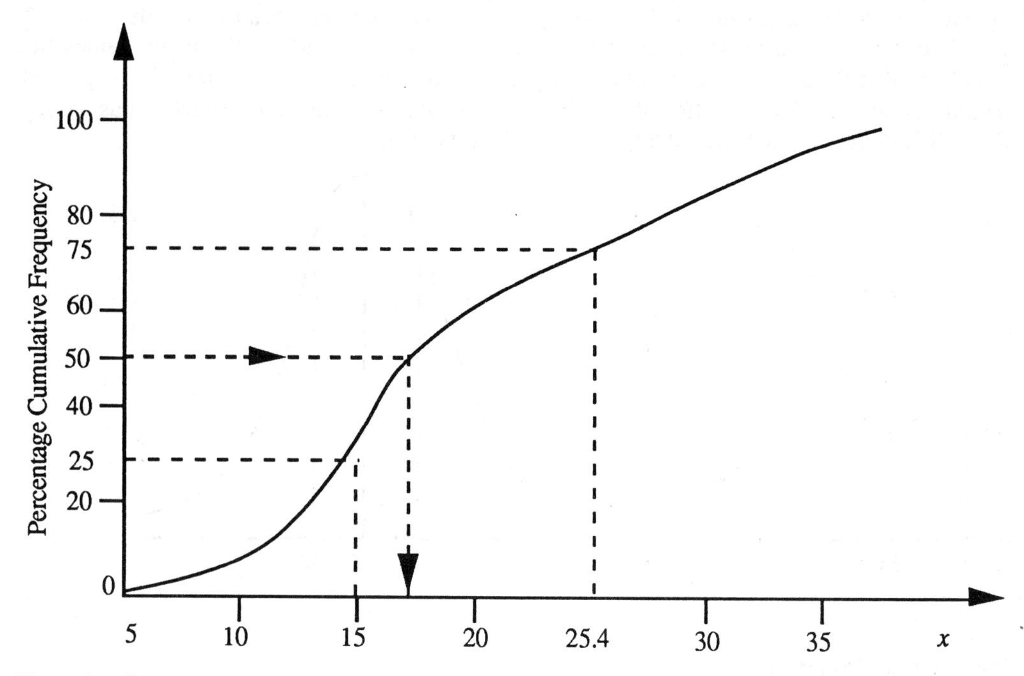

From the diagram, we see that:

$Q_1 = 15$ and $Q_3 = 25.4$

Thus, $IR = Q_3 - Q_1 = 25.4 - 15 = 10.4$

and $SIR = \frac{Q_3 - Q_1}{2} = \frac{10.4}{2} = 5.2$

But why is the SIR (or IR) a measure of dispersion? Consider the diagram below:

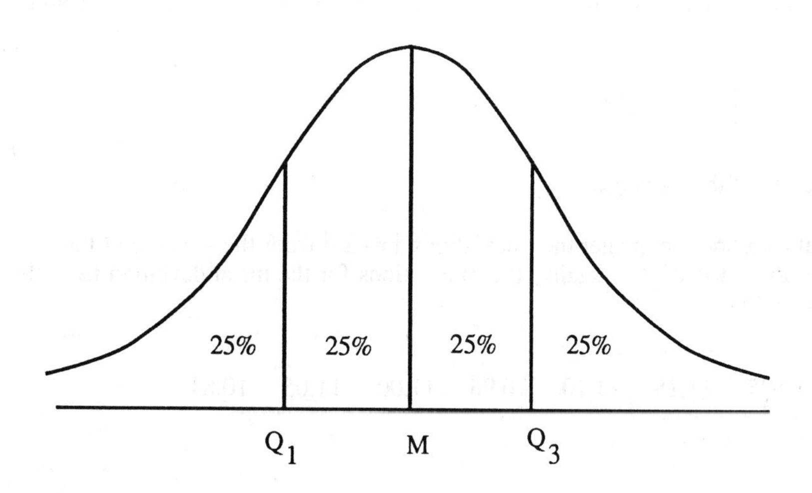

Note: Remember that the *area* of *all* the bars of a histogram represent the total frequency of the distribution; the curve in the diagram above is the shape of the histogram of a distribution, and thus the area under the curve represents the total frequency.

Here we see the two quartiles, Q_1 and Q_3, and the median M (which is really Q_2) dividing up the area under the curve (which represents the total frequency into four equal parts (25% of the values in each part). Thus, between the two quartiles Q_1 and Q_3, we see that the middle 50% of the values lie. Clearly, the closer that Q_1 and Q_3 are together, the more tightly those 50% of values will be packed together, and the smaller the IR and SIR will be (this is illustrated in the diagram below); thus we see that the IR and SIR do reflect, to some extent, how spread out the values are.

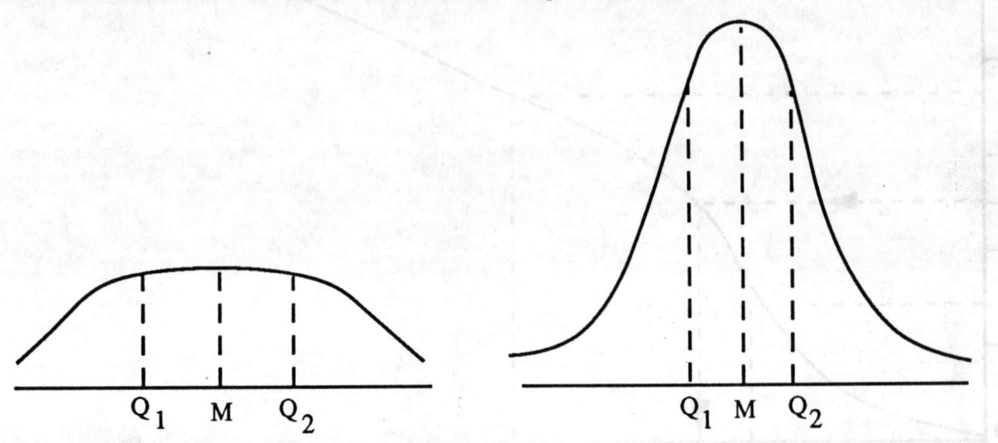

4.4 The mean deviation

When a set of values has a small dispersion, ie is tightly bunched together, the values cluster tightly about the 'centre'. Now, for the following, let us assume that the mean, $\bar{x}$, is a good measure of the centre of the set of values, as would be the case with symmetrically distributed data. We can then see that the distance, or *deviation*, of each value x from the centre $\bar{x}$ will be small, on average if the values are mostly tightly clustered about the centre. If we let $|x - \bar{x}|$ denote the *positive* distance (*deviation*) of x from $\bar{x}$, then the mean of all such deviations, called the *mean deviation from the mean* and denoted MD, is given by the formula:

$$\text{MD} = \frac{\Sigma |x - \bar{x}|}{n}$$

where n is the number of values of the variate x.

The more spread out the values x are, the greater their deviations $|x - \bar{x}|$ from the mean, and thus the larger the MD is. Consider the set A of data again; the calculations for the mean deviation from the mean for this data are set out below:

Set A: 11.10 10.90 10.95 11.15 11.10 10.96 11.00 11.05 10.83

For the data, $\bar{x} = 11.00$. Calculating the mean deviations, $|x - \bar{x}|$, of each value x from $\bar{x}$, we have:
$|11.10 - 11.00| = 0.10$; $|10.90 - 11.00| = 0.10$; $|10.95 - 11.00| = 0.05$;
$|11.15 - 11.00| = 0.15$; $|11.10 - 11.00| = 0.10$; $|10.96 - 11.00| = 0.04$;
$|11.00 - 11.00| = 0.00$; $|11.05 - 11.00| = 0.05$; $|10.83 - 11.00| = 0.17$

Thus, MD $= \dfrac{\Sigma |x - \bar{x}|}{n}$

$$= \frac{0.10 + 0.10 + 0.05 + 0.15 + 0.10 + 0.04 + 0.00 + 0.05 + 0.17}{9}$$

$$= \frac{0.76}{9} = \underline{0.084}$$

These calculations can be set out more economically as demonstrated below for the values of x in Set B above.

x	3	5	6	9	12	13	15	17	19
$\lvert x - \bar{x} \rvert$	8	6	5	2	1	2	4	6	8

Then MD $= \dfrac{\Sigma |x - \bar{x}|}{n}$

$$= \frac{8 + 6 + 5 + 2 + 1 + 2 + 4 + 6 + 8}{9}$$

$$= \frac{42}{9} = \underline{4.67}$$

Again, we see that the mean deviation from the mean does indeed reflect the dispersion of these two sets of data.

Now, instead of using the *mean*, $\bar{x}$, as the measure of the 'centre' from which the x-values deviate, we *could* use the median, M, or mode. As an example, we now calculate the mean deviation from the *median*, M, for set A.

Set A: 10.8 10.9 10.95 10.96 11.00 11.05 11.10 11.10 11.15

Now, M $= \dfrac{10.96 + 11.00}{2} = 10.98$

Thus, we have:

x	10.8	10.9	10.95	10.96	11.00	11.05	11.10	11.10	11.15
$\lvert x - M \rvert$	0.18	0.08	0.03	0.02	0.02	0.07	0.12	0.12	0.17

So, mean deviation from the median M

$$= \frac{\Sigma \lvert x - M \rvert}{n} = \frac{0.18 + 0.08 + 0.03 + 0.02 + 0.02 + 0.07 + 0.12 + 0.12 + 0.17}{9}$$

$$= 0.09$$

The calculation of the mean deviation from the mode for this set of data is left as an exercise for the reader.

4.5 The variance (S^2)

The variance is calculated in a similar manner to the mean deviation in as much as it is based upon the deviations of the x values from the mean $\bar{x}$. (*Note*: *only* the mean can be used as the measure of centre here); however, before averaging these deviations from the mean, we first *square* them, to obtain $(x - \bar{x})^2$. The reason for this will be given later in the chapter.

Thus, we obtain the formula below:

$$\text{Variance } (S^2) = \frac{\Sigma(x - \bar{x})^2}{n}$$

For the set of data A above we have:

x	11.10	10.90	10.95	11.15	11.10	10.96	11.00	11.05	10.83
$x - \bar{x}$	0.10	-0.10	-0.05	0.15	0.10	-0.04	0.00	0.05	-0.17
$(x - \bar{x})^2$	0.01	0.01	0.0025	0.0225	0.01	0.016	0.00	0.0025	0.0289

Thus, $S^2 = \frac{\Sigma(x - \bar{x})^2}{n}$

$$= \frac{0.01 + 0.01 + 0.0025 + 0.0225 + 0.01 + 0.016 + 0.00 + 0.0025 + 0.0289}{9}$$

$= 0.0114$

Similar calculations for Set B yield the variance:

$S^2 = 27.8$

Thus we see that the variance too reflects the dispersions of the two sets of data.

4.6 The standard deviation (S)

This is simply the square root of the variance, and is thus denoted S; the formula for calculating S is therefore:

$$\text{Standard Deviation (S)} = \sqrt{\frac{\Sigma(x - \bar{x})^2}{n}}$$

The process of taking the square root of the variances 'counteracts' the squaring of the deviation, thereby giving a measure of dispersion of the same *order* as the original data. For the data in sets A and B, we have:

Set A: $S = \sqrt{0.0114} = 0.11$

Set B: $S = \sqrt{27.8} = 5.27$

As for the measures of central tendency in the last chapter, we now consider the calculation of the above measures of dispersion for data given in the form of frequency distributions, beginning firstly with *ungrouped* distributions; naturally, this will entail modifying the formula given above.

4.7 Ungrouped frequency distributions

Range

Consider the distribution below.

x	1	2	3	4	5	6	7	8	Total
f	6	8	15	20	19	13	4	5	90

We can see that the smallest value of x is 1 (there are, in fact, 6 of these) and the largest value of x is 8 (there are 5 of these); thus

Range (R) = 8 - 1 = 7.

Semi-interquartile range

Here, since there are 90 values, the median, M, lies mid-way between the 45th and 46th values; below and above the median, therefore, are 45 values. Q_1 is the middle of the lower 45 values, and is therefore the 23rd value, which is 3. Also, Q_3 is the middle of these 45 values and is, therefore, the 68th value, which is 5.

Thus we have:

$$SIR = \frac{Q_3 - Q_1}{2} = \frac{5 - 3}{2} = 1$$

Mean deviation

We will only calculate the mean deviation from the mean, $\bar{x}$, and leave the calculation of the mean deviation from the median and mode to the reader.

Here from p.34 we have:

$$\bar{x} = \frac{\Sigma fx}{n} = \frac{6x1}{90} + \frac{8x2}{90} + \frac{15x3}{90} + \frac{20x4}{90} + \frac{19x5}{90} + \frac{13x6}{90} + \frac{4x7}{90} + \frac{5x8}{90} = 4.31$$

Thus, we have:

x	1	2	3	4	5	6	7	8	
$\lvert x - \bar{x} \rvert$	3.31	2.31	1.31	0.31	0.69	1.69	2.69	3.69	Total
f	6	8	15	20	19	13	4	5	90

Now, each deviation $x - \bar{x}$ occurs with a frequency f; thus, in order to calculate the mean of all 90 deviations we must use the modified formula below:

$$MD = \frac{\Sigma f . \lvert x - \bar{x} \rvert}{n}$$

giving us:

MD =

$$\frac{6 \times 3.31 + 8 \times 2.31 + 15 \times 1.31 + 20 \times 0.31 + 19 \times 0.69 + 13 \times 1.69 + 4 \times 2.69 + 5 \times 3.69}{90}$$

= <u>1.43</u>

Variance

This is calculated in a similar manner to the mean deviation from the mean, $\bar{x}$, above, using the modified formula:

$$S^2 = \frac{\Sigma f(x - \bar{x})^2}{n}$$

Using the same data, we have:

x	1	2	3	4	5	6	7	8	
$(x - \bar{x})$	-3.31	-2.31	-1.31	-0.31	0.69	1.69	2.69	3.69	
$(x - \bar{x})^2$	10.96	5.34	1.72	0.10	0.48	2.86	7.24	13.62	Total
f	6	8	15	20	19	13	4	5	90

Then we have:

$$S^2 = \frac{\Sigma f(x - \bar{x})^2}{n}$$

$$= \frac{6 \times 10.96 + 8 \times 5.34 + 15 \times 1.72 + 20 \times 0.10 + 19 \times 0.48 + 13 \times 2.86 + 4 \times 7.24 + 5 \times 13.62}{90}$$

$= \underline{3.10}$

> *Note*: The calculator, in the 'statistics' mode, can be used to calculate S^2 and S by inputting the data as described on page 35, and then pressing $\boxed{\sigma_n}$ for S, and squaring it for S^2.

Standard deviation

Again, this is simply the square root of the variance; thus, the formula for S becomes:

$$S = \sqrt{\frac{\Sigma f(x - \bar{x})^2}{n}}$$

In the example above, we have:

$$S = \sqrt{\frac{\Sigma f(x - \bar{x})^2}{n}} = \sqrt{3.10} = \underline{1.76}$$

4.8 Grouped frequency distributions

Range

Consider the data below.

x	5-10	10-15	15-20	20-30	30-35	Total
f	1	6	8	7	3	25

What is the smallest value? We know that there is one value in the class 5-10 but do not know its exact value; it could be as small as 5 or as large as 10 (almost). The same argument applies to the largest value; it could be as large as 35 or as small as 30. Thus, we cannot state the range with any certainty; all we can do is give two numbers that it lies between, viz the smallest and largest *possible* ranges.

Now, the largest possible range = largest possible value - smallest possible value. In this case,

largest possible range = 35 - 5 = 30

Also, the smallest possible range = smallest possible largest value - largest possible smallest value. In this case,

smallest possible range = 30 - 10 = 20

Thus, we can say that the range lies between 20 and 30, ie

$20 < r < 30$

Semi-interquartile range

The calculation of the quartiles Q_1 and Q_3 is similar to the calculation of the median, M, on page 41. Consider the data above. The lower quartile, Q_1, lies between the 6th and 7th values, which both lie somewhere in the class 10-15, illustrated below.

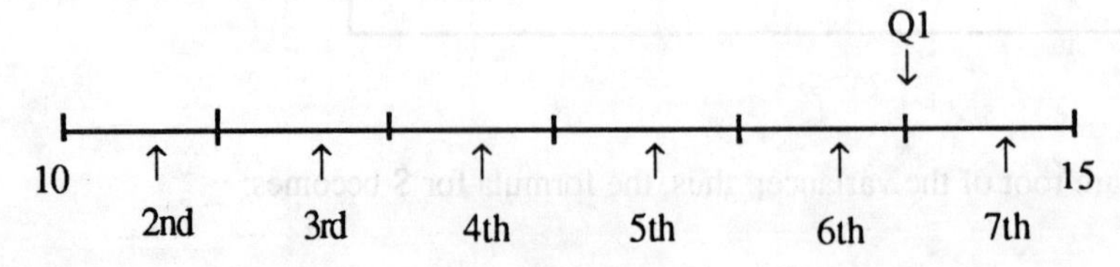

Each small interval has width $\frac{1}{6}(15 - 10) = \frac{5}{6}$.

Also, Q_1 is a distance 5 of these intervals into the class; thus,

$$Q_1 = 10 + 5 \times \frac{5}{6} = 14.17$$

Similarly, Q_3 lies between the 19th and 20th values, which both lie in the class 20-30, as illustrated below.

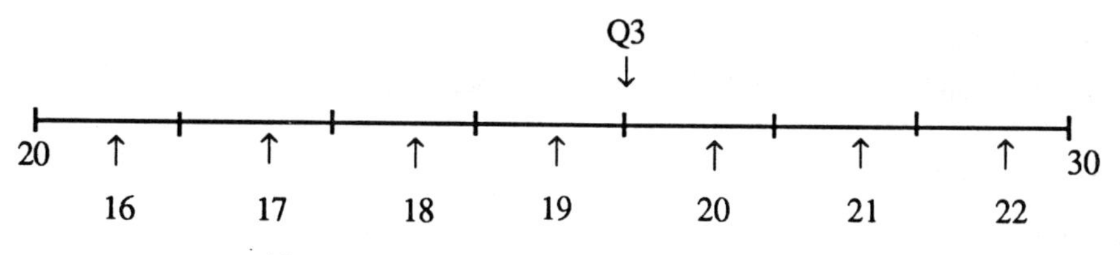

Thus, $Q_3 = 20 + 4 \times \frac{10}{7} = 25.71$

Hence, $SIR = \frac{Q_3 - Q_1}{2} = \frac{25.71 - 14.17}{2} = \underline{5.67}$

We could, of course, plot the percentage cumulative frequency diagram for the distribution and read off estimates for Q_1 and Q_3; this is left as an exercise for the reader.

Mean deviation

Again, we will only consider the mean deviation from the mean, $\bar{x}$. As for the calculation of the mean for grouped data, we use the MCV of each class, x_m, in place of x, assuming that the x-values of each class are evenly spread throughout each class. The formula for MD thus becomes:

$$MD = \frac{\Sigma f. |x_m - \bar{x}|}{n}$$

In the above example, $\bar{x} = 19.8$; thus we have:

x	5-10	10-15	15-20	20-30	30-35	
MCV (x_m)	7.5	12.5	17.5	25	32.5	
$\|x_m - \bar{x}\|$	12.3	7.3	2.3	5.2	12.7	Total
f	1	6	8	7	3	25

Hence, $MD = \frac{\Sigma f. |x_m - \bar{x}|}{n}$

$$= \frac{1 \times 12.3 + 6 \times 7.3 + 8 \times 2.3 + 7 \times 5.2 + 3 \times 12.7}{25}$$

$= \underline{5.96}$

Variance

Replacing x by the MCV x_m, the formula for the variance, S^2, becomes:

$$S^2 = \frac{\Sigma f.(x_m - \bar{x})^2}{n}$$

In the above example, we have:

x	5-10	10-15	15-20	20-30	30-35	
$(x - \bar{x})$	-12.3	-7.3	-2.3	5.2	12.7	
$(x - \bar{x})^2$	151.29	53.29	5.29	27.04	161.29	Total
f	1	6	8	7	3	25

Thus, $S^2 = \frac{\Sigma f.(x_m - \bar{x})^2}{n}$

$= \frac{1 \times 151.29 + 6 \times 53.29 + 8 \times 5.29 + 7 \times 27.04 + 3 \times 161.29}{25}$

$= \underline{47.46}$

Standard deviation

Taking the square root of the variance, the formula for the standard deviation, S, becomes:

$$S = \sqrt{\frac{\Sigma f(x_m - \bar{x})^2}{n}}$$

In this case, $S = \sqrt{47.46} = \underline{6.89}$

Note: If the grouped frequency distribution has discrete classes, they must first be converted to continuous classes, as described on page 9.

If we are calculating the SIR; however, this is unnecessary for the calculation of the other measures.

Advantages and disadvantages of the measures of dispersion

In a similar way as for the measures of central tendency, the choice of measure of dispersion to be used for a given set of data depends upon the nature of the data, ie how well the measure reflects the dispersion of the data. We now consider the advantages and disadvantages of the above measures in view of the criterion.

Range

Advantage

Quick and easy to calculate.

Disadvantage

Only uses two values in its calculation, viz the largest and the smallest. This could result in a high range when there is, in fact, low dispersion; consider the data below.

1 12.1 12.2 12.05 11.99 12.05 12.19 12.3 99

The range is 99-1 = 98, which clearly does not reflect the tightly clustered nature of *most* of the values.

Semi-interquartile range

Advantage

Not sensitive to extreme values. This follows since it only measures the dispersion of the middle 50% of the values (see page 59), and thus takes no heed of extremely small or large values lying outside this middle range.

Disadvantage

Based on only part of the data. Because it only measures the dispersion of the middle 50% of values, the remaining 50% of the values are not taken into consideration. The diagram below shows two distributions with the *same* SIR and yet one has a much higher dispersion than the other.

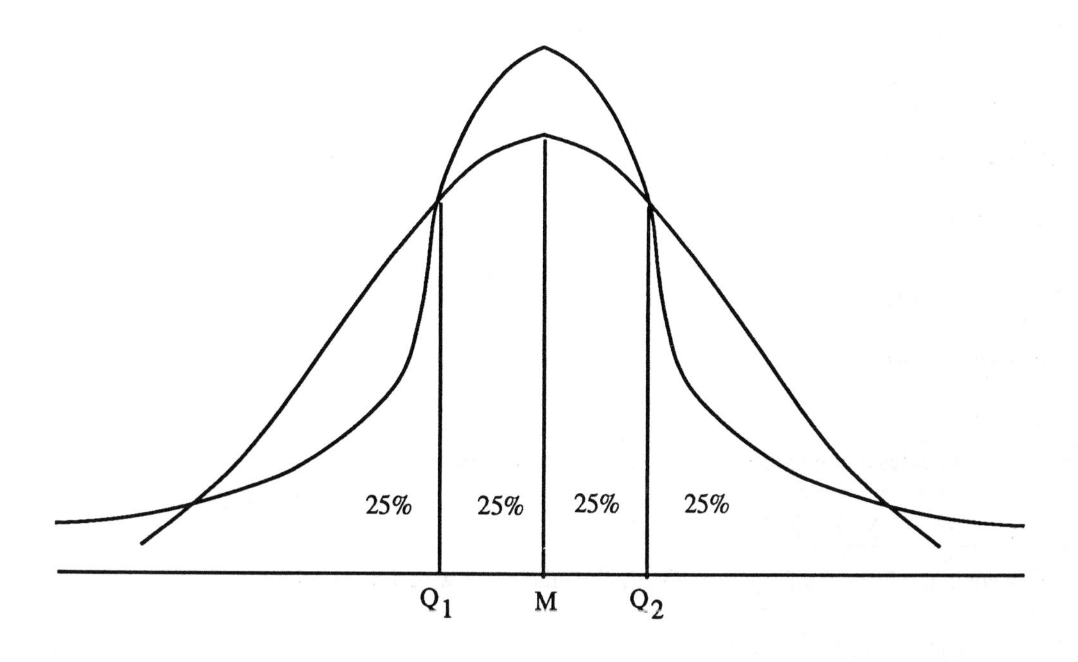

Mean deviation

Advantage

Uses all of the data in its calculation and thus summarises the individual deviations of all the values.

Disadvantages

Sensitive to extreme values. This is because, like the mean, any extremely large or small values will have extremely large or small deviations and therefore drag up or down the value of the mean deviation.

Variance

Advantages

a) Like the mean deviation, it uses all of the data in its calculation.

b) It also plays an important role in the theory of statistics (like the mean) because of its mathematical properties; the other measures of dispersion do not possess these.

Standard deviation: As for the variance.

Thus, if the data is *symmetrically* distributed, the most suitable measure of *dispersion* is the variance (or standard deviation).

However, if the data has a *skewed* distribution, thereby containing *extreme* values, the SIR is the most appropriate measure of *dispersion*.

Alternative formula for the variance

We now derive an alternative formula for the variance which will be of considerable use later on, and also immediately in the analogue to the 'special problem' at the end of the last chapter (see page 49). Recall the original formula for the variance, S^2, of a set of values of the variate x:

$$S^2 = \frac{\Sigma(x - \bar{x})^2}{n}$$

Now, $\Sigma(x - \bar{x})^2 = \Sigma(x^2 - 2x.\bar{x} + \bar{x}^2)$

$$= \Sigma x^2 - \Sigma 2x\,\bar{x} + \Sigma\bar{x}^2$$

$$= \Sigma x^2 - 2\bar{x}\,\Sigma x + n\bar{x}^2$$

$$= \Sigma x^2 - 2\bar{x}n\left(\frac{\Sigma x}{n}\right) + n\bar{x}^2$$

$$= \Sigma x^2 - 2\bar{x}.n.\bar{x} + n\bar{x}^2$$

$$= \Sigma x^2 - 2n.\bar{x}^2 + n\bar{x}^2$$

$$= \Sigma x^2 - n\bar{x}^2$$

This result is sometimes required to be proved, and so is stated below:

$$\boxed{\Sigma\,(x - \bar{x})^2 = \Sigma x^2 - n\bar{x}^2}$$

If the data were in the form of a frequency distribution, the result would be as below:

$$\boxed{\Sigma\,f(x - \bar{x})^2 = \Sigma\,f.x^2 - n\bar{x}^2}$$

We can now replace the $\Sigma(x - \bar{x})^2$ part of the variance formula to obtain a new formula:

$$S^2 = \frac{\Sigma(x - \bar{x})^2}{n} = \frac{\Sigma x^2 - n\bar{x}^2}{n}$$

$$= \frac{\Sigma x^2}{n} - \bar{x}^2$$

Re-writing this we have:

$$\boxed{S^2 = \frac{\Sigma x^2}{n} - \bar{x}^2}$$

Special problem involving the variance

The nature of this problem is similar to the problem at the end of the last chapter, and is in fact, an extension of the same problem.

Example

50 girls and 30 boys were weighed during a routine physical examination at a school. The weights of the girls had a mean of 55 kg and a standard deviation of 5 kg. The weights of the boys had a mean of 41 kg with a standard deviation of 4 kg. Calculate the mean weight of all 80 chldren, and the standard deviation of their weights.

From page 49 we see that the mean, $\bar{x}_T$, of all 80 children is 52 kg. From the data in the question, we know the variances of the girls' weights and boys' weights, $S_G{}^2$ and $S_B{}^2$, say:

$$S_G{}^2 = 5^2 = 25 \text{ kg}^2 \text{ and } S_B{}^2 = 4^2 = 16 \text{ kg}^2$$

(Notice that variances are measured in square units.)

If we let the means of the girls' and boys' weights be $\bar{x}_G$ and $\bar{x}_B$, we also have:

$$\bar{x}_G = 55 \qquad \text{and } \bar{x}_B = 47$$

Finally, let n_G and n_B denote the number of girls and boys, respectively; then $n_G = 50$ and $n_B = 30$.

Now, we wish to calculate $S_T{}^2$; using the new formula above, we have:

$$S_T{}^2 = \frac{\Sigma_T x^2}{n_T} - \bar{x}_T{}^2$$

where n_T is the total number of children,, and $\Sigma_T x^2$ is the 'sum of the squares' of all their weights.

Since we know n_T and $\bar{x}_T{}^2$, we have:

$$S_T{}^2 = \frac{\Sigma_T x^2}{80} - (52)^2$$

Thus, all we need to find is $\Sigma_T x^2$.

Now, using the new formula for the girls' and boys' variances, we have:

$$S_G^{\ 2} = \frac{\Sigma_G x^2}{n_G} - \bar{x}_G^{\ 2} \text{ and } S_B^{\ 2} = \frac{\Sigma_B x^2}{n_B} - \bar{x}_B^{\ 2}$$

where $\Sigma_G x^2$ and $\Sigma_B x^2$ are the sums of squares of the girls' and boys' weights respectively.

Again, we know $S_G^{\ 2}$, n_G and $\bar{x}_G 2$, and $S_B^{\ 2}$, n_B and $\bar{x}_B^{\ 2}$, and so we can write:

$$25 = \frac{\Sigma_G x^2}{50} - (55)^2 \text{ and } 16 = \frac{\Sigma_B x^2}{30} - (47)^2$$

ie $\Sigma_G x^2 = 50 \times 25 + 50 \times 55^2$ and $\Sigma_B x^2 = 30 \times 16 + 30 \times 47^2$

$= 152{,}500$ $= 66{,}750$

We can now add $\Sigma_G x^2$ and $\Sigma_B x^2$ together to give us $\Sigma_T x^2$:

$$\Sigma_T x^2 = 152{,}500 + 66{,}750 = 219{,}250$$

Thus, $S_T^{\ 2} = \dfrac{219{,}250}{80} - (52)^2 = 36.625$

$\therefore$ $S_T = \sqrt{36.625} = \underline{6.05 \text{ kg}}$

Exercise 4

1. Calculate the range, semi-interquartile range, mean deviation from the mean, median and mode, variance, and standard deviation of the following sets of values:

   a) 2, 5, 6, 3, 1, 2, 8, 9, 4, 8, 3, 2, 5, 1

   b) 0.3, 0.35, 0.34, 0.45, 0.57, 0.20, 0.35, 0.22, 0.17

   c) 28.3, 39.4, 26.1, 27.0, 27.5, 29.2, 29.8, 28.3

   d) 141, 100, 106, 129, 152, 138, 133, 129, 125, 135.

2. Calculate the range, semi-interquartile range, mean deviation from the mean, median and mode, variance and standard deviation of the data in Exercise 2, No. 3.

3. a)-g) Use the cumulative frequency diagrams of Exercise 2, Nos. 10-16 to estimate the upper and lower quartiles and semi-interquartile range of the data.

4. Find the mean deviation of the following set of observations.

   5 8 11 2 4 6

5. Calculate the variance for the data below in which the variable is discrete.

Value of variable	1	2	3	4
Frequency	11	0	7	2

6. Calculate the mean deviation of

4 7 7 7 7 10

7. The examination marks obtained by 100 candidates are distributed as follows:

Mark	0-19	20-29	30-39	40-49	50-59	60-69	70-79	80-89
No. of candidates	8	7	14	23	26	12	6	4

Make use of an assumed mean to estimate the mean and the standard deviation. A further group of 200 candidates obtained a mean mark of 52.3 with a standard deviation of 14.5 marks. Calculate the mean and the standard deviation of the whole group of 300 candidates.

8. Calculate the standard deviation of the distribution below.

Variable	1	2	3	4
Frequency	1	2	3	4

9. An electronic device is made from three components, one each of type A, B and C. Components of each type are produced by several manufacturers, the prices being shown below.

Component	Type A	£3, £4 £4 £5
	Type B	£12, £14, £15
	Type C	£1, £1, £1, £2

Draw up a frequency table of the cost of the 48 possible ways of purchasing the necessary components. Use this table to obtain for this distribution

a) the mean,

b) the mean deviation from the mean,

c) the standard deviation.

10. Recorded below are the goals scored by 60 teams on a particular Saturday.

0	5	2	3	2	2	0	1	1	0
1	1	1	0	1	0	1	1	1	2
1	0	2	3	0	1	1	1	3	2
0	1	2	0	1	5	1	1	3	1
2	1	1	3	1	3	0	2	1	2
1	2	0	2	0	1	1	0	0	1

a) Tabulate these scores and state the type of frequency distribution to which they approximate.

b) Use your table in order to calculate the mean number of goals scored, the standard deviation, and the mean deviation from the mean.

c) A further 32 teams played on the same Saturday as the first 60 teams. The mean number of goals scored by all 92 teams was 1.25. Of the further 32 teams, calculate the number scoring two goals given that the remainder scored 27 goals between them.

11. Find the range of the following distribution.

No. of persons in family	1	2	3	4	5	6	9
No. of families	7	12	10	6	4	1	1

12. Calculate the variance of the following distribution.

Value of variable	4	6	8	10
Frequency	240	340	200	20

13. The table below gives the distribution of scores obtained when a die was thrown 40 times

Score	1	2	3	4	5	6
Frequency	5	8	7	6	4	10

If a series of fives is then thrown, how many would be required for the overall arithmetic mean to be 3.8?

14. The standard deviation of a set of 5 numbers is 7.0, and the standard deviation of another set of 15 numbers is 6.0. Given that the two sets of numbers have the same mean, calculate the standard deviation of the combined set of 20 numbers.

15. The following frequency table gives the heights in inches of 100 male students, classified in intervals of 2 in, with the corresponding class frequencies.

Height in inches	61-63	64-66	67-69	70-72	73-75	
Observed class frequencies	5	20	39	28	8	Total 100

Show that the mean and the standard deviation of the distribution are 68.42 in. and 2.97 in. respectively. Fit a normal curve to the above data. Compare the theoretical class frequencies with the observed values.

16. The Much Booking village football team, in 200 matches, had scores given in the following frequency table. The number of goals scored by the team in one match is x, and the frequency of each value of x is f.

x	0	1	2	3	4	5	6	7	8	9	10	11	12
f	38	35	28	21	18	19	17	15	5	1	2	0	1

Write down the modal score and find the arithmetic mean, median, variance and standard deviation of x.

17. Prove the formula $\Sigma(x - \bar{x})^2 = \Sigma x^2 - n\bar{x}^2$.

In a Middle School there are 253 girls whose ages have a mean 11.8 yr and a standard deviation 1.7 yr. There are also 312 boys whose ages have a mean 12.3 yr and a standard deviation 1.9 yr. Calculate the mean and standard deviation of the ages of all the 565 pupils.

18. In an agricultural experiment the gains in mass, in kilograms, of 100 pigs during a certain period were recorded as follows:

Gains in mass (kilograms)	5-9	10-14	15-19	20-24	25-29	30-34
Frequency	2	29	37	16	14	2

Construct a histogram and a relative cumulative frequency polygon of these data. Obtain

i) the median and the semi-interquartile range,

ii) the mean and the standard deviation.

Which of these pairs of statistics do you consider more appropriate in this case, and why?

19. The number of eggs laid per week by 25 hens was found to have a mean of 5.8 and standard deviation of 1.5. When the performance of a further 50 hens was calculated, it was found that the mean of all 75 hens was 6.3 with a standard deviation of 1.7. Calculate the mean and standard deviation of the number of eggs laid per week by these other 50 hens.

20. State which measure of dispersion is most appropriate for the data in Exercise 2, No. 3.

5 PROBABILITY

5.1 Introduction

This chapter will form the basis of the further development of the theory of statistics in the remaining chapters. Probability is also the most difficult topic to master and, as such, should be persevered with (to this end, a large number of examples are given at the end of the chapter).

Probability is concerned with *prediction*, an important part of our lives. Suppose we choose a card from an ordinary pack of 52 cards; what will it be? What is the likelihood that it will be an Ace, for instance? We know that there are 4 Aces in the pack of 52 cards, and that each of the 52 cards is *equally likely* to be chosen (this is an important assumption for the present). Thus, there is a *relatively* small chance that an Ace will be chosen. In fact, if we calculate the so-called 'relative frequency' of an Ace being drawn, as below, we see that this *number* reflects this small likelihood:

$$\text{relative frequency of an Ace} = \frac{\text{No. of Aces}}{\text{No. of cards in pack}} = \frac{4}{52} = \frac{1}{13}$$

In a similar manner, we can calculate the relative frequency of a Heart being drawn:

$$\text{relative frequency of a Heart} = \frac{\text{No. of Hearts}}{\text{No. of cards in pack}} = \frac{13}{52} = \frac{1}{4}$$

This is a much larger number* and reflects the far greater chance of a heart being drawn (since there are more of them!)

[*Notice that *all* relative frequencies are numbers lying between 0 and 1; it is for this reason that $\frac{1}{4}$ is much larger than $\frac{1}{13}$]

Thus, we can see that the relative frequency of an 'event' occuring (eg choosing a Heart; choosing an Ace) is given by:

$$\frac{\text{Number of different ways the event } can \text{ occur}}{\text{Total number of possible ways}}$$

For example, if the event is 'choosing an Ace', then there are 4 different ways that this event can occur, viz Ace of Spades, Ace of Hearts, Ace of Diamonds, Ace of Clubs.

We now have a method of calculating a number (the relative frequency) which reflects the likelihood of a given 'event' occuring, and hence, a way of *predicting* (numerically) what is likely to happen. Thus, for the time being, we make the following 'working' definition of the *probability* of a given event occuring:

> Probability of an event occuring = relative frequency of the event occuring
>
> $$= \frac{\text{No. of different ways the event can occur}}{\text{Total number of possible ways}}$$
>
> If there are r different ways of the event occuring out of a total possible number of n *equally likely* ways, then,
>
> Probability of the event occuring $= \frac{r}{n}$

Consider the following examples illustrating the use of this definition of probability.

Example:

What is the probability of throwing a 6 with a fair, six-sided die?

First, there is a total of six possible 'ways' of throwing the die, viz 1, 2, 3, 4, 5, 6, each of which is *equally likely* to occur since the die is *fair*. The *event* we are trying to calculate the probability of occuring is 'throwing a 6', which can only occur in one way, since there is only one face of the die with a 6 on it.

Thus, probability of a 6 occuring

$$= \frac{\text{No. of ways of throwing a 6}}{\text{Total possible number of ways of throwing a die}}$$

$$= \frac{1}{6}$$

Example:

Two fair coins are flipped.

What is the probability that one is heads and one is tails?

Now, if we let the two coins be A and B then the possible ways of flipping them are shown below:

A	B
H	H
H	T
T	H
T	T

Thus, we see that there are 4 possible ways of flipping two coins, each equally likely to occur, 2 of which 'satisfy' the event 'one heads, the other tails'. From the definition of probability we have:

Probability of one head and one tail $= \frac{2}{4} = \frac{1}{2}$

[*Note*: From now on, we will write P(Event) to mean the probability of the event occuring.]

Example:

Two fair, six-sided dice are tossed.

What is the probability that the sum of their uppermost faces will be 7?

Since each number of one die can occur with each number of the other, there are 36 equally likely ways of throwing two dice, eg (1, 1), (1, 2), ..., (2, 1), (2, 2), ..., (6, 5), (6, 6).

Of these, there are 6 ways which yield a total of 7, viz (2, 5), (5, 2), (3, 4), (4, 3), (1, 6), (6, 1).

Thus we have:

$$\text{P(Total of 7)} = \frac{6}{36} = \frac{1}{6}$$

Having introduced the idea of probability with these few examples, we can now develop the theory formally. We begin with a few definitions.

A *trial* is a repeatable act, eg 'flipping a coin', 'choosing a card from a pack'.

Any possible 'result' of a trial is called an *outcome*, eg the two 'outcomes' of the trial 'flipping a coin' are H and T. The set of *all* such outcomes is called the possibility space, and is denoted S; thus, for example, the *possibility space* for the trial 'tossing a die' is:

$$S = \{ 1, 2, 3, 4, 5, 6 \}$$

An *event*, usually denoted E, is any subset of the possibility space S. For example, for the above trial 'tossing a die', the event E = 'an odd number' is the set: $E = \{ 1, 3, 5 \}$. An event then, is simply a certain subset of possible *outcomes* which we are interested in. The *number* of outcomes in an event, E, is denoted n(E), just as in ordinary set theory (with which we assume the reader is conversant). Thus, for the above event, n(E) = 3. Also, in the above example, n(S) = 6.

We are now in a position to define formally the probability of an event, E, occuring.

> If the outcomes of a trial are equally likely to occur and form the possibility space S, then, the probability of the event E occuring is given by:
>
> $$P(E) = \frac{n(E)}{n(S)}$$

Thus, in the above example,

$$P(\text{an odd number}) = P(E) = \frac{n(E)}{n(S)} = \frac{3}{6} = \frac{1}{2}$$

We can see that this definition of probability is really the same as the definition given earlier, since n(E) is just the number of different ways that the event E can occur, and n(S) is the total number of possible ways.

We can illustrate events and possibility spaces using Venn diagrams. The above event, E, and possibility space, S, are depicted in the Venn diagram below.

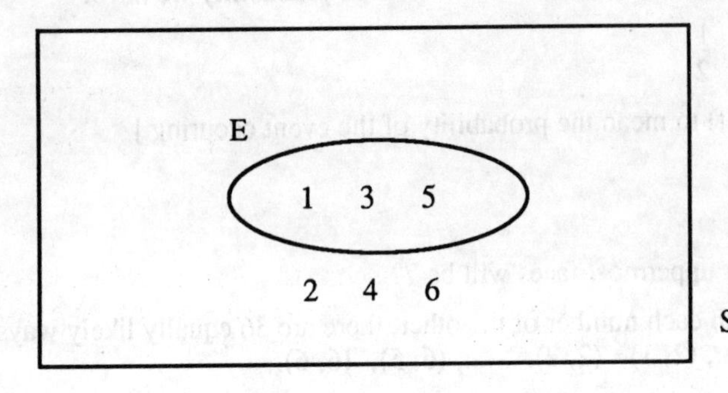

Example

A coin and a die are tossed together. Find the set of all possible outcomes. Illustrate the following events on a Venn diagram:

E_1 = 'coin shows a H and die shows an odd number'

E_2 = 'coin shows a H and die shows a number greater than 3'

E_3 = 'coin shows a T and die shows a 1 or 6'

Find $P(E_1)$ and $P(E_2)$ and $P(E_3)$.

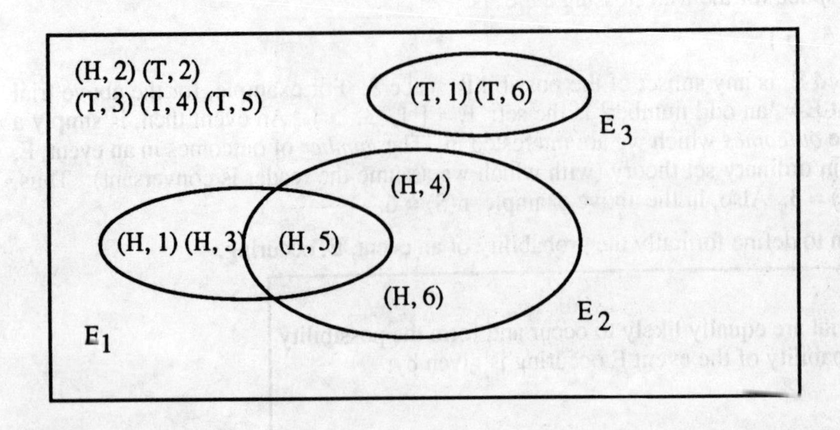

We see that: N(S) = 12, $n(E_1) = 3$, $n(E_2) = 3$, $n(E_3) = 2$.

Thus we have:

$$P(E_1) = \frac{n(E_1)}{n(S)} = \frac{3}{12} = \frac{1}{4}$$

$$P(E_2) = \frac{n(E_2)}{n(S)} = \frac{3}{12} = \frac{1}{4}$$

$$P(E_3) = \frac{N(E_3)}{n(S)} = \frac{2}{12} = \frac{1}{6}$$

Notes:

i) If there are *no* possible ways for an event, E, to occur, then obviously n(E) = 0, and so,

$P(E) = \frac{n(E)}{n(S)} = 0$; thus, zero probability indicates an *impossibility*, eg for the trial tossing an ordinary, fair, six-sided die the event E = 'die shows a 7' is impossible, since there are no outcomes in the possibility space

S = { 1, 2, 3, 4, 5, 6 } which satisfy this event, ie n(E) = 0 and so P(E) = 0.

ii) If *all* of the possible outcomes lead to a certain event, E, occuring, then P(E) = 1, ie a probability of 1 indicates an *absolute certainty* of the event occuring, eg for the trial 'tossing a die' the event 'uppermost face shows a number' is satisfied by *every* possible outcome, since a die can *only* show a number on its uppermost face (providing it is an ordinary die, of course),

thus, E = { 1, 2, 3, 4, 5, 6 } and so $P(E) = \frac{n(E)}{n(S)} = \frac{6}{6} = 1$

5.2 Counting methods

In order to be able to calculate more complicated probabilities, we must first consider ways of 'counting' up the numbers of possible outcomes in E and S, which are sometimes very large numbers (think of how many different 5 card hands can be dealt from a pack of 52, for instance).

Multiplication

We have in fact already used the following rule:

> If there are m outcomes of one trial, and n outcomes of another, then there are m x n outcomes when both are performed together.

This can be seen in the example below.

Example: A fair coin and a fair, six-sided die are tossed together.

How many possible outcomes are there?

There are 2 possible outcomes of tossing a coin; H, T.

There are 6 possible outcomes of tossing a die: 1, 2, 3, 4, 5, 6.

Thus, there are 2 x 6 = 12 outcomes of tossing both coin and die;

these are illustrated in the example above on page 76.

5.3 Permutations

Suppose we wish to *arrange* five *different* flowerpots on a windowsill. Beginning at the left hand space on the windowsill:

There are 5 flowerpots to choose from to fill the first space with. This leaves 4 flowerpots to choose from to fill the second space with. Continuing in this manner we have:

this leaves 3 flowerpots to choose from to fill the third space with,

this leaves 2 flowerpots to choose from to fill the fourth space with,

this leaves 1 flowerpot to fill the fifth space with.

Now, in filling the five spaces we are, in effect, performing five trials together; thus, from the multiplication rule above, we have:

Number of ways of arranging 5 flowerpots = 5 x 4 x 3 x 2 x 1.

We can write the product 5 x 4 x 3 x 2 x 1 using the so-called *factorial* notation of 5!

Thus, number of arrangements of the 5 flowerpots = 5!

Another name for an arrangement is *permutation*. In general, we have the rule below:

> The number of permutations of n different objects is n!,
>
> where n! = n(n-1). (n-2) ..... 3.2.1.

Now, suppose we have 9 flowerpots from which we must choose 5 to put on the windowsill; in a similar way to above, we have:

9 flowerpots to choose from to fill the first space with,

this leaves 8 flowerpots to choose from to fill the second space with,

this leaves 7 flowerpots to choose from to fill the third space with,

this leaves 6 flowerpots to choose from to fill the fourth space with,

this leaves 5 flowerpots to choose from to fill the fifth space with.

Thus, using the multiplication rule again, we see that:

No. of ways of arranging 5 flowerpots chosen from 9 flowerpots

$$= 9 \times 8 \times 7 \times 6 \times 5$$

We can write the product 9 x 8 x 7 x 6 x 5 more briefly using factorial notation again, ie

$$9 \times 8 \times 7 \times 6 \times 5 = \frac{9 \times 8 \times 7 \times 6 \times 5 \times 4 \times 3 \times 2 \times 1}{4 \times 3 \times 2 \times 1}$$

$$= \frac{9!}{4!}$$

$$= \frac{9!}{(9-5)!}$$

A further abbreviation for this number is ${}^{9}P_{5}$, ie ${}^{9}P_{5} = \frac{9!}{(9-5)!}$

Thus, we have:

No. of permutations of 5 flowerpots drawn from 9 flowerpots = ${}^{9}P_{5}$.

In general, we have the following rule:

> The number of permutations of r objects drawn from n different objects is ${}^{n}P_{r}$, where
>
> $${}^{n}P_{r} = \frac{n!}{(n-r)!}$$

Example: A postman has 15 different letters to deliver, and 10 letterboxes to deliver them to, only one letter being allowed to go into any one letterbox.

a) How many ways can all 10 letterboxes receive a letter?

b) How many ways can the letters be delivered such that the first letterbox remains empty?

c) How many ways can the letters be delivered such that a particular letter goes into a particular letterbox?

The 10 letterboxes are like the 5 spaces on the windowsill.

a) Thus, number of ways of delivering letters $= {}^{15}P_{10}$

$$= \frac{15!}{(15-10)!}$$

(using a calculator) $= \frac{15!}{5!}$

$= \underline{108972860000}$

b) Now, if the first letterbox does *not* receive a letter, this leaves 9 letterboxes to be delivered to; thus,

Number of ways of delivering letters $= {}^{15}P_{9}$

$$= \frac{15!}{(15-9)!}$$

$$= \frac{15!}{6!}$$

$= \underline{1816214400}$

c) Now, if one of the letters is delivered to one of the letterboxes, this leaves 14 letters to be delivered to 9 letterboxes, thus:

Number of ways of delivering letters $= {}^{14}P_9$

$= \frac{14!}{(14-9)!}$

$= \frac{14!}{5!}$

$= \underline{726485760}$

Suppose now that the n objects are *not* all *different*; for example, suppose we have 7 coloured balls, 3 of which are Red, the other 4 being Green, Black, White and Yellow. We could call these 7 balls:

R_1 R_2 R_3 G B W Y

where the subscripts 1, 2 and 3 on the Red balls indicate that they are, in fact, *different* balls.

Now, if we wish to arrange all 7 balls in a row, as above, clearly we will be unable to distinguish between the 3 Red balls, since they are identical; thus swapping, say, R_1 with R_2 has no visible effect on the arrangement. We know that there are 3! = 6 different arrangements of the 3 Red balls, each of which looks the same as the other. Thus, the 6 arrangements of the 7 balls below all *look the same.*

R_1 R_2 R_3 G B W Y

R_1 R_3 R_2 G B W Y

R_2 R_3 R_1 G B W Y

R_2 R_1 R_3 G B W Y

R_3 R_1 R_2 G B W Y

R_3 R_2 R_1 G B W Y

If we assumed that the 7 balls all looked *different* (imagine the numbers 1, 2 and 3 on the Red balls) then we know that there would be 7! permutations of them, including the above 6 permutations; also, for every 6 permutations like those above, we have one *apparent* permutation (because all the Red balls appear the same).

Thus, number of (apparently) different permutations $= \frac{7!}{6}$

But 6 = 3! and so,

number of (apparently) different permutations of

7 balls, 3 of which are identical, $= \frac{7!}{3!}$

In general, we have the result:

The number of permutations of n objects, s of which are the same, is $\frac{n!}{s!}$

Furthermore, by a similar argument, it can be shown that:

> The number of permutations of n objects, s of which are of one kind, t of which are another kind, etc, is:
>
> $$\frac{n!}{s!\ t!\ \ldots\ldots}$$

This latter result is illustrated in the following example.

Example: A bag contains 3 orange, 2 lemon, 5 blackcurrant and 4 lime tasting sweets. How many different ways could I eat them, one at a time?

Here, there are 3 + 2 + 5 + 4 = 14 sweets, 3 of which are of one kind (orange), 2 of which are another kind (lemon), etc. Thus, from the above result we have:

$$\text{No. of different ways of eating sweets} = \frac{14!}{3!\ 2!\ 5!\ 4!}$$

$$= \underline{1522520}$$

5.4 Combinations

Now, suppose there are 5 coloured balls: A, B, C, D and E, in a bag, and we draw out 3 of them (either one at a time or all at once, it makes no difference); we will have a set of 3 balls, say A, B, E, which we will call a *combination* (of 3 balls). Notice that we are not concerned with the *order* in which they were selected, but just in *which* 3 balls were selected - this is what differentiates a *combination* from a *permutation*, where order (arrangement) *is* important. How many different combinations can be drawn?

In order to answer this question, let us firstly *assume* that the order in which the 3 balls are selected *is* important; then, from the above rule, we have:

No. of permutations of 3 balls drawn from 5 balls = ${}^5P_3$.

Thus, there are ${}^5P_3$ different ways that 3 balls can be drawn from the bag, *taking the order* (permutation) *in which they were drawn into account*. For example, the above 3 balls A, B and E could be drawn in the following ways:

1st	2nd	3rd
A	B	E
A	E	B
B	A	E
B	E	A
E	B	A
E	A	B

In fact, this is simply the number of permutations (orderings) of the 3 balls A, B and E, which we know, from the rule on page 80 is 3!

If we now forget about the *order* in which the 3 balls are chosen, we see that all of the *6 permutations* above are equivalent to the *single combination* A, B, E, ie for every 6 permutations there is 1 combination.

But there are ${}^5P_3$ permutations altogether, therefore,

$$\text{No. of combinations of 3 balls drawn from 5} = \frac{{}^5P_3}{6}$$

Now, we calculated the number 6 by 3! using the rule on page 78; thus, we have:

$$\text{No. of combinations of 3 balls drawn from 5} = \frac{{}^5P_3}{3!}$$

$$= \frac{5!}{3!(5-3)!}$$

Thus number can be written using the so-called *Binomial Coefficient* notation: ${}^5C_3$

Thus, we have:

$$\text{No. of combinations of 3 balls drawn from 5} = {}^5C_3$$

In general, we have the rule:

> The number of combinations of r objects drawn from n different objects is ${}^nC_r$, where
>
> $${}^nC_r = \frac{n!}{r!(n-r)!}$$

We can now use the above counting methods in calculating further probabilities, as demonstrated in the following examples.

Example: A bag contains 7 differently coloured balls:

White, Black, Green, Yellow, Blue, Orange, Red.

a) If one ball is drawn from the bag what is the probability that it is (i) White, (ii) Green or Red;

b) If 3 balls are drawn from the bag, what is the probability that

i) the White ball is amongst them,

ii) the Green and Red balls are amongst them,

iii) the three balls are Red, Orange and Black;

c) Four balls are drawn out of the bag one at a time, without replacement. What is the probability that

i) the first is Green,

ii) the second and third are Red and Orange, respectively,

iii) the balls are Green, Red, White, Black in that order.

a) i) P(ball is White) $= \frac{\text{No. of White balls}}{\text{Total No. of balls}} = \frac{1}{7}$

ii) P(ball is Red or Green) $= \frac{\text{No. of Red and Green balls}}{\text{Total No. of balls}} = \frac{2}{7}$

b) Now, total number of ways (combinations) of drawing 3 balls from a bag of 7 is ${}^7C_3$.

i) If the White is one of the 3 balls, then the other 2 can be any of the remaining 6, and can therefore be chosen from these in ${}^6C_2$ ways. This means that there are ${}^6C_2$ ways; this means that there are ${}^6C_2$ possible selections that include White.

Thus,

P(White is one of them) $= \frac{\text{No. of selections including White}}{\text{Total No. of possible selections}}$

$$= \frac{{}^6C_2}{{}^7C_3} = 0.43$$

ii) If 2 of the 3 balls are Green and Red, then 3rd ball must be any one of the remaining 5 balls; this means that there are 5 possible selections of 3 balls containing Green and Red. Thus,

P(Green and Red included) $= \frac{5}{{}^7C_3} = \underline{0.143}$

c) There are ${}^7P_4$ different ways that 4 balls can be drawn from the bag of 7, if *order* is taken into considerations.

i) If the first is Green, the 2nd, 3rd and 4th balls must be chosen from the remaining 6 balls, which can be done in ${}^6P_3$ different ways, including order; this gives ${}^6P_3$ different ways of selecting the 4 balls such that the 1st is Green.

Thus, P(first is Green) $= \frac{{}^6P_3}{{}^7P_4} = \underline{0.143}$

ii) The 2nd and 3rd balls are Red and Orange, respectively, which means that the 1st and 4th can be any two (in order) of the remaining 5 balls, which can be chosen, therefore, in ${}^5P_2$ different ways. Thus:

P(2nd and 3rd are Red and Orange) $= \frac{{}^5P_2}{{}^7P_4} = 0.024$

iii) There is only one way that the 1st, 2nd, 3rd and 4th balls are Green, Red, White and Black, respectively.

Thus,

P(Green, Red, White and Black) $= \frac{1}{{}^7C_4} = \underline{\frac{1}{5040}}$

Example: Five ordinary, fair, six-sided dice are thrown together.

What is the probability that:

i) they all show the same number,

ii) they all show odd numbers,

iii) they total 10,

iv) the product of their uppermost faces is 12.

The total number of different ways of throwing the 5 dice is $6 \times 6 \times 6 \times 6 \times 6 = 6^5$, by the multiplication rule.

i) If they all show the same number, this number could be 1, 2, 3, 4, 5 or 6; thus, there are 6 ways that this could happen.

Therefore, $P(\text{all show same number}) = \frac{6}{6^5} = \frac{1}{6^4} = 6^{-4}$

ii) There are three odd numbers on each die, 1, 3, 5 and so the number of ways that all 5 dice can show odd numbers is $3 \times 3 \times 3 \times 3 \times 3 = 3^5$, by the multiplication rule.

Thus, $P(\text{all odd}) = \frac{3^5}{6^5} = \left(\frac{1}{2}\right)^5 = \underline{\frac{1}{32}}$

iii) A typical example of the 5 dice totalling 10 is (1, 1, 2, 3, 3), the *order* of the numbers being important, of course, since each position in the brackets represents a different die; thus, a different way they can total 10 is (1, 3, 2, 3, 1), even though the same digits are used. Thus, for each such *combination* totalling 10, there are 5! permutations giving 5! different ways the dice can total 10. All we have to do then is find all such combinations and multiply their number by 5!; the various combinations are listed below.

(1, 1, 1, 1, 6) (1, 1, 1, 2, 5) (1, 1, 1, 3, 4)
(1, 1, 2, 3, 3) (1, 2, 2, 2, 3) (2, 2, 2, 2, 2)

Thus, there are 6 x 5! ways that the dice total 10.

Hence, $P(\text{total of } 10) = \frac{6 \times 5!}{6^5} = \underline{0.015}$

iv) In a similar way, we must find all the combinations whose products are 12; these are listed below.

(1, 1, 1, 2, 6) (1, 1, 1, 3, 4) (1, 1, 2, 2, 3)

Thus, there are 3 x 5! ways of obtaining a product of 12.

Hence, $P(\text{product of } 12) = \frac{3 \times 5!}{6^5} = \underline{0.0077}$

Example: A machine producing electrical components occasionally makes faulty ones.

In a batch of 20 such components, 5 are found to be faulty. A sample of 3 of this batch is taken and examined. What is the probability that:

i) it contains no faulty components,

ii) it contains exactly one faulty component,

iii) exactly two components are faulty.

The number of different ways 3 components can be chosen from 20 is ${}^{20}C_3$.

i) If the sample contains no faulty components, then they must all come from the other 17 good components, which means that they can be any of the ${}^{17}C_3$ possible combinations.

$$\text{Thus, P(no faulty components)} = \frac{{}^{17}C_3}{{}^{20}C_3} = \underline{0.60}$$

ii) If it contains one faulty component, this could be any of the 5 in the sample. The other 2 components come from the remaining 17 good components, which can be any of ${}^{17}C_2$ combinations.

Thus, there are 5 x ${}^{17}C_2$ possible ways that the sample contains one faulty component.

$$\text{Therefore, P(one faulty component)} = \frac{5 \times {}^{17}C_2}{{}^{20}C_3} = \underline{0.596}$$

iii) If exactly two components are faulty, these two could come from the five faulty components in the batch, and could therefore be any of ${}^{5}C_2$ combinations. The other one could be any of the remaining 17 components. Thus, there are 17 x ${}^{5}C_2$ possible ways that the sample contains exactly two faulty components.

$$\text{Therefore, P(exactly two faulty)} = \frac{17 \times {}^{5}C_2}{{}^{20}C_3} = \underline{0.447}$$

5.5 Compound events

We now consider how to calculate the probabilities of events which are in themselves compounds of other events, eg the probability of 'throwing a 6 with a die' *and* 'flipping a head with a coin' at the same time.

5.6 Intersection of two events

Suppose A and B are two events, and S is the possibility space, as in the diagram below.

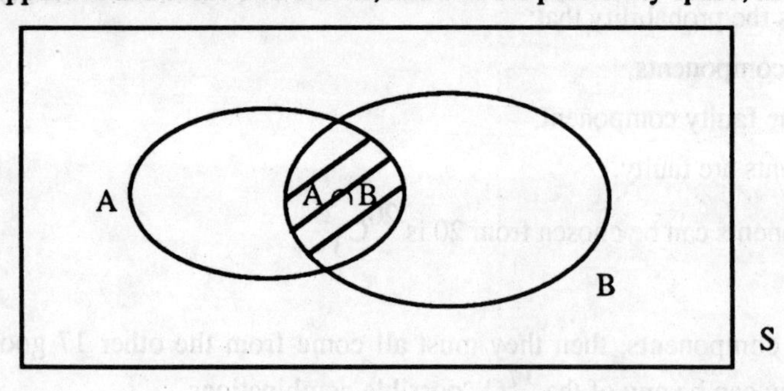

We define $A \cap B$ to be the intersection of the two events A and B, comprising all those outcomes occuring in *both* A and B.

For example, let S be the possibility space of the trial 'tossing two fair, six-sided dice', and A be the event 'uppermost faces total 7' and B be the event 'product of uppermost faces is 12'.

Then, A = { (1, 6), (6, 1), (2, 5), (5, 2), (3, 4), (4, 3) } and
B = { (6, 2), (2, 6), (3, 4), (4, 3) }

Thus, $A \cap B$ = { (3, 4), (4, 3) }, ie if one die is 3 and the other is 4 then their total will be 7 *and* their product will be 12; ie $A \cap B$ means 'a total of 7 *and* a product of 12'.

So we can read the event $A \cap B$ as meaning 'A *and* B', ie replace the $\cap$ sign by *and.*

Furthermore, the probability of $A \cap B$ is given by:

$$P(A \cap B) = \frac{n(A \cap B)}{n(S)}$$

In this case, $n(A \cap B) = 2$ and $n(S) = 36$;

thus, $P(A \cap B) = \frac{2}{36} = \frac{1}{18}$

We say that the two events A and B are *mutually exclusive* if $A \cap B = \varnothing$, ie if their intersection is empty, as illustrated in the Venn diagram below.

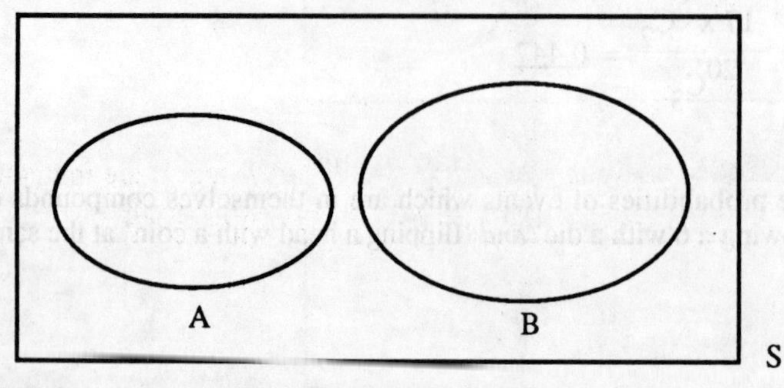

In this case, obviously $n(A \cap B) = 0$, and so $P(A \cap B) = 0$. An example of this are the two events A = 'a total of 2' and b = 'a product of 36' in the above possibility space S; clearly they have no intersection (there is no way that two dice can total 2 and have a product of 36). Thus, A and B are mutually exclusive, ie $A \cap B = \varnothing$. Hence, P(total of 2 *and* product of 36) = 0.

5.7 Union of two events

The *union* of the two events A and B is written $A \cup B$ and is the event comprising all of the outcomes in *both* A and B, and is illustrated in the diagram below.

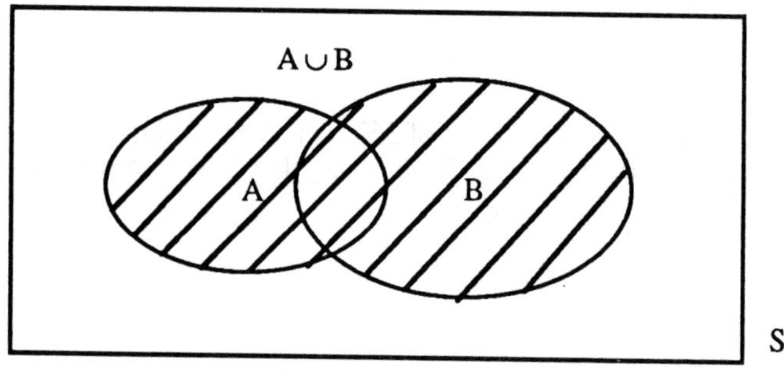

Using the same two original events, A and B, above we see that

$A \cup B = \{ (6, 1), (1,6), (6, 2), (2, 6), (3, 4), (4, 3), (2, 5), (5, 2) \}$

We see, therefore, that $A \cup B$, in this case, is satisfied by *either* a total of 7 *or* a product of 12.

Thus, $A \cup B$ means A *or* B, ie we can replace $\cup$ by *or*.

Now, $P(A \cup B) = \dfrac{n(A \cup B)}{n(S)}$. But, from a simple result of set theory, we know that:

$$n(A \cup B) = n(A) + n(B) - n(A \cap B)$$

Thus, $$P(A \cup B) = \frac{n(A) + n(B) - n(A \cap B)}{n(S)}$$

$$= \frac{n(A)}{n(S)} + \frac{n(B)}{n(S)} - \frac{n(A \cap B)}{n(S)}$$

$$= P(A) + P(B) - P(A \cap B)$$

Summarising this result, we have:

> If A and B are any two events, then,
>
> $P(A \cup B) = P(A) + P(B) - P(A \cap B)$

An immediate corollary of this is the so-called addition rule; if the two events A and B are mutually exclusive we have already seen above that $P(A \cap B) = 0$. Thus, we have the following result:

> Addition Rule: If two events A and B are mutually exclusive, then:
>
> $P(A \text{ or } B) = P(A \cup B) = P(A) + P(B)$

For example, if A and B are as above, then,

P(total of 7 *or* product of 12) = P(A *or* B)

$= P(A \cup B) = P(A) + P(B) - P(A \cap B)$

$= \frac{6}{36} + \frac{4}{36} - \frac{2}{36} = \frac{2}{9}$

However, if A is the event 'total of 2' and B is the event 'product of 36', then A = {(1, 1)} and B = {(6, 6)}; also, as mentioned above, A and B are mutually exclusive. Thus, using the above corollary:

P(total of 2 *or* product of 36) = P(A *or* B)

$= P(A \cup B)$

$= P(A) + P(B) = \frac{1}{36} + \frac{1}{36} = \frac{1}{18}$

We now consider an important corollary of the *addition rule* concerning events and their 'opposites'. Consider the event E = 'throwing a 6 with a fair, six-sided die'; the opposite of this event is '*not* throwing a 6', which we therefore call 'not E'. Clearly, E and 'not E' are mutually exclusive (you cannot throw a 6 and 'not a 6' at the same time) - this is illustrated in the diagram below.

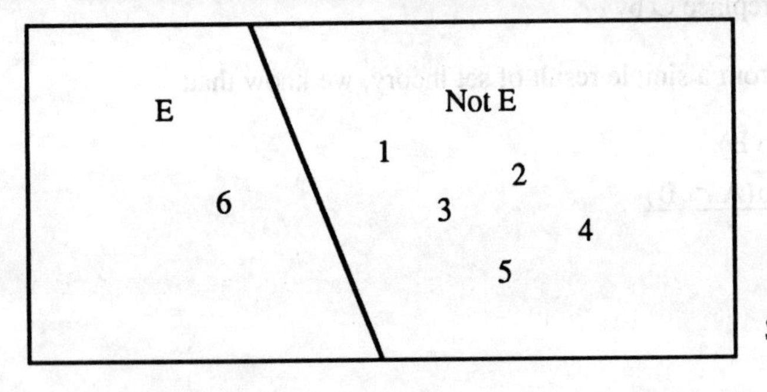

Furthermore, E ∪ 'not E' is simply S. Thus, using the addition rule above, we have:

$P(S) = P(E \cup \text{'not E'}) = P(E) + P(\text{not } E)$

But, $P(S) = \frac{n(S)}{n(S)} = 1$, and so:

$P(E) + P(\text{not } E) = 1$

or, as in more common, we have the result:

$$\boxed{P(E) = 1 - P(\text{not } E)}$$

This is a very useful result since it is often easier to calculate the probability of an event *not* occuring than it is to find the probability of it occuring. Consider the following examples.

Example: What is the probability of the uppermost faces of two ordinary fair, six-sided dice totalling 10 or less?

We *could* find all of the ways that two dice add up to 10 or less. However, it is much easier to find the probability of the opposite event, viz 'the dice total 11 or 12'; this can be done in 3 ways:

(6, 5), (5, 6), (6, 6).

Thus, $\quad P(\text{total of 11 or 12}) = \dfrac{3}{36} = \dfrac{1}{12}$

Here, we have E = 'total of 10 or less' not E = 'total of 11 or 12'

and so, using the above rule, we have:

$$P(\text{total of 10 or less}) = P(E) = 1 - P(\text{not } E)$$

$$= 1 - \frac{1}{12} = \underline{\frac{11}{12}}$$

Example: If a hand of 6 cards is dealt from an ordinary pack of 52 cards, what is the probability that it contains at least one king?

It is far easier to consider the opposite event 'contains no kings'; if the hand contains no kings then all 6 cards must come from the remaining 48 cards, which can be done in ${}^{48}C_6$ ways.

The total number of hands that can be dealt is ${}^{52}C_6$.

Thus, $P(\text{no kings}) = \dfrac{{}^{48}C_6}{{}^{52}C_6} = 0.603.$

Hence, using the above rules, we have:

$P(\text{at least one king}) = 1 - P(\text{no kings})$

$= 1 - 0.603 = \underline{0.397}$

A generalisation of the above result involves the concept of *exhaustive* events. The events A, B, C, ... are said to be *exhaustive* if their union equals the possibility space S, ie if

$$A \cup B \cup C \cup \; ..... = S$$

For example, suppose that we toss a fair, six-sided die, and define the events A = 'an odd number', B = 'a prime number', C = 'a multiple of 3', D = 'a power of 2'; these events are depicted in the Venn diagram below.

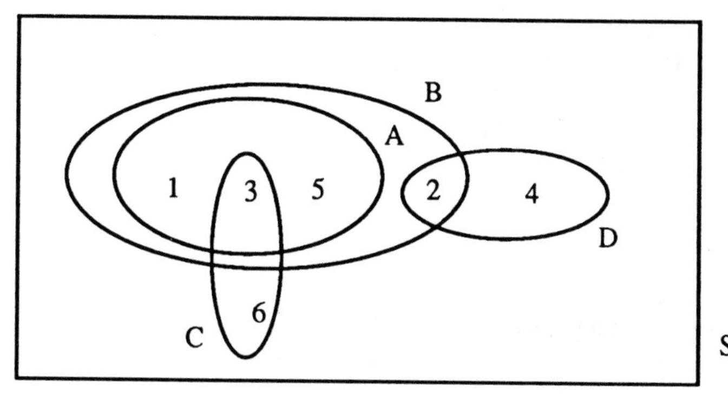

As can be seen, their union includes all of the outcomes of S, ie

$$A \cup B \cup C \cup D = \{1, 3, 5\} \cup \{1, 2, 3, 5\} \cup \{3, 6\} \cup \{2, 4\}$$

$$= \{1, 2, 3, 4, 5, 6\} = S$$

and so A, B, C and D are *exhaustive*.

Now, if these events are also *mutually exclusive* (see page 74), then, by the addition rule, we have:

$P(A \cup B \cup C \cup \ldots\ldots) = P(S)$

ie $P(A) + P(B) + P(C) + \ldots\ldots = P(S)$

But, $P(S) = \dfrac{n(S)}{n(S)} = 1$, and so we have the result:

> **If A, B, C, ... are mutually exclusive and exhaustive, then,**
>
> $P(A) + P(B) + P(C) + \ldots\ldots = 1$

For example, suppose we toss a fair, six-sided die again, and let A = 'an odd number', B = 'a power of 2', and C = 'a six'; these are depicted in the diagram below.

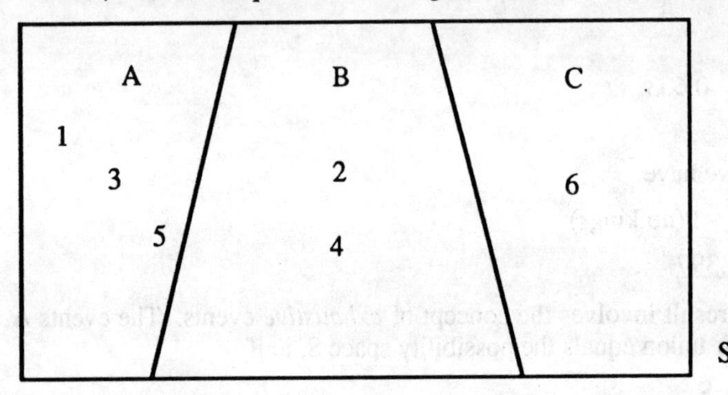

As can be seen, this time the event A, B and C are mutually exclusive (have no intersection) but are *also* exhaustive (because their union is S), ie

$A \cup B \cup C = \{1, 3, 5\} \cup \{2, 4\} \cup \{6\} = \{1, 2, 3, 4, 5, 6\} = S$

Also, $P(A \cup B \cup C) = P(A) + P(B) + P(C)$, by the addition rule,

$= \dfrac{3}{6} + \dfrac{2}{6} + \dfrac{1}{6} = \dfrac{6}{6} = 1 = P(S)$,

thus demonstrating the above rule.

5.8 Independent events

Suppose the following two events are performed together:

A = 'flip a head with a fair coin'

B = 'toss a 6 with a fair, six sided die'

It is obvious that the occurence of one event is *independent* of the other, since the two trials are physically separate from one another.

The following definition of the independence of two events embodies this intuitive 'feeling' of independence:

> The two events A and B are said to be independent if $P(A \cap B) = P(A)\,P(B)$.

Now, for the above two events, we have:

$P(A) = \frac{1}{2}$ and $P(B) = \frac{1}{6}$

Also, $P(A \cap B)$ = P(A and B)

= P(Head and a 6)

$= \frac{1}{12}$ (see Venn diagram on page 76)

Thus, we see that:

$$P(A)\,P(B) = \frac{1}{2} \times \frac{1}{6} = \frac{1}{12} = P(A \cap B)$$

and so the definition is consistent with the intuitive independence of A and B.

Example: A card is drawn from an ordinary pack of 52 cards.

Are the following three events independent of each other?

A = 'choosing a Heart'

B = 'choosing a King'

C = 'choosing the King of Clubs'

Now, P(Heart) $= \frac{13}{52} = \frac{1}{4}$ and P(King) $= \frac{4}{52} = \frac{1}{13}$ and P(King of Clubs) $= \frac{1}{52}$

Also, the event $A \cap B$ = A and B = 'choosing a King *and* it being a Heart'

= 'choosing the King of Hearts'

Thus, $P(A \cap B) = \frac{1}{52}$

Now, $P(A)\,P(B) = \frac{1}{4} \times \frac{1}{13} = \frac{1}{52} = P(A \cap B)$; thus, A and B are independent.

Also, $B \cap C$ = B and C = 'choosing a King *and* it being the King of Clubs'

= 'choosing the King of Clubs'

Thus, $P(B \cap C) = \frac{1}{52}$

Now, $P(B)\,P(C) = \frac{1}{13} \times \frac{1}{52} \neq \frac{1}{52} = P(B \cap C)$

Hence, B and C are *not* independent. (This stands to reason since choosing the King of Clubs certainly implies choosing a King.)

Finally, $A \cap C$ = A and C = 'choosing a Heart *and* the King of Clubs'.

This is clearly impossible ie A and C are mutually exclusive; thus, $P(A \cap C) = O$.

However, $P(A)\,P(C) = \frac{1}{4} \times \frac{1}{52} \neq O = P(A \cap C)$

Thus, A and C are *not* independent, simply because they *cannot* occur together.

5.9 More than two events

We can generalise the addition rule and the multiplication of independent events to more than just two events, as stated below:

> If A, B, C, ... are mutually exclusive events, then,
>
> P(A *or* B *or* C *or* ...) = $P(A \cup B \cup C \cup \ldots) = P(A) + P(B) + P(C) + \ldots$

and

> If A, B, C, ... are mutually independent then we have:
>
> P(A and B and C ....) = $P(A \cap B \cap C \cap \ldots)$
>
> $= P(A) \times P(B) \times P(C) \ldots$

Example: One card is dealt from the top of three well shuffled ordinary packs of 52 cards in order.

What is the probability that:

a) the three cards are a Heart, King and a Club, respectively;

b) two cards are Clubs or two cards are King of Diamonds or two cards are Hearts;

c) first card is the King of Spades and the other two are Diamonds or all three are below 10 (Ace counts as 1);

d) at least two are spades or all three are fives or at least two are Hearts.

Since the three cards are dealt from three *separate* packs, they are all *independent* of each other (so we can multiply probabilities as above). Also, the total number of ways that three cards can be dealt from three separate packs is 52 x 52 x 52, by the multiplication rule above.

a) P(Heart and King and Club) = P(Heart) P(King) P(Club)

$$= \frac{13}{52} \cdot \frac{4}{52} \cdot \frac{13}{52} \cdot = \frac{1}{208}$$

since all three events are independent of each other.

b) P(Two Clubs *or* Two Kings of Diamonds *or* Two Hearts)

= P(Two Clubs) + P(Two King of Diamonds) + P(Two Hearts), by the addition rule above,

since the three events are mutually exclusive.

Now, if two of the three cards are Clubs, the other can be any of the 39 non-Clubs in that pack. There are $^3C_2 = 3$ ways that the two Clubs come from two of the three packs. Finally, each of these two packs has 13 Clubs. Thus, the number of ways that the three cards can contain two clubs is 3 x 13 x 13 x 39; and so,

$$\text{P(Two Clubs)} = \frac{3 \times 13 \times 13 \times 39}{52 \times 52 \times 52}.$$

The other two probabilities are calculated in a similar manner; thus, the required probability

$$= \frac{3 \times 39 \times 13 \times 13}{52 \times 52 \times 52} + \frac{3 \times 51 \times 1 \times 1}{52 \times 52 \times 52} + \frac{3 \times 39 \times 13 \times 13}{52 \times 52 \times 52}$$

= 0.141 + 0.001 + 0.141 = 0.283

c) P(1st is King of Spades *and* Two Diamonds *or* All three below 10)

= P(lst is King of Spades *and* Two Diamonds) + P(All three below 10)

since these two events are mutually exclusive,

= P(1st is King of Spades *and* 2nd is a Diamd *and* 3rd is a Diamond) +

P(1st is under 10 *and* 2nd is under 10 *and* 3rd is under 10)

= P(1st is King of Spades). $[\text{P(Diamond)}]^2 + [\text{P(Under 10)}]^3$

$$= \frac{4}{52} \cdot \frac{13}{52} \cdot \frac{13}{52} + \frac{36}{52} \cdot \frac{36}{52} \cdot \frac{36}{52} = 0.34$$

d) P(At least two Spades *or* All Fives *or* At least two Hearts)

= P(At least two Spades) + P(All Fives) + P(At least two Hearts),

since all three events are mutually exclusive.

Now, P(At least two Spades) = 1 - P(No Spades) - P(One Spade)

$$= 1 - \left(\frac{39}{52}\right)^3 - 3\left(\frac{39}{52}\right)\left(\frac{13}{52}\right)^2 = \underline{0.292}$$

and P(All Fives) = $\left(\frac{4}{52}\right)^3 = 0.0005$

and P(At least two Hearts) = P(two *or* three Hearts)

= P(two Hearts) + P(three Hearts)

$$= \frac{3 \times 13 \times 13}{52 \times 52 \times 52} + \left(\frac{13}{52}\right)^3$$

= 0.0192

Thus, required probability = 0.297 + 0.0005 + 0.0192

= 0.317

More applications of the above two results can be seen at the end of the chapter under the section headed Tree diagrams.

5.10 Conditional probability

Suppose a bag contains 20 sweets, of which 7 are toffees wrapped in green paper, 4 are barley sugar wrapped in red paper, 3 are toffees wrapped in red paper, and 6 are barley sugar wrapped in green paper. Now, suppose a sweet is taken from the bag.

Let G be the event 'it has a green wrapper', and R be the event 'has a red wrapper' and B be the event 'is a barley sugar' and T be the event 'it is a toffee'.

The Venn diagram below illustrates these events.

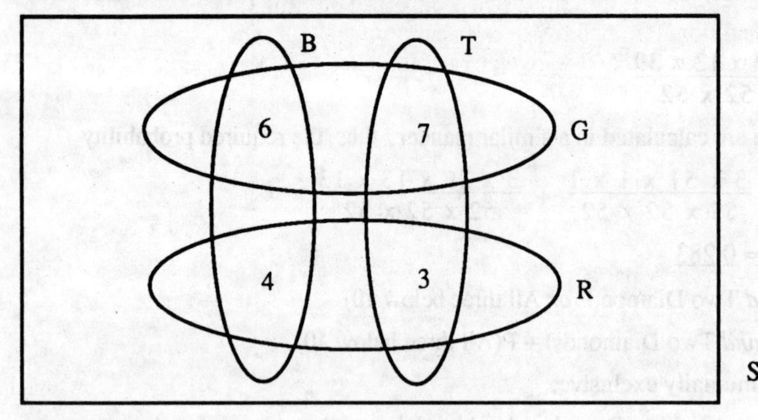

The probabilities of these two events occuring are:

$$P(G) = \frac{13}{20} \quad \text{and } P(T) = \frac{10}{20}.$$

Now, suppose we take out a sweet and see that it has a green wrapper; what is the probability that a toffee is inside? We are *given* the information that the sweet has a green wrapper which we can use to calculate the probability that a toffee is inside. Since we know it has a green wrapper, we know it is one of the 13 green wrapped sweets, thus, we can restrict our attention to this set of sweets, ie the set G becomes our *new* possibility space.

$$\text{Thus,} \quad P(\text{Toffee}) = \frac{\text{No. of toffees in G}}{\text{Total No. of sweets in G}} = \frac{7}{13}$$

We call this probability a *conditional probability*, since its calculation is conditional upon the sweet having a green wrapper; we use the notation $P(T|G)$ to denote this, which reads 'the probability of T given that G has occurred'. Thus, we have:

$$P(T|G) = \frac{7}{13}$$

Now, dividing top and bottom of this fraction by 20 (the total number of sweets), we have:

$$P(T|G) = \frac{7/20}{13/20}$$

But, $P(T \cap G) = \frac{7}{20}$ (see Venn diagram above)

and $P(G) = \frac{13}{20}$, and so we have:

$$P(T|G) = \frac{P(T \cap G)}{P(G)}$$

This is a general formula for calculating conditional probabilities, and is stated below:

> If A and B are any two events, then, the conditional probability that A occurs given that B has occured is given by the formula:
>
> $$P(A \mid B) = \frac{P(A \cap B)}{P(B)}$$

An immediate corollary of this formula follows if A and B are independent events; this implies that $P(A \cap B) = P(A) \cdot P(B)$, by definition. Thus, the above formula becomes:

$$P(A \mid B) = \frac{P(A) \cdot P(B)}{P(B)} = P(A)$$

This says that the probability of A *given that B has occured* is the *same* as the probability of A without being given this information, which is exactly what you would expect if A and B are independent of (ie do not affect) each other.

The following example shows how the above formula can also be used to calculate $P(A \cap B)$.

Example: Suppose the probability of it raining on a day in September is $\frac{1}{4}$; also if there is a red sky in the morning, suppose the probability of it raining is $\frac{1}{3}$. What is the probability that, on a typical day in September, there is a red sky and it rains that day?

Here, A = 'it rains', B = 'red sky'.

Also, $P(A) = \frac{1}{4}$ and $P(A \mid B) = \frac{1}{3}$; thus, we have:

$$\frac{1}{3} = \frac{P(A \cap B)}{\frac{1}{4}}$$

Hence, P(red sky *and* raining) $= P(A \cap B) = \frac{1}{3} \times \frac{1}{4} = \underline{\frac{1}{12}}$

This can be generalised into the following multiplication rule:

> <u>Multiplication Rule</u>
>
> If A and B are any two events then:
>
> $P(A \text{ and } B) = P(A \cap B) = P(A \mid B) \cdot P(B)$

5.11 Tree diagrams

These are sometime useful in calculating conditional probabilities, and make use of the multiplication rule. The diagram below illustrates the probabilities in the sweets example.

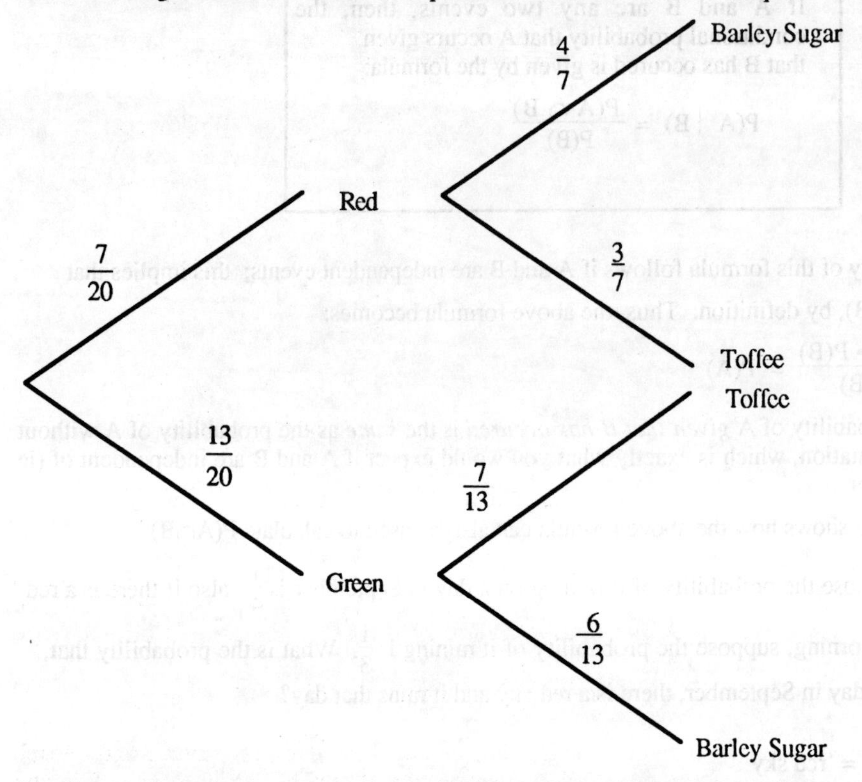

The first two branches represent the two different colours the sweet wrappers can be with their associated probabilities of occuring. The second set of branches represent the *conditional* probabilities of the sweet being a toffee or a barley sugar, *given* the colour of paper it is wrapped in (we calculated one of these above). Notice that the probabilities of either set of branches total 1; this is because one of each pair of branches must occur, ie the two events they represent are opposites. This is a good way of calculating the probabilities of the branches, eg knowing the conditional probability of a toffee given that it has a green wrapper is $\frac{7}{13}$ (as we calculated above), we know that:

$$P(\text{Barley Sugar} \mid G) = 1 - \frac{7}{13} = \frac{6}{13}$$

since the opposite of it being a toffee is it being a barley sugar.

Example: A bag contains 9 balls: 2 White, 3 Green and 4 Yellow.

Two balls are drawn, one after the other, *without* replacement.

Draw a tree diagram to represent the possible choices.

Calculate the probability of:

a) the 1st being White, and the 2nd Yellow;

b) both being Yellow;

c) the 2nd ball is Green.

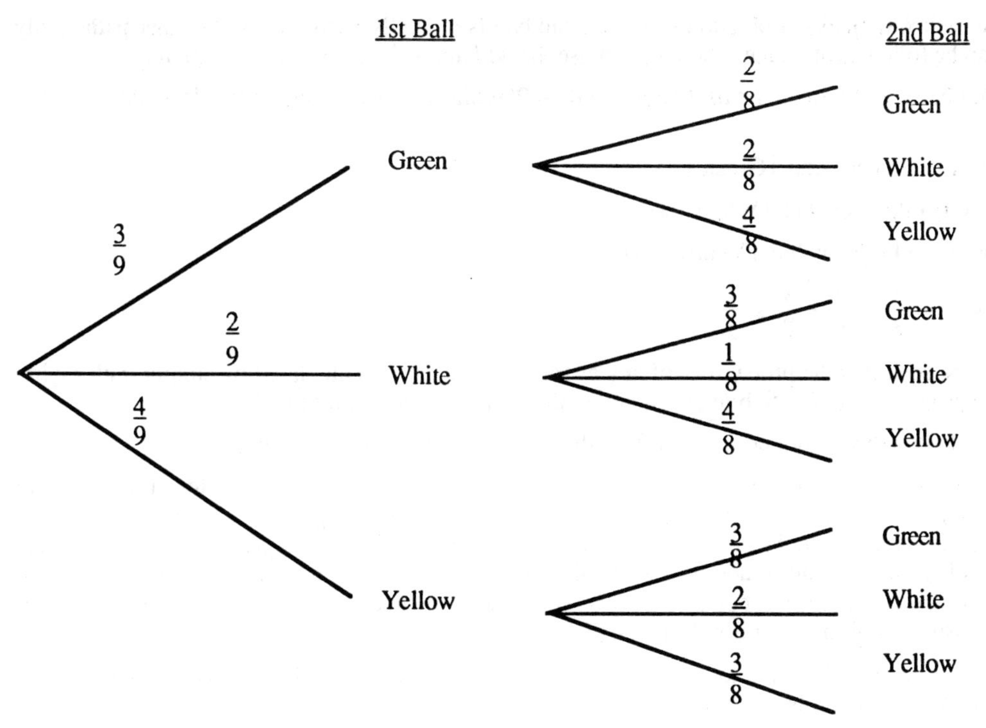

The probabilities for the 1st ball are:

$$P(W) = \frac{2}{9}; \qquad P(G) = \frac{3}{9} = \frac{1}{3}; \qquad P(Y) = \frac{4}{9}.$$

After the 1st ball is taken out, there remains 8 balls; the colours of these are conditional upon which one was taken out. Thus, *if* the 1st ball was Yellow, then there will only be 3 Yellow balls left, and so:

$$P(W) = \frac{2}{8} = \frac{1}{4}; \qquad P(G) = \frac{3}{8}; \qquad P(Y) = \frac{3}{8}.$$

These and the other conditional probabilities for the 2nd ball are shown in the tree diagram above.

a) From the above multiplication rule, we have:

P(1st White and 2nd Yellow) = P(1st White) • P(2nd Yellow | 1st White)

$$= \frac{2}{9} \cdot \frac{4}{8} = \underline{\frac{1}{9}}$$

b) P(Both Yellow) = P(1st Yellow *and* 2nd Yellow)

= P(1st Yellow) • P(2nd Yellow | 1st Yellow)

$$= \frac{4}{9} \cdot \frac{3}{8} = \underline{\frac{1}{6}}$$

c) Now, the 2nd ball can be Green, following a Green, White or Yellow 1st ball; the tree diagram shows these three 'paths' leading to a Green 2nd ball (pointed to by arrows).

Thus, P(Green 2nd ball) = P(Green 1st *and* Green 2nd *or* White 1st *and* Green 2nd *or* Yellow 1st *and* Green 2nd).

Now, since each way (path) of obtaining a Green 2nd ball is mutually exclusive of the other paths (only one path can be followed at a time), then we can use the *addition rule* to replace *or* by giving:

P(Green 2nd ball) = P(Green 1st *and* Green 2nd) + P(White 1st *and* Green 2nd) + P(Yellow 1st *and* Green 2nd)

= P(Green 1st) • P(Green 2nd | Green 1st)

+ P(White 1st) • P(Green 2nd | White 1st)

+ P(Yellow 1st) • P(Green 2nd | Yellow 1st)

$= \frac{3}{9} \bullet \frac{2}{8} + \frac{2}{9} \bullet \frac{3}{8} + \frac{4}{9} \bullet \frac{3}{8} = \frac{1}{3}$

Thus, it can be seen that the probability of any event occuring can be calculated by simply multiplying together the probabilities of each branch leading to the event and then summing these.

The example below shows further uses of a Tree diagram in a more complicated example.

Example: An oil company instructs a seismic exploration team to make soundings for oil in a certain area on the basis of past findings, the probability of oil being discovered is 0.2. If oil is discovered, it could be of three grades: high, medium or low. The probability of high grade is 0.2 and medium grade is 0.5. If oil is *not* discovered in the seismic survey, a second survey is carried out, but with only a 0.05 chance of success this time. If both surveys are unsuccessful then no oil will be found.

Draw a tree diagram to represent the above probabilities, filling in any missing ones. Calculate the probability of:

a) first survey being successful with high grade subsequently found;

b) finding no oil;

c) finding medium grade oil after a second survey;

d) finding low grade oil.

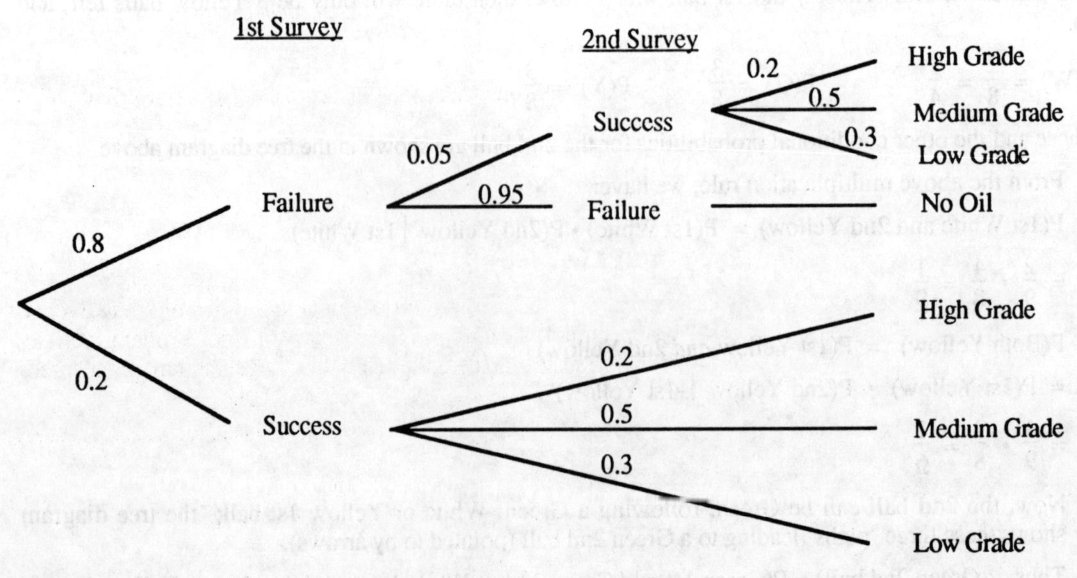

The above tree diagram shows the possible sequences of events. Firstly, the probability of finding low grade oil after a successful survey is 1 - 0.5 - 0.2 = 0.3, since the events: high grade, medium grade and low grade are mutually exhaustive (ie one must occur), and so their probabilities must add up to 1 (see page 88). Also, the probability of the 1st survey being unsuccessful is 1 - 0.2 = 0.8, and the 2nd survey being unsuccessful is 1 - 0.05 = 0.95.

These probabilities are shown on the diagram above.

a) P(1st survey a success *and* high grade found)

= P(1st survey success) • P(high grade 1st survey success) (by the multiplication rule)

= 0.2 x 0.2 = 0.04

ie multiply the probabilities along the path whose branches are '1st survey a success' and 'high grade oil found'.

b) P(finding no oil) = P(1st survey unsuccessful *and* 2nd survey unsuccessful)

= P(1st survey unsuccessful) x P(2nd survey unsuccessful)

= 0.8 x 0.95 = 0.760

Notice that the event '2nd survey unsuccessful' is conditional upon the 1st survey being unsuccessful.

c) P(medium grade found after 2nd survey) = P(medium grade *and* 2nd survey successful)

= 0.5

We are *given* that the 2nd survey was necessary, and so do not need to consider the 1st survey (which has *already happened*).

d) Now, low grade oil can be found in two ways (ie is at the end of two paths): either after a successful first survey or after a successful second survey.

Thus P(low grade oil found)

= P(successful 1st survey *and* low grade found *or* unsuccessful 1st survey *and* successful 2nd survey and low grade found)

= P(successful 1st survey) • P(low grade found) + P(unsuccessful 1st survey) • P(successful 2nd survey) • P(low grade found)

= 0.2 x 0.3 + 0.8 x 0.05 x 0.3 = 0.072

Notice the use of the words *and* and *or* ultimately producing the x and +.

5.12 Posterior probabilities

So far, the discussion has only been concerned with the assessment of the probabilities of some events occuring before they actually happen. Another important use is in the revision of Prior Probabilities (ie assessed before the event) into Posterior Probabilities (ie assessed in the light of an event having occured or deemed to have occured).

In general,

$$\text{posterior probability} = \frac{\text{Joint probability}}{\text{Marginal probability}}$$

Example: On the basis of past experience a manufacturer has found that a machine which automatically produces a given component is 'set up' correctly 90% of the time. If the machine is set up correctly then 90% good parts are expected. On the other hand, if the machine is not 'set up' correctly then the probability of a good part is only 30%.

The machine is 'set up', started, and the first part is inspected and found to be good. What is the probability that the machine is 'set up' correctly at this point in time?

Let:

1. Prob (correct set-up) = P(C) = .90
2. Prob (incorrect set-up) = $P(C_1)$ = .10
3. Prob (good part if machine is correctly set-up) = P(P(G/C) = .90
4. Prob (defective part if machine is correctly set-up) = P(D/C) = .10
5. Prob (good part if machine is incorrectly set-up) = $P(D/C_1)$ = .30
6. Prob (defective part if machine is incorrectly set-up) = $P(D/C_1)$ = .70

Tree diagram:

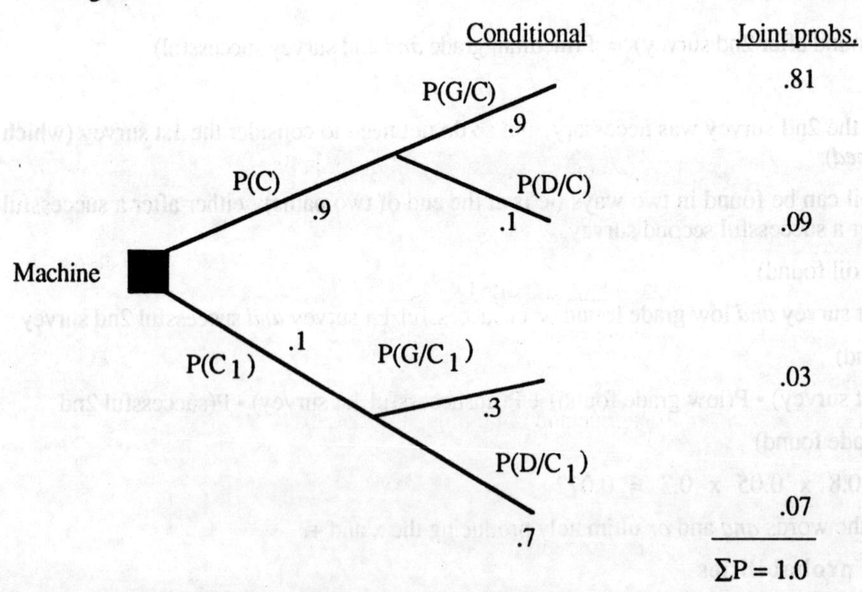

The posterior probability required is P(C/G) and is obtained from:

$$P(C/G) = \frac{P(C) \times P(G/C)}{P(G)}$$

$$\text{ie} = \frac{\text{Joint}}{\text{Marginal}}$$

$$= \frac{.9 \times .9}{(.9 \times .9) + (.1 \times .3)}$$

$$= \frac{.81}{.84}$$

$$= .96$$

Exercise 5

1. Write down all the possible outcomes of the following trials:
   a) throwing a fair, six-sided die;
   b) flipping a coin;
   c) flipping a coin twice;
   d) throwing two fair, six-sided dice;
   e) choosing a digit at random;
   f) choosing a ball from a bag containing 2 red and 5 yellow balls;
   g) choosing two balls from a bag containing 1 white and 2 black balls.

2. Illustrate the following events on a Venn diagram:
   a) A = 'throwing a 5 or a 6 with a fair, six-sided die';
   b) A = 'choosing a heart from an ordinary pack of 52 cards';
   c) A = 'flipping a head on a coin';
   d) A = 'throwing two fair, six-sided dice and obtaining a total of 7;
   e) A = 'choosing a card less than 6'
      B = 'choosing a heart'
      from an ordinary pack of 52 cards (Ace low);
   f) A = 'throwing a 5'
      B = 'throwing an even number'
      with a fair, six-sided die.
   g) A = 'throwing a total of 5'
      B = 'throwing a product of 6'
      with two fair, six-sided dice.

3. Find the probabilities of each of the above events in number 2. Also, in parts f) and g) find $P(A \cap B)$, $P(A \cup B)$, $P(A')$, $P(B')$, $P(A \cap B')$, $P(A' \cap B)$.

4. Two fair, six-sided dice are thrown together. If A is the event 'their total is even', B is the event 'their product is odd' and C is the event 'the highest number showing is 4', calculate the following probabilities.

i) $P(A \cap B)$;	ii) $P(A \cup B)$;	iii) $P(A \cap C)$;
iv) $P(A \cup C)$;	v) $P(C \cap B)$;	vi) $P(C \cup B)$;
vii) $P(A \cup B \cup C)$;	viii) $P(A')$;	ix) P (not B);
x) $P(A \cap B \cap C)$;	xi) $P(A \cap B \cup C)$;	xii) $P(A' \cap B)$;
xiii) $P(A' \cap B' \cup C')$;	xiv) $P(C' \cap A')$	xv) $P(A \mid B)$;
xvi) $P(B \mid A)$;	xvii) $P(A \mid C)$;	xviii) $P(C \mid B)$;
xix) $P(B \mid C)$;	xx) $P(A \mid B')$;	xxi) $P(A' \mid B')$;
xxii) $P(A' \mid C')$;	xxiii) $P(A \mid C')$;	xxiv) $P(B \mid C')$;
xxv) $P(C \mid B')$.		

5. In question 4 above, which of the following statements are true. Give reasons.
   i) A and B are independent
   ii) A and B are mutually exclusive
   iii) A, B and C are exhaustive
   iv) B and C are independent
   v) A and C are independent
   vi) A and C are mutually exclusive
   vii) A, B and C' are exhaustive.

6. A bag contains 2 green, 2 white, 1 black and 4 yellow balls; one ball is drawn from the bag. What is the probability that:
   a) the ball is black or white
   b) the ball is white or yellow

   Are the two events in a) and b) independent?

7. How many different arrangements can be made using the letters of the word TEMPERATURE?

8. How many different ways can 5 people be seated at a round table with 5 chairs? If another chair is added, how many ways can they now be seated. If two of the 5 people wish to sit next to each other, how many ways can this be arranged? If there are 3 men and 2 women, what is the probability that two women sit together?

9. Five poker dice have 6 faces representing the following cards of an ordinary pack: 9, 10, J, Q, K, Ace, each of the *same* suit. If all five dice are thrown together, what is the probability of obtaining:

   a) three Queens;

   b) all five 'higher than J' (Ace is high);

   c) a 'straight', ie a run of 5 consecutive cards, eg 10, J, Q, K, Ace (highest card of a straight can be an Ace);

   d) two pairs, eg 9, 9, Q, Q;

   e) a 'full house', ie a pair and a treble eg J, J, J, Ace, Ace?

   If the five dice are thrown, and two Aces appear what is the probability that, on throwing the remaining three dice:

   f) another Ace shows;

   g) a full house is obtained?

10. A box contains a large number of three different sizes of washers in the ratio 2 : 5 : 3 of large, medium, small washers. What is the probability of picking a medium size washer out of the box in one go. What is the probability that three washers need to be withdrawn in order to find a medium size washer. If there are 50 washers in the box and a sample of 10 washers is removed, what is the probability that:

   a) they are all of one kind;

   b) there are 5 medium and 5 large;

   c) there are three times as many small as there are large washers?

11. The probability of it raining in a day in September is 0.2. If it has rained the day before, the probability of it raining is 0.15. Draw a tree diagram to illustrate the weather over three consecutive days. What is the probability that:

   a) it rains on three consecutive days;

   b) it rains on only one of the three days.

   If it has rained on one of the first two days, what is the probability of it raining on the third day?

12. How many different arrangements are there of all the letters of the word SYNONYM?

   Counting Y as a vowel, how many of the arrangements have consonants and vowels alternately?

   If an arrangement is taken at random, what is the probability that the third letter is an N?

   If we are told that the arrangement has consonants and vowels alternatively and the first letter is an S, what is the probability that the third letter is an N?

13. Four boys are taken at random. What is the probability that they were

   i) all born on the same day of the week,

   ii) all born on different days of the week?

   It may be assumed that births are equally likely on the seven days of the week.

14. In a quiz programme, there are five competitors, each of whom makes a random choice out of ten prizes, three of which are booby prizes. No two competitors can have the same prize. What is the probability

    i) that none of the booby prizes is selected;

    ii) that all three booby prizes are chosen?

15. a) A typist types four letters and four corresponding envelopes. She then puts one letter into each envelope. Find the probability that every letter is in the wrong envelope.

    b) Find the number of different ways of choosing six letters from the word DISPIRITEDLY, each selection to contain at least one letter of the word PREY. No account is taken of the order in a selection of six. What is the probability that a random selection of six letters will contain at least one letter of the word PREY?

16. A box contains 20 blue and 30 red counters. One counter is taken at random, and then a second. What is the probability that the second counter is of the same colour as the first

    i) when the first counter is not replaced,

    ii) when the first counter is replaced before the second is drawn?

17. The events A and B are such that:

$$P(A) = \frac{1}{2}$$

$$P(A \mid B') = \frac{2}{3}$$

$$P(A \mid B) = \frac{3}{7}$$

where B' is the event 'B does not occur'.

Find $P(A \cap B)$, $P(A \cup B)$, $P(B)$, $P(B \mid A)$.

State, with reasons, whether A and B are

i) independent,

ii) mutually exclusive.

18. Twenty men took equal parts in a rescue operation, and a medal is to be awarded to one of them, chosen at random. Comment on the fairness of the following selection procedures:

    a) Twenty pieces of paper, one of which bears a cross, are folded and put in a hat. The men take one at a time, without replacement, until one of them takes the paper with the cross, and this man receives the medal.

    b) Each man is allotted a different number from 1 to 20. Two sets of cards numbered from 1 to 10 are shuffled and one card is drawn from each. The man having a number equal to the sum of the two numbers on the cards receives the medal.

    c) Each man is allotted a number from 1 to 20 and an independent person is asked to choose a number from 1 to 20 at random. The man with the corresponding number is given the medal.

    d) The names of the men are written on twenty similar pieces of paper which are put in a hat and shaken. An independent person takes one of the papers out, and the man named on it receives the medal.

19. Three bags each contain ten balls which are identical apart from colour. Bag A contains 4 red and 6 white balls, bag B contains 5 red and 5 white balls, and bag C contains 8 red and 2 white balls.

    a) From bag A, two balls are drawn at random and without replacement. Find the probability that they are both of the same colour.

    b) The original contents of bag A having been restored, the following procedure is carried out. One ball is drawn at random from bag A:

       if it is red, a ball is then drawn at random from bag B;

       if it is white, a ball is then drawn at random from bag C.

       The event that the ball from A is red is denoted by X and the event that the second ball (drawn from bag B or C) is red is denoted by Y. Find the probabilities

       i) $P(Y)$,

       ii) $P(X \cup Y)$,

       iii) $P(X \mid Y)$.

20. Two red urns each contain 2 black balls and 1 white ball; 2 yellow urns each contain 8 black balls and a green urn contains 3 black balls and 2 white balls.

    i) If a ball is chosen at random, find the probability that it is white.

    ii) If an urn is chosen at random and a ball drawn from it randomly, find the probability that it is white.

    iii) If a ball, drawn at random from one of the urns is black, find the probability that it was drawn from a red urn.

21. a) Evaluate ${}^{12}P_3$ and ${}^{8}C_5$.

b) Four cards are selected at random from a normal pack of 52.

i) How many possible selections of four cards are there?

ii) What is the probability that all four cards are diamonds?

iii) What is the probability that the four cards will differ only in suit?

iv) In how many different ways can four aces be arranged when held in the hand?

c) A player deals two cards at random from a normal pack of 52. What is the probability that they are an ace and a picture card given that the first card is

i) an ace,

ii) a picture card?

22. a) Six people are going on a motoring holiday in a six-seater car. In how many ways can they be seated if

i) all six are able to drive,

ii) the owner of the car insists on driving,

iii) the owner of the car is to drive and his wife is to sit next to him?

When the party stops for refreshments, two people are required to go to purchase them. In how many ways can these two people be selected?

b) In the directory of a certain large telephone exchange, 15 per cent of the subscribers are named Smith, 12 per cent Jones, 8 per cent Brown and 7 per cent Thomas. If, at the exchange, a connection is set up at random, calculate the probability that

i) both subscribers would be called Smith,

ii) just one of the subscribers would be called Jones,

iii) at least one of the subscribers would be called Brown,

iv) neither of the subscribers would be called Thomas,

v) A Smith would be connected to a Brown.

23. The probability that a golfer sinks a putt is $\frac{2}{3}$. Find the probability that, of three similar putts, the golfer sinks

a) all three,

b) less than three,

c) at least two.

24. A die, with faces numbers 1 to 6, is biased in such a way that on throwing it the number shown on the uppermost face is twice as likely to be an even number as an odd number. Find the probability:

    a) that, on throwing the die once, an odd number appears on the uppermost face,

    b) that, on throwing the die twice, the sum of the two numbers appearing on the uppermost face is odd.

25. A bag contains 100 beads of which 30 are white. If two beads are drawn simultaneously from the bag, calculate the probability that at least one bead is white.

26. In a certain game of chance, two common six-sided dice are thrown simultaneously and the score obtained is the sum of the two numbers shown by the dice.

    a) Assuming the two dice to be unbiased, calculate the probability of each of the possible scores, giving your answers clearly in a table.

    b) If the two dice are thrown once only, what is the most probable score?

    c) If the two dice are thrown twice, what is the probability that both scores will be 5?

    d) If the two dice are thrown three times, what is the probability that none of the scores will be 7?

27. For a certain card game, a normal pack of 52 cards is taken and all the twos, threes, and fours are removed, leaving a reduced pack of 40 cards. An Ace scores 11, the picture cards (King, Queen and Jack) score 10, while the other cards score their face value. Each player receives two cards, the total score being the sum of the scores of the two cards.

    a) A player is dealt two cards, a heart and a spade, from this reduced pack. Produce a table showing the total scores of all the possible pairs of cards which the player could have received and hence find:

        i) the most probable score,

        ii) the probability that the player has two picture cards,

        iii) the probability that his score is exactly 15,

        iv) the probability that his score is 15 or less.

    b) If instead the player received two hearts, use the appropriate portion of your table to find

        i) the most probable score,

        ii) the probability that the player has two picture cards,

        iii) the probability that his score is exactly 15,

        iv) the probability that his score is 15 or less.

28. In 18 games of chess between A and B, A wins 8, B wins 6 and 4 are drawn. A and B play a tournament of 3 games.

    On the basis of the above data, estimate the probability that:

    a) A wins all three games

    b) A and B win alternately

    c) two games are drawn

    d) A wins at least one game.

29. a) What is the probability that there are not more than 3 boys in a family of 5 children? (Assume that the probability of a male or a female birth is $\frac{1}{2}$.)

    b) Six balls are thrown into three boxes so that each ball falls into a box and is equally likely to fall into any one of the boxes. What is the probability that the first box will contain exactly 3 balls?

30. Three fair cubical dice are thrown. Find the probability that:

    i) the sum of the scores is 18,

    ii) the sum of the scores is 5,

    iii) none of the three dice shows a 6,

    iv) the product of the scores is 90.

31. a) In the Who, What or Where game, three contestants each choose one of the three categories ofquestion. Assuming that the contestants choose independently and that they are equally likely to select any of the categories, find the probability that

    i) all will choose the same category

    ii) all will choose different categories

    iii) two will be alike and the third different

    b) From a pack of 52 cards, seven are taken at random, examined, and replaced. The cards are shuffled and another seven are drawn at random. Find the probability that at least one card will be drawn twice.

32. a) Two cards are drawn at random, without replacement, from an ordinary pack of 52 cards.

    Find the probability that they are:

    i) of the same suit,

    ii) of the same value (both aces, both kings, etc.)

    iii) either of the same suit or the same value

b) Two cards are drawn at random, one from each of two ordinary packs. Find the probability that they are

   i) of the same suit,

   ii) of the same value,

   iii) either of the same suit, the same value, or both.

c) Three fair cubical dice are thrown. Find the probability that the sum of the numbers of spots on the upper face is a perfect square.

33. a) A biased die is such that the probability that any face is shown is proportional to the numbershown on that face. Find the probability with which each of the numbers 1, 2, 3, 4, 5, 6 appears. Two independent tosses of this die are made. What is the probability that the sum of the faces is ten?

   b) Consider a young man waiting for his young lady who is late. To amuse himself while waiting, he decides to take a walk under the following set of rules. He tosses a fair coin. If the coin falls heads he walks 10 metres north; if the coin falls tails he walks 10 metres south. He repeats this process every 10 metres and thus executes what is called a 'random walk'. What is the probability that after walking 100 metres he will be

   i) back at his starting point,

   ii) no more than 10 metres from his starting point,

   iii) exactly 20 metres away from his starting point?

34. The following are three of the classical problems in probability.

   a) Compare the probability of a total of 9 with the probability of a total of 10 when three fair dice are tossed once (*Galileo and Duke of Tuscany*).

   b) Compare the probability of at least one six in 4 tosses of a fair die with the probability of at least one double-six in 24 tosses of two fair dice (*Chevalier de Mere*).

   c) Compare the probability of at least one six when 6 dice are rolled with the probability of at least two sixes when 12 dice are rolled (*Pepys to Newton*).

   Solve each of these problems.

35. a) An electric circuit contains 20 components, each of which has a probability of 0.01 of beingdefective.

   Calculate the probability that none of the components is defective.

   b) A bag contains 3 red counters, 4 white counters and 5 blue counters. Two counters are drawn at random. Find the probability that they are of different colours.

36. a) Ruby Welloff, the daughter of a wealthy jeweller, is about to get married. Her father decidesthat as a wedding present she can select one of two similar boxes. Each box contains three stones. In one box two of the stones are real diamonds, and the other is a worthless imitation; and in the other box one is a real diamond, and the other two are worthless imitations. She has no idea which box is which. If the daughter were to choose randomly between the two boxes, her chance of getting two real diamonds would be 1/2. Mr Welloff, being a sporting type, allows his daughter to draw one stone from one of the boxes and to examine it to see if it is a real diamond. The daughter decides to take the box that the stone she tested came from if the tested stone is real, and to take the other box otherwise. Now what is the probability that the daughter will get two real diamonds as her wedding present?

b) A fair die is cast; then n fair coins are tossed, where n is the number shown on the die. What is the probability of exactly two heads?

37. a) Three naval officers have an annual reunion dinner and this year they have agreed to meet atthe Castle Inn, Dartmouth. Unknown to these men there are three inns with this name in Dartmouth. Assuming that each officer is equally likely to choose any of the three Castle Inns, what is the probability that

i) all three officers go to different Castle Inns,

ii) all three officers go to the same Castle Inn?

b) The hamlet of Chatton has 25 residents and is prone to rumours. Rumours spread when one person tells a story to someone else, who then repeats it to another person and so on. The rumours spread in such a way that at each stage the person receiving the rumour is chosen at random from the 24 residents available. If a rumour is told five times what is the probability that six different residents are involved? Calculate the probability that a rumour will be told four times without getting back to the originator.

38. a) Out of 50 patients being treated at a clinic for a severe allergy, 10 are chosen at random toreceive a new dietary treatment as opposed to the standard drug treatment given to the other 40 patients. It is known that the probability of a cure with the standard treatment is 0.6 whereas the probability of a cure with the new treatment is 0.9. Sometime later, one of these patients returns to the clinic to thank the doctors for his cure. What is the probability that he has been given the new treatment?

b) A small, but very select, restaurant has 6 tables and provides for one sitting only at 8.30 pm each evening. The number of tables in use on any given evening is equally likely to be 0, 1, 2, 3, 4, 5 or 6. The waiter serving these tables knows that he has a probability of 0.5 of receiving a tip from any table, and that tables tip or not independently. What is the probabilities that on a particular evening he receives at least two tips.

6 DISCRETE RANDOM VARIABLES

6.1 Introduction

This chapter brings together the concepts of the variate and probability by combining them to form the so-called *random variable* (in particular, the *discrete* random variable; chapter 8 introduces the *continuous* random variable.)

Consider the trial of throwing an ordinary, fair, six-sided die; the possible outcomes are 1, 2, 3, 4, 5, 6. Let r denote this *variate*. We know, from chapter 5, that the probability of each number occuring is $\frac{1}{6}$;

thus, we can ascribe to each value of the variate r a probability, viz $\frac{1}{6}$, and write them as below:

r	1	2	3	4	5	6
P(r)	$\frac{1}{6}$	$\frac{1}{6}$	$\frac{1}{6}$	$\frac{1}{6}$	$\frac{1}{6}$	$\frac{1}{6}$

Notice that $\Sigma P(r) = 1$, ie the probabilities add up to 1; this follows from the result on page 77, since the values of the variate r are *mutually exclusive* (only one number can turn up at once) and *exhaustive* (one of the numbers 1 to 6 must turn up).

We call this combination of the values of the *variate* r and their associated *probabilities* P(r) a *discrete random variable*, and denote it by R (the capital of the variate r). More generally, we have the definition below:

> If every value of the discrete variate r has an associated probability P(r) such that Σ P(r) = 1, then we refer to this arrangement as a *discrete random variable*, and denote it by R, the capital of the variate r.

Notice in the above definition that it is required that Σ P(r) = 1, ie the probabilities add up to 1; this is necessary since we require that the values of the variate constitute a set of mutually exclusive and exhaustive events, as shown in the above example.

Just as a variate can be specified by a *frequency distribution* a discrete random variable can be totally specified by a *probability distribution*; the probability distribution for the variate r in the above example is:

r	1	2	3	4	5	6
P(r)	$\frac{1}{6}$	$\frac{1}{6}$	$\frac{1}{6}$	$\frac{1}{6}$	$\frac{1}{6}$	$\frac{1}{6}$

Returning to the above example, suppose we now consider actually throwing the die 120 times, say; if, as we are assuming, the die is *fair*, then we would *expect* 20 of each number to occur since each number has an equal chance of occuring, as shown in the *frequency distribution* below.

r	1	2	3	4	5	6	Total
f	20	20	20	20	20	20	120

As can be seen, the relative frequency, $\frac{f}{120}$, of each value of r is $\frac{20}{120} = \frac{1}{6}$, the same as the probability of each value of r above; thus, we could obtain these *expected* frequencies by the formula: f = 120.P(r). In this sense, the random variable R is a (statistical) *model* of the variate r, in that its probabilities can be used to predict (see beginning of chapter 5) the frequencies of the values of r in, say, 120 outcomes. This can be generalised to the following result:

> The discrete random variable, R, with variate r and probabilities P(r), is a *model* of the variate r; if the total frequency of the values of r is n, then this model predicts the frequency, f, of each value r using the formula:
>
> f = n.P(r)

Returning again to the above example, it is, of course, unlikely that each number will occur *exactly* 20 times; the frequency distribution below gives the frequencies obtained in a typical experiment involving throwing the die 120 times.

r	1	2	3	4	5	6	Total
f	18	21	22	19	23	17	120

The relative frequencies, $\frac{f}{120}$, are all close to $\frac{1}{6}$ and, as would be expected, the larger the number of throws of the die, the closer they would become. We therefore state the following relationship between the observed *relative frequencies*, $\frac{f}{n}$, of values of the variate r, and the *probabilities* P(r) of the *discrete random variable* R which models this variate:

> If R is the discrete random variable which models the variate r, and if the total frequency of the observed values of this variate is n, then, the relative frequencies, $\frac{f}{n}$, of the values of r become closer to their probabilities, P(r), the larger n becomes.

In the above example we could actually *calculate* the probability of each outcome using the probability theory of chapter 5, since we *knew* that the die was *fair*, ie each outcome was *equally likely* (see page 64); however, suppose we do not have such information. Suppose we flip a coin 100 times and obtain the frequencies below:

r	H	T	Total
f	30	70	100

Now, looking at these observed frequencies, it appears that the coin is *biased* ie each outcome is *not* equally likely; what is a good *model* for this variate then? Since this is the only available data and bearing in mind the above result (n is large here), a good model will be obtained by putting $P(r) = \frac{f}{100}$, to give the probability distribution:

r	H	T
P(r)	$\frac{30}{100}$	$\frac{70}{100}$

or

r	H	T
P(r)	0.3	0.7

We now consider some special discrete random variables which are likely to occur in the examination.

6.2 The discrete uniform distribution

The first example considered in this chapter is a *uniformly distributed* discrete random variable; the reason for this is that all of its probabilities are equal, ie *uniform*. More generally, however, we have the following definition:

> The discrete random variable R is said to be uniformly distributed if the probabilities P(r) are given by the formula:
>
> $$P(r) = \frac{1}{k}, \text{ for all } r,$$
>
> where k is the number of values of R.

For the example mentioned above, we have:

$$P(r) = \frac{1}{6}, \ r = 1, 2, 3, 4, 5, 6. \text{ (thus, } k = 6 \text{ here)}$$

Such a 'formula' which calculates the probabilities for the discrete random variable R is called its *probability mass function.*

Notice that the probability mass function for the discrete uniform distribution is well-defined, ie $\Sigma P(r) = 1$, since

$$\Sigma P(r) = \frac{1}{k} + \frac{1}{k} + \dots\dots\dots + \frac{1}{k}$$

$$= k \cdot \frac{1}{k} = 1$$

6.3 The Bernoulli distribution

This is a very simple probability distribution, an example of which is the model for the biased coin considered above; this discrete random variable has only *two* values, viz H and T. Furthermore, we have:

$$P(H) = 0.3 \quad \text{and } P(T) = 0.7$$

More generally, we have the following definition:

> The discrete random variable R is said to follow a *Bernoulli distribution* if its only values are r = 0 and r = 1 with probabilities:
>
> $P(r = 1) = \beta$ and $P(r = 0) = 1 - \beta$

In the above example, it makes no difference if we let 0 represent H and 1 represent T; then, we would have

$$P(r = 0) = P(H) = 0.3 \text{ and } P(r = 1) = P(T) = 0.7 = 1 - 0.3$$

and so $\beta = 0.3$.

6.4 The binomial distribution

This is the most important probability distribution in statistics.

Suppose we have a bag containing 5 balls, 3 of which are Red and the other 2 being Green; suppose further that we take out one ball, look at it, and then *put it back in the bag*. We know that the probabilities of it being Green or Red are:

$$P(R) = \frac{3}{5} \quad \text{and} \quad P(G) = \frac{2}{5}.$$

Now, suppose we repeat the above process 7 times, say. Let r be the number of Red balls drawn out of the bag; then, clearly r can take the values 0, 1, 2, 3, 4, 5, 6, 7. Consider the selection of 7 balls below that might be chosen:

1st	2nd	3rd	4th	5th	6th	7th
R	R	R	G	G	G	R

The probability of this particular selection occuring is:

$$\frac{3}{5} \times \frac{3}{5} \times \frac{3}{5} \times \frac{2}{5} \times \frac{2}{5} \times \frac{2}{5} \times \frac{3}{5} = \left(\frac{3}{5}\right)^4 \left(\frac{2}{5}\right)^3$$

However, we could have selected 4 Red balls and 3 Green balls in a different way, eg

1st	2nd	3rd	4th	5th	6th	7th
R	G	R	R	G	R	G

In fact, using the rule on page 69 of result under heading 'Permutations' in chapter 5, with the formula:

$$\frac{n!}{s!\ t!\ \ldots}$$

$$\frac{7!}{4!\ 3!} = {}^{7}C_{4}$$

different permutations of these 4 Red and 3 Green balls, each of which has a probability of $\left(\frac{3}{4}\right)^4 \left(\frac{2}{5}\right)^3$ of occuring. Thus, we have:

$$P(4 \text{ Red balls}) = P(r = 4) = {}^{7}C_{4} \left(\frac{3}{5}\right)^4 \left(\frac{2}{5}\right)^3.$$

We can perform similar calculations to find the probability of any number of Red balls occuring using the formula:

$$P(r \text{ Red balls}) - P(r) = {}^{7}C_{2} \left(\frac{3}{5}\right)^r \cdot \left(\frac{2}{5}\right)^{7-r}$$

Thus,

$$P(2 \text{ Red balls}) = P(r = 2) = {}^{7}C_{2} \left(\frac{3}{5}\right)^2 \left(\frac{2}{5}\right)^{7-2}$$

$$= \underline{0.077}$$

We can now define the Binomial Distribution as follows:

> The discrete random variable R is said to be *Binomially distributed* if it takes the values $r = 0, 1, \ldots, n$, for some n, and has probabilities P(r) given by the probability mass function:
>
> $$P(r) = {}^{n}C_{r}\, \beta^{r} (1 - \beta)^{n-r},\ r = 0, 1, 2, \ldots, n,$$
>
> for some constant β such that $0 < \beta < 1$.

In the above example,

$$n = 7,\ \beta = \frac{3}{5},\ \text{and } 1 - \beta = 1 - \frac{3}{5} = \frac{2}{5}$$

We generally refer to the variate r as the number of *successes* in n trials; thus, in the above example, we are interested in calculating the probabilities of r Red balls out of 7, a *Red ball* here being counted as a *success*. Thus, we have:

$$P(\text{Success}) = P(\text{Red ball}) = \frac{3}{5}$$

More generally, as in the above definition, we have:

$$P(\text{Success}) = \beta$$

This implies that:

P(Failure) = P(Not a Success) = 1 - P(Success) = $1 - \beta$.

In the above example, a 'Failure' is a Green ball.

Now, P(r) is well-defined since:

$$\sum_{r=0}^{n} P(r) = \sum_{r=0}^{n} {}^{n}C_{r}\, \beta^{r} (1-\beta)^{n-r}$$

$= [\beta + (1 - \beta)]^n$, from the Binomial Theorem on page 47 of the Pure Mathematics Textbook.

$= 1^n = 1$

Let us consider another example.

Example: It has been found in the past that 20% of all electronic components manufactured by a certainprocess are faulty. In a routine quality control check, a sample of 10 components is taken off the production line.

What is the probability that:

a) none of the components are faulty;

b) one of the 10 is faulty;

c) at least 3 of the components are faulty?

Here, a 'Success' is a faulty component.

So, P(a component is faulty) = 20% = 0.2.

Thus, $\beta = 0.2$; also, n = 10. Hence, if r is the number of faulty components, we have:

$$P(r) = {}^{10}C_{r}\, (0.2)^{r}\, (1 - 0.2)^{10-r}$$

ie $P(r) = {}^{10}C_{r}\, (0.2)^{r}\, (0.8)^{10-r}$, r = 0, 1, 2, ..., 10

a) Then, P(no faulty component)

$= P(r = 0) = {}^{10}C_{0}\, (0.2)^{0}\, (0.8)^{10-0}$

$= 0.8^{10} = \underline{0.107}$

b) Also, P(one faulty component)

$= P(r = 1) = {}^{10}C_{1}\, (0.2)^{1}\, (0.8)^{10-1}$

= 0.268

c) Now, P(at least three faulty components)

= 1 - P(less than three faulty components)

= 1 - P(r = 0, 1 or 2)

= 1 - P(r = 0) - P(r = 1) - P(r = 2)

= 1 - 0.107 - 0.268 - P(r = 2)

Now, $P(r = 2) = {}^{10}C_2\,(0.2)^2\,(0.8)^{10-2}$

$= 0.302$

Thus, P(at least three faulty components)

$= 1 - 0.107 - 0.268 - 0.302 = \underline{0.32}$

Conditions for the binomial model

When we derived the probability mass function for P(r) above, we made certain assumptions, which are now stated; these must be satisfied by any variate, r, which is to be modelled by a binomial distribution.

1. The n repeated trials must be *independent* of each other. In the first example considered above, the trial was 'taking a ball from a bag, and then replacing it'; clearly, the colour of one ball drawn from the bag has no influence over the colour of the next ball drawn, and so the 7 repeated trials are independent of each other.

   In the last example, the quality of one electrical component taken from the production line is assumed to be unrelated to whether or not the next component is faulty, ie it is assumed that the production process does not produce 'runs' of faulty components.

2. The probability, β, of a success must remain *constant* over all the n trials.

   Clearly, in the first example, the probability that a ball will be Red is always $\frac{3}{5}$, since the balls in the bag are unchanging.

   In the second example, we make this assumption on the basis of past production performance.

3. The number of trials is finite (ie is same number n). In the first example, this is 7; in the second example this is 10.

4. Each trial has only two possible outcomes: 'Success' or 'Failure'. In the first example, a ball could be Green or Red. In the second example, an electrical component could be faulty or not faulty.

Consider one last example.

Example: A marksman has a probability of 0.25 of hitting the bullseye of a certain target.

What is the probability that, out of 5 shots, he obtains:

a) no bullseye,

b) all bullseyes,

c) less than 4 bullseyes?

Here, a 'success' is a bullseye; also, P(Bullseye) = 0.25.

Thus, $\beta = 0.25$; and n = 5 (the number of shots).

Hence,

$$P(r) = {}^{5}C_r\,(0.25)^r\,(0.75)^{5-r}, \quad r = 0, 1, 2, 3, 4, 5.$$

a) P(no bullseye) = P(r = 0) $= {}^{5}C_0\,(0.25)^0\,(0.75)^{5-0}$

$= (0.75)^5 = \underline{0.237}$

If we let r = no. of days for it to rain, then:

β = P(Success) = P(Rains) = 0.2.

Thus, $P(r) = 0.2\ (1 - 0.2)^{r-1}$,

ie $P(r) = 0.2\ (0.8)^{r-1}$, r = 1, 2, 3, ...

a) So, P(it doesn't rain for 4 days)

= P(it rains on the 5th day)

$= P(r = 5) = 0.2\ (0.8)^{5-1} = 0.2\ (0.8)^4 = \underline{0.08192}$

b) Here, we shall count a Success as a dry day;

Then, β = P(Success) = P(dry day)

= 1 - P(raining)

= 1 - 0.2 = 0.8

So, $P(S) = 0.8\ (1 - 0.8)^{-1}$, s = 1, 2, 3, ...

where s = no. of days for a dry day to occur.

Thus, P(rains on 3 successive days) = P(it is dry on the 4th day)

= P(s = 4) = 0.8 (1 - 0.8)4-1 = 0.8 (0.2)3 = 0.0064

c) P(it doesn't rain in September) = P(dry on all 30 days)

$= P(\text{rains on 31st day}) = P(r = 31) = 0.2\ (0.8)^{30}$

$= \underline{0.00025}$

The conditions for a geometric model to be appropriate are as follows:

1. Each trial has *only two possible outcomes*. In the first example, the trial is 'tossing a die', and the two outcomes were 6 and Not 6. In the second example, the trial was the next day in September, and the two outcomes were Rains and Dry.

2. The trials are *independent* of each other.

   In the first example, one toss of the die cannot affect another; in the second example, it is assumed that the weather on one day is not related to the weather on other days - however, this is not so likely, and forms a criticism of the geometric model used in calculating the above probabilities.

3. The probability, β, of a Success is constant throughout all the trials.

   In the first example, the probability of 6 occuring remains constant since the die remains the same; in the second example, this is not so likely, since as September wears on it is more likely to rain - this is another criticism of the geometric distribution used above.

4. The number, r, of the trials for a Success to occur is (theoretically) unlimited.

   In the first example, we *could*, although it would be most unlikely, go on tossing the die 'ad infinitum' without a 6 turning up. In the second example, again, this is *not* true, since September has only a *finite* number of days (30), which limits the values r can take; in part (c) of the above example, we had to 'extend' the model into October (one might argue, however, that there is a small possibility that it *never* rains again!).

6.6 The mean (or expectation) of a discrete random variable

Just as we can find the *mean*, $\bar{x}$, of the discrete *variate* x, we can also find the mean, or rather the *expectation*, E[X], of the *discrete random variable* X.

Recall the definition of the *mean*, $\bar{x}$, of the *discrete variate* x given in the form of a frequency distribution (see page 40)

$$\text{mean} = \bar{x} = \Sigma \frac{fx}{n}$$

Now, we saw on page 63 above that the relative frequency, $\frac{f}{n}$, of the value x is the same as the probability P(x); thus, replacing $\frac{f}{n}$ by P(x), we obtain the *expectation* E[X] of the *discrete random variable* X as below:

> The expectation, E [X] of the discrete random variable X is given by:
>
> $$E[X] = \Sigma\, x.P(x)$$

Thus, if X has the probability distribution below:

x	0	1	2	3	4
P(x)	0.1	0.2	0.4	0.2	0.1

then, the mean or expectation of X is given by:

$$\begin{aligned} E[X] &= \Sigma\, x.P(x) \\ &= 0 \times 0.1 + 1 \times 0.2 + 2 \times 0.4 + 3 \times 0.2 + 4 \times 0.1 \\ &= 2.1 \end{aligned}$$

Example: It has been found that in the production of a certain garment the number, r, of stitching faults per garment has the probability distribution below; find the mean number of faults per garment.

No. of faults (r)	0	1	2	3	4
P(r)	0.7	0.2	0.05	0.03	0.02

Mean numbers of faults = E [R] $= \Sigma r.P(r)$

$= 0 \times 0.7 + 1 \times 0.2 + 2 \times 0.05 + 3 \times 0.03 + 4 \times 0.02$

$= 0.47$

6.7 Winnings

Here we use expectation to calculate the average winnings from playing games of chance for money, ie gambling.

Suppose a fairside stall charges 25p to play a certain game involving throwing balls onto a table with a number of coloured round holes in it. There are 20 holes altogether, 10 Red, 5 Blue, 3 Green and 2 Black. You are given one ball to throw, there being an equally likely chance of it going down any one of the 20 holes (it must go down one every throw); if it goes down a Red hole, you lose (ie win nothing); a Blue hole you win 20p; a Green hole, you win 50p; a Black hole, you win £1.00. What would be your average winnings? Is the game 'fair', ie on average, do you and the fairground come out quits?

To answer these questions, we must firstly calculate the probabilities of winning the above prizes.

$$P(\text{Red}) = \frac{10}{20} = \frac{1}{2}$$

$$P(\text{Blue}) = \frac{5}{20} = \frac{1}{4}$$

$$P(\text{Green}) = \frac{3}{20}$$

$$P(\text{Black}) = \frac{2}{20} = \frac{1}{10}$$

Thus, we can construct the *winnings table* below, denoting the discrete variate 'winnings' by W.

Ball	Red	Blue	Green	Black
Winnings (W)	0	20	50	100
P(W)	$\frac{1}{2}$	$\frac{1}{4}$	$\frac{3}{20}$	$\frac{1}{10}$

We can now calculate the average or expected winnings using the above definition:

$$E[W] = \Sigma\, WP(W) = 0 \times \frac{1}{2} + 20 \times \frac{1}{4} + 50 \times \frac{3}{20} + 100 \times \frac{1}{10}$$

$$= 22.5\text{p}$$

But remember that it costs 25p to play the game; therefore, on average, you lose 25p - 22.5p per game, and the fairground wins 2.5p per game; thus, the game is *not* fair.

Example: How would you modify the prize for throwing a ball down the Black hole so that the game was fair?

Let the prize for the Black hole be x; then the winnings table is as below:

Ball	Red	Blue	Green	Black
Winnings (W)	0	20	50	x
P(W)	$\frac{1}{2}$	$\frac{1}{4}$	$\frac{3}{20}$	$\frac{1}{10}$

We require that the average winnings equal the game fee, ie we require that:

$E[W] = 25$

ie $\quad 0 \times \frac{1}{2} + 20 \times \frac{1}{4} + 50 \times \frac{3}{20} + x \times \frac{1}{10} = 25$

ie $\quad 12.5 + \frac{x}{10} = 25$

ie $\quad x = 10 \times 12.5 = 125p$

We now derive the means, or expectation, of the Uniform, Geometric and Binomial distributions.

Mean of the Discrete Uniform Distribution

Here, we consider the special discrete uniform distribution with probability mass function:

$$P(r) = \frac{1}{n}, \quad r = 1, 2, 3, \ldots\ldots, n$$

Then,
$$E[R] = \sum_{r=1}^{n} r.P(r)$$
$$= \sum_{r=1}^{n} r.\frac{1}{n}$$
$$= \frac{1}{n}\sum_{r=1}^{n} r$$
$$= \frac{1}{n} \bullet \frac{n(n+1)}{2}, \text{ see formula book; page 5.}$$
$$= \frac{n+1}{2}$$

Mean of the geometric distribution

Recall the probability mass function for a geometrically distributed variable R,

$$P(r) = \beta(1-\beta)^{r-1}, \quad r = 1, 2, \ldots.$$

$$\text{Then, } E[R] = \sum_{r=1}^{\infty} r.\ P(r)$$

$$= \sum_{r=1}^{\infty} r.\beta(1-\beta)^{r-1}$$

$$= \beta \sum_{r=1}^{\infty} r\,(1-\beta)^{r-1}$$

Now, using the binomial theorem of Pure Mathematics (see page 47 of the Pure Mathematics textbook), we have:

$$\frac{1}{(1-\theta)^2} = (1-\theta)^{-2} = 1 + 2\theta + 3\theta^2 + \ldots..$$

$$= \sum_{r=1}^{\infty} r\,\theta^{r-1}$$

Replacing θ by $(1 - \beta)$, we have:

$$\sum_{r=1}^{\infty} r.\theta^{r-1} = \sum_{r=1}^{\infty} r.(1-\beta)^{r-1} = \frac{1}{[1-(1-\beta)]^2} = \frac{1}{\beta^2}$$

$$\text{Thus, } E[R] = \beta \bullet \frac{1}{\beta^2} = \frac{1}{\beta}$$

Example: It is known that 20% of cars driven in the UK are Fords. If I stand by the roadside observingthe passing cars, what is the average number of cars that will pass for a Ford to pass me?

Here, β = P(Success) = P(Ford) $= 0.2$

Thus, average number of cars $= E[R]$

$$= \frac{1}{\beta}$$

$$= \frac{1}{0.2}$$

$= 5$ cars

Mean of the binomial distribution

Recall the probability mass function of the binomial distribution:

$$P(r) = {}^nC_r\,\beta^r (1-\beta)^{n-r},\ r = 0, 1, \ldots.., n.$$

Then, we have:

$$E[R] = \sum_{r=0}^{n} r.\ P(r)$$

$$= \sum_{r=0}^{n} r.\ {}^{n}C_{r}\ \beta^{r}(1-\beta)^{n-r}$$

$$= \sum_{r=0}^{n} r.\ \frac{n!}{r!(n-r)!}\ \beta^{r}(1-\beta)^{n-r}$$

Now, when r = 0, the term is 0, and so we can begin the sum at r = 1,

$$\therefore \quad E[R] = \sum_{r=1}^{n} r\ \frac{n!}{r!(n-r)!}\ \beta^{r}(1-\beta)^{n-r}$$

$$= \sum_{r=1}^{n} \frac{n!}{(r-1)!(n-r)!}\ \beta^{r}(1-\beta)^{n-r}$$

We now attempt to change r into r-1, and n into n-1, as follows:

$$E[R] = \sum_{r=1}^{n} \frac{n.(n-1)!}{(r-1)![(n-1)-(r-1)]!}\ \beta.\beta^{r-1}(1-\beta)^{[(n-1)-(r-1)]}$$

$$= n\beta \sum_{r=1}^{n} {}^{n-1}C_{r-1}\ \beta^{r-1}(1-\beta)^{[(n-1)-(r-1)]}$$

Putting s = r - 1, we have: when r = 1, s = 0

and when r = n, s = n-1,

and

$$E[R] = n\beta \sum_{s=0}^{n-1} {}^{n-1}C_{s}\ \beta^{s}(1-\beta)^{(n-1)-s}$$

Now, from the binomial theorem of Pure Mathematics (see page 47 of the Pure Mathematics textbook).

$$\sum_{s=0}^{(n-1)} {}^{n-1}C_{s}\ \beta^{s}(1-\beta)^{n-s} = [\beta + (1-\beta)]^{n-1} = 1$$

Thus, $E[R] = n\beta \cdot 1 = \underline{n\beta}$

Example: In the production of radio sets it is found that, on average, 10% require fixing before leaving the factory. A sample of 8 radios is taken from the production line; how many would you expect need fixing?

Here, n = 8, $\beta = 0.1$, and so:

expected number needing to be fixed = $n\beta = 8 \times 0.1 = \underline{0.8}$

We now consider how to find the median and mode of a discrete random variable.

6.8 Median of a discrete random variable

Consider the following discrete frequency distribution:

x	1	2	3	4	5	Total
f	3	5	10	9	3	30

From the definition of the median, M, of a frequency distribution, M is the value of x below which $\frac{1}{2}$ of the values lie; in this case, M is the mean of the 15th and 16th largest values, which are both 3. Hence, the median M is 3.

Now, suppose we convert the above frequency distribution to a *probability distribution*, by calculating the relative frequencies $\frac{f}{30}$ to get:

x	1	2	3	4	5
P(x)	$\frac{3}{30}$	$\frac{5}{30}$	$\frac{10}{30}$	$\frac{9}{30}$	$\frac{3}{30}$

We need to find a method that yields the same median of 3; the following definition also gives a median of 3.

<u>Definition</u>

The median, M, of the discrete random variables, X, is defined as the value of x such that:

$$P(x \leq M) \geq \frac{1}{2}$$

and $$P(x \geq M) \geq \frac{1}{2}$$

If there are two such values, then, we take the mean of these to be the median.

Applying the definition to the above probability distribution, we see that:

$$\begin{aligned} P(x \geq 3) &= P(x = 3 \text{ or } x = 4 \text{ or } x = 5) \\ &= P(x = 3) + P(x = 4) + P(x = 5) \\ &= \frac{10}{30} + \frac{9}{3} + \frac{3}{30} \\ &= \frac{22}{30} \geq \frac{1}{2} \end{aligned}$$

$$\begin{aligned} \text{and } P(x \leq 3) &= P(x = 1 \text{ or } x = 2 \text{ or } x = 3) \\ &= P(x = 1) + P(x = 2) + P(x = 3) \\ &= \frac{3}{30} + \frac{5}{30} + \frac{10}{30} \\ &= \frac{18}{30} \geq \frac{1}{2} \end{aligned}$$

Thus, M = 3 satisfies the definition and so is the median. The reason why this complicated looking definition works can be seen by returning to the corresponding *frequency* distribution, which is written out long-hand below.

$$\overbrace{111}^{3} \quad \overbrace{22222}^{5} \quad \overbrace{3333333333}^{10} \quad \overbrace{444444444}^{9} \quad \overbrace{555}^{3}$$

$x \leq 3$ (left bracket: 111 22222 3333333333)

$x \geq 3$ (right bracket: 3333333333 444444444 555)

Now, $x \geq 3$ includes *all ten 3s*, nine 4s and three 5s, as the right hand square bracket shows, making 22 values in all; thus, estimating $P(x \geq 3)$ we see that:

$$P(x \geq 3) = \frac{22}{30}$$

Similarly, $x \leq 3$ includes the three 1s, five 2s, and all ten 3s again, as the left hand square bracket shows, making 18 values in all; thus, the estimate for $P(x \leq 3)$ is:

$$P(x \leq 3) = \frac{18}{30}$$

Now, the reason why each of these probabilities *exceed* $\frac{1}{2}$ is that *all* of the ten 3s are included in each calculation (instead of only the first seven, as for the *frequency* distribution definition); this is necessary, however, since we have to add the entire probability of 3 occuring, viz $\frac{10}{30}$, in the above definition.

The following example illustrates the case where *two* values satisfy the above conditions, requiring their mean to be taken.

Example: Find the median of the discrete uniform random variable

X, where $P(x) = \frac{1}{6}$, $x = 1, 2, 3, 4, 5, 6$

x	1	2	3	4	5	6
$P(x)$	$\frac{1}{6}$	$\frac{1}{6}$	$\frac{1}{6}$	$\frac{1}{6}$	$\frac{1}{6}$	$\frac{1}{6}$

Here, we see that the values 3 and 4 satisfy the conditions:

$$P(x \geq M) \geq \frac{1}{2} \quad \text{and} \quad P(x \leq M) \geq \frac{1}{2}$$

since

$$P(x \geq 3) = \frac{4}{6} \geq \frac{1}{2} \quad \text{and} \quad P(x \leq 3) = \frac{3}{6} \geq \frac{1}{2}$$

and $$P(x \geq 4) = \frac{3}{6} \geq \frac{1}{2} \quad \text{and} \quad P(x \leq 4) = \frac{4}{6} \geq \frac{1}{2}$$

Thus, the median is the mean of these,

ie $$M = \frac{1}{2}(3 + 4) = \underline{3.5}$$

Hence, the method used in finding the median is to accumulate the probabilities from left to right, stopping at the value M (the median) where the accumulated probability exceeds or equals $\frac{1}{2}$; a check should then be made by accumulating the probabilities from right to left, to see if the same value of M is obtained (if not, the mean of the two values must be calculated).

Example: Find the median of the binomial distribution with $n = 5$ and $\beta = \frac{1}{3}$.

The probability mass function for this binomial distribution is:

$$P(r) = {}^5C_r \, \frac{1}{3}^r \left(1 - \frac{1}{3}\right)^{5-r}, \quad r = 0, \ldots\ldots, 5$$

yielding the probability distribution:

r	0	1	2	3	4	5
P(r)	0.13	0.33	0.33	0.16	0.04	0.004

Accumulating the probabilities from left to right, we see that $\frac{1}{2}$ is exceeded when r = 2.

ie $\quad 0.13 + 0.33 + 0.33 = 0.79 \geq \frac{1}{2}$.

From right to left, we have:

$$0.004 + 0.04 + 0.16 + 0.33 = 0.534 \geq \frac{1}{2}$$

Hence, M = 2 is the median.

6.9 Mode of a discrete random variable

Considering the frequency distribution

x	1	2	3	4	5	Total
f	3	5	10	9	3	30

again, we see that the mode is 3, since it has the highest frequency (10).

x	1	2	3	4	5
P(x)	$\frac{3}{30}$	$\frac{5}{30}$	$\frac{10}{30}$	$\frac{9}{30}$	$\frac{3}{30}$

and it can be seen that the value 3 has the highest probability of occuring.

We thus have the following definition.

> Definition:
>
> The mode of discrete random variable X, is the value (s) of x with the greatest corresponding probability P(x)

Clearly, there can be more than one mode. In the above discrete uniform distribution the modes are 1, 2, 3, 4, 5, 6, since their corresponding probabilities are all equally great.

In the above binomial distribution, the modes are 1 and 2.

Finally, the same relationships hold between the mean, median and mode of a probability distribution as does for a skewed frequency distribution, viz for a positively skewed distribution (see page 41).

Derived random variables

Here we consider how a different random variable can be derived from the random variable R, say. Suppose R has the probability distribution below:

r	2	3	4	5	6
P(r)	0.1	0.4	0.2	0.2	0.1

We can derive the new discrete random variable S, say, by doubling each value of R, to obtain the probability distribution below:

s	4	6	8	10	12
P(s)	0.1	0.4	0.2	0.2	0.1

and we can write $S = 2R$.

Notice that the probabilities remain *unchanged*; this follows since, for example:

$P(s = 4) = P(2r = 4) = P(r = 2) = 0.1.$

Now, the mean of S is as follows:

$E[S] = \Sigma\, s.P(s) = 4 \times 0.1 + 6 \times 0.4 + 8 \times 0.2 + 10 \times 0.2 + 12 \times 0.1$

$= 7.6$

Also, the mean of R is:

$E[R] = \Sigma\, r.P(s) = 2 \times 0.1 + 3 \times 0.4 + 4 \times 0.2 + 5 \times 0.2 + 6 \times 0.1$

$= 3.8$

which is double E [S] .

And so we see that,

$E[2R] = E[S] = 2.E[R]$

Suppose now we add 5 on to each value of S, to obtain the new discrete random variable T, say;

then $T = S + 5$ and has the probability distribution below:

t	9	11	13	15	17
P(t)	0.1	0.4	0.2	0.2	0.1

Furthermore,

$E[T] = \Sigma\, t.P(t) = 9 \times 0.1 + 11 \times 0.4 + 13 \times 0.2 + 15 \times 0.2 + 17 \times 0.1$

$= 12.6$

which is 5 more than E [S] , and so we see that:

$E[T] = E[S] + 5$

Finally, since T = S + 5 and S = 2R, we have:

T = 2R + 5 and so:

$E[2R + 5) = E[T] = E[S] + 5 = 2.E[R] + 5$

ie $E[2R + 5] = 2.E[R] + 5$

Generalising, we obtain the following result:

> If R is a discrete random variable, and a and b are constants, we have:
>
> $E[a.R + b] = a.E[R] + b$

(Thus, in the above example, a = 2 and b = 5.)

Example: If the mean of R is 5, find the mean of

i) 2R - 1,

ii) -5R + 3

Here, E [R] = 5.

Using the above result:

i) $E[2R - 1] = 2E[R] - 1 = 2 \times 5 - 1 = 9$

ii) $E[-5R + 3] = -5.E[R] + 3 = -5 \times 5 + 3 = -22$

Example: At a fairground stall, it costs 20p to play a certain game involving 10 cards numbered 1 to 10.

Each player chooses a card, and is paid out the number of pence equal to double the number on the card plus 5. Is the game fair?

If we let r be the number on the card, then R is uniformly distributed, with:

$$P(r) = \frac{1}{10}, \quad r = 1, \ldots\ldots, 10$$

Let W be the winnings; then W = 2R + 5.

Using the result on page 109 above, we know that:

$$E[R] = \frac{10 + 1}{2} = 5.5$$

Thus, E [W] = E [2R + 5] = 2E [R] + 5 = 2 x 5.5 + 5 = 16p

Hence, the game is not fair, with the player losing 20p - 16p = 4p on average.

We can derive further random variables from the discrete random variable R by using more complicated functions of R; for example, suppose R has the probability distributed below:

r	-1	0	1	2	3
P(r)	0.1	0.4	0.2	0.2	0.1

and we now square each value of r. We thereby obtain the new discrete random U, say, where $U = R^2$, which takes the values

U = 0, 1, 4, 9, and has the probability distribution:

U	0	1	4	9
P(U)	0.4	0.3	0.2	0.1

Notice that the probabilities of r = -1 and r = 1 have been added together to give the probability of U = 1; this follows since:

$$P(U = 1) = P(r^2 = 1) = P(r = 1 \text{ or } r = -1)$$
$$= P(r = 1) + P(r = -1) = 0.1 + 0.2 = 0.3$$

We can find the mean of U as follows:

$$E[U] = \Sigma U.P(U)$$
$$= 0 \times 0.4 + 1 \times 0.3 + 4 \times 0.2 + 9 \times 0.1$$
$$= 2.0$$

Thus, $E[R^2] = 2.0$

Also, the mean of R is:

$$E[R] = \Sigma r.P(r) = -1 \times 0.1 + 0 \times 0.4 + 1 \times 0.2 + 2 \times 0.2 + 3 \times 0.1$$
$$= 0.8$$

Notice therefore that:

$E[R^2] \neq (E[R])^2$

Generally speaking, it is *not* true that

$E[f(R)] = f(E[R])$

where f(R) is any function of R.

However, we do have the following result:

> If R is a discrete random variable and f(R) is a function of R, then:
>
> $E[f(R)] = \Sigma f(r) P(r)$

We can demonstrate this using the example above, where

$f(R) = R^2$; we must show that $E[R^2] = \Sigma r^2.P(r)$.

$$\begin{aligned} \text{Now, } \Sigma r^2 P(r) &= (-1)^2 \times 0.1 + 0^2 \times 0.4 + 1^2 \times 0.2 + 2^2 \times 0.2 + 3^2 \times 0.1 \\ &= 1^2 \times (0.1 + 0.2) + 0^2 \times 0.4 + 2^2 \times 0.2 + 3^2 \times 0.1 \\ &= 1^2 \times 0.3 + 0^2 \times 0.4 + 2^2 \times 0.2 + 3^2 \times 0.1 \\ &= \Sigma u.P(u) \\ &= E[U] \\ &= E[R^2] \end{aligned}$$

Thus, in this case: $E[f(R)] = \Sigma f(r) P(r)$

6.10 Variance of a discrete random variable

Recall from page 66 for 'alternative formula for variance' in chapter 4.7 the equivalent formula for the variance of the discrete variate *x*:

$$\text{variance of X} = \frac{\Sigma fx^2}{n} - (\bar{x})^2$$

We can rewrite $\frac{\Sigma fx^2}{n}$ as $\Sigma \frac{f}{n} x^2$, by a similar argument, see chapter 3.4 page 40;

thus, we have:

$$\text{variance of } x = \Sigma \frac{f}{n} x^2 - (\bar{x})^2$$

Now, we can convert the variate x into the random variable X by replacing the relative frequency $\frac{f}{n}$ of *x*

by the probability P(*x*) of *x*; also, replacing the mean $\bar{x}$ by the expectation E [X] , we have,

$$\text{variance of X} = \Sigma x^2.P(x) - (E[X])^2$$

But, we saw above that $E[R^2] = \Sigma r^2 P(r)$, and so we have:

$$\text{variance of X} = E[X^2] - (E[X])^2$$

If we let V [X] denote the variance of X, we have the result:

> The variance, V [X] , of the discrete random variable X is given by:
>
> $$V[X] = E[X^2] - (E[X])^2$$

Example: Calculate the variance of the discrete random variable R with probability distribution:

r	2	3	4	5	6
P(r)	0.1	0.2	0.3	0.2	0.2

Here, $E[R] = \sum r.P(r) = \sum r.P(r) = 2 \times 0.1 + 3 \times 0.2 + 4 \times 0.3 + 5 \times 0.2 + 6 \times 0.2$

$= 4.2$

Also, $E[R^2] = \sum r^2.P(r)$

$= 2^2 \times 0.1 + 3^2 \times 0.2 + 4^2 \times 0.3 + 5^2 \times 0.2 + 6^2 \times 0.2$

$= 19.2$

Thus, $V[X] = E[X^2] - (E[X])^2$

$= 19.2 - (4.2)^2 = \underline{1.36}$

We now derive the variances of the uniform, and binomial distributions (the variance of the geometric distribution is too complicated to be considered here).

Variance of the discrete uniform distribution

Now, $P(r) = \frac{1}{n}$, $r = 1, 2, \ldots\ldots, n$.

Thus, $\sum_{r=1}^{n} r^2 .P(r) = \sum_{r=1}^{n} r^2 . \frac{1}{n}$

$= \frac{1}{n} \sum_{r=1}^{n} r^2$

$= \frac{1}{n} \bullet \frac{1}{6} n(n+1)(2n+1)$, from page 5 of the formula book

$= \frac{(n+1)(2n+1)}{6}$

We have already shown $E[R] = \frac{(n+1)}{2}$;

thus, $V[R] = E[R^2] - (E[R])^2$

$$= \frac{(n+1)(2n+1)}{6} - \left[\frac{(n+1)}{2}\right]^2$$

$$= \frac{(n+1)[2(2n+1) - 3(n+1)]}{12}$$

$$= \frac{(n+1)(4n+2 - 3n - 3)}{12}$$

$$= \frac{n^2 - 1}{12}$$

Variance of the binomial distribution

Now, $P(r) = {}^nC_r \beta^r (1 - \beta)^{n-r}$, $r = 0, 1, \ldots, n$

We have already shown that $E[R] = n\beta$.

Also, $E[R^2] = E[R(R-1) + R)$

$= E[R(R-1)] + E[R]$

$= E[R(R-1)] + n\beta$

And, $E[R(R-1)] = \sum_{r=p}^{n} r(r-1).P(r)$, see result on page 107.

Again, when r = 0 or r = 1, these terms are 0, and so we can begin the sum with r = 2;

$\therefore \quad E[R(R-1)] = \sum_{r=2}^{n} r(r-1)\, {}^nC_r \beta^r (1-\beta)^{n-r}$

$$= \sum_{r=2}^{n} r(r-1) \frac{n!}{r!(n-r)!} \beta^r (1-\beta)^{n-r}$$

$$= \sum_{r=2}^{n} \frac{n!}{(r-2)!(n-r)!} \beta^r (1-\beta)^{n-r}$$

We now change r into r-2 and n into n-2, as follows

$$E[R(r-1)] = \sum_{r=2}^{n} \frac{n(n-1)(n-2)!}{(r-2)!\,[(n-2)-(r-2)]!} \beta^2.\beta^{r-2}(1-\beta)^{[(n-2)-(r-2)]}$$

$$= n(n-1)\beta^2 \sum_{r=2}^{n} {}^{n-2}C_{r-2}\, \beta^{r-2} (1-\beta)^{[(n-2)-(r-2)]}$$

Now, putting $s = r-2$, we have: when $r = 2$, $s = 0$ and when $r = n$, $s = n-2$, and so:

$$E[R(R-1)] = n(n-1)\beta^2 \sum_{s=0}^{n-2} {}^{n-2}C_s \, \beta^s (1-\beta)^{(n-2)-s}$$

Again, from the binomial theorem, we have:

$$\sum_{s=0}^{n-2} {}^{n-2}C_s \, \beta^s (1-\beta)^{(n-2)-s} = [\beta + (1-\beta)]^{n-2} = 1$$

and so:

$$E[R(R-1)] = n(n-1)\,\beta^2.1 = n(n-1)\,\beta^2$$

Hence $E[R^2] = n(n-1)\,\beta^2 + n\beta$, and so:

$$V[R] = E[R^2] - (E[R])^2$$

$$= n(n-1)\beta^2 + n\beta - (n\beta)^2$$

$$= n^2\beta^2 - n\beta^2 + n\beta - n^2\beta^2$$

$$= n\beta(1-\beta)$$

Summarising the mean and variance, we have:

> If R is binomially distributed with parameters n and β, the mean and variance of R are:
>
> $E[R] = n\beta$ and $V[R] = n\beta\,(1-\beta)$

Example: Two ordinary fair, six-sided dice are thrown together ten times, the total of their uppermostfaces being recorded each time; if r is the number of times they total 7, calculate the mean and variance of the discrete random variable R.

Here, $\beta = P(\text{success}) = P(\text{total of } 7) = \frac{6}{36} = \frac{1}{6}$.

Also, $n = 10$, and so:

$$E[R] = n\beta = 10 \times \frac{1}{6} = \frac{5}{3}$$

and $V[R] = n\beta(1-\beta) = 10 \times \frac{1}{6} \times \left(1 - \frac{1}{6}\right) = \frac{50}{36} = \underline{\frac{17}{18}}$

6.11 Variance of derived random variables

We can prove a result similar to that for the expectation of a discrete random variable on page 116; viz if a and b are constants, then:

$$V[aR + b] = a^2 . V[R]$$

For simplicity, we show the proof when b = 0, ie we will show that:

$$V[aR] = a^2 . V[R]$$

Now, $V[aR] = E[(aR)^2] - (E\{aR])^2$

$= E[a^2R^2] - (a.E[R])^2$

$= a^2 E[R^2] - a^2(E[R])^2$

$= a^2 (E[R^2] - (E[R])^2)$

$= a^2.V[R]$

The proof of the following general result is left to the reader:

> If R is a discrete random variable, and a and b are constants then,
>
> $$V[aR + b] = a^2.V[R]$$

This is a particularly useful result, and will be used extensively later on.

Example: If R is binomially distributed with n = 5 and $\beta = \frac{1}{4}$, calculate the mean and variance of

i) R

ii) 5R - 3

i) From the above results (on page 126) we have:

$$E[R] = n\beta = 5 \text{ x } \frac{1}{4} = \frac{5}{4}$$

and $V[R] = n\beta(1-\beta) = 5 \text{ x } \frac{1}{4} \text{ x } \left(1 - \frac{1}{4}\right) = \underline{\frac{15}{16}}$

ii) From the result on page 132, we have:

$$E[5R - 3] = 5.E[R] - 3 = 5 \text{ x } \frac{5}{4} - 3 = \frac{13}{14}$$

From the last result above, we have:

$$V[5R - 3] = 5^2.V[R] = 25 \text{ x } \frac{15}{16} = \underline{\frac{375}{16}}$$

6.12 Choosing a random sample

Suppose we wish to choose a random sample of size 10, say, from the discrete probability distribution below.

x	1	2	3	4	5
P(x)	0.2	0.2	0.3	0.1	0.2

We would *expect* to obtain a sample whose *frequency* distribution is as below (see page 112).

x	1	2	3	4	5	Total
f	2	2	3	1	2	10

ie we would expect the frequency of each value to reflect its corresponding probability of being chosen. Thus, for example, we would expect twice as many 5s than 4s.

One way of ensuring this is to use random number tables in the following way.

Firstly, decide on the degree of accuracy you wish to work to, eg 3 decimal places.

Secondly, choose 10 three digit numbers (probabilities) lying between 0 and 1 from the random number tables (see chapter 1.5).

Thirdly, construct the *cumulative* probability distribution below:

x	1	2	3	4	5
Cumulative Probability	0.000 -0.199	0.200 -0.399	0.400 -0.699	0.700 -0.799	0.800 -0.999

Finally, allocate each 'probability' to its appropriate class, thereby obtaining the required sample, eg if the 10 probabilities chosen from the random number tables are:

0.110 0.793 0.241 0.222 0.589 0.941 0.322
0.941 0.620 0.810

then their respective values are:

1 4 2 2 3 5 2

5 3 5

(is it possible to put each set of numbers on one line, one above the other?)

The reasoning behind this method is that, the larger the probability a value has of occuring, the larger its corresponding cumulative probability class and, hence, attracts correspondingly larger proportion of the three-digit probabilities (since each three digit random number occurs with equal probability).

Example: Choose a random sample of size 5 from the binomial distribution with $n = 4$ and $\beta = 0.2$.

Using the probability mass function:

$$P(r) = {}^4C_r\ (0.2)^r\ (0.8)^{4-r},\quad r = 0, 1, 2, 3, 4$$

we obtain the probability distribution below:

r	0	1	2	3	4
P(r)	0.4096	0.4096	0.1536	0.0256	0.0016

The corresponding cumulative probability distribution is:

r	0	1	2	3	4
Cumulative Probability	0.0000 -0.4095	0.4096 -0.8191	0.8192 -0.9727	0.9728 -0.9983	0.9984 -0.9999

From the random number tables, five 4-digit probabilities are:

0.2941 0.7488 0.5520 0.0197 0.3347

This yields the following random sample

0 1 1 0 0

Exercise 6

1. Construct the probability distribution of the discrete random variable R which models the variate below:

r	1	2	3	4	5	Total
f	20	50	20	8	2	100

2. The number of defectives, r, in samples of size 200 taken from a production line has the probability distribution below. Write down the frequency distribution of the number of defectives found in 20 such samples.

r	0	1	2	3	4
P(r)	0.2	0.4	0.2	0.1	0.1

3. Write down the probability distribution of

   a) the binomial variable, R, with n = 5 and $\beta = \frac{1}{3}$

   b) the geometrically distributed variable, R, with $\beta = \frac{1}{4}$

   c) the uniformly distributed variable, R, where r = 1, ..., 9.

4. A bag contains 3 Red, 2 Black and 4 Yellow balls. One ball is drawn from the bag, its colour noted, and then returned to the bag; this is repeated 5 times.

   What is the probability that:

   a) 2 are Red,

   b) all 5 are Yellow,

   c) none are Black,

   d) at least 2 are Yellow.

5. In an ESP experiment, a person has to guess which of 5 cards has a star on it; on past experience, he has a success rate of 70%. Calculate the probability that, in 10 guesses, he guesses:

   a) none correctly,

   b) at least 3 correctly,

   c) all 10 correctly.

6. A marksman shoots at a target until a hit is obtained. In the past she has a probability of success of 0.3. What is the probability that she hits a target.
   a) with the first shot,
   b) with the 4th shot?

7. In the production of a certain kind of chocolate, 5% have to be discarded due to imperfection. If each chocolate travelling along a conveyor belt is examined, what is the probability that:
   a) the 5th chocolate examined has to be discarded,
   b) the first 25 chocolates are perfect,
   c) the first 5 chocolates are imperfect?

8. If a biased coin turns up heads 70 times out of 100, what is the probability that, if the coin is repeatedly tossed,
   a) the first toss shows a tail,
   b) it takes 8 tosses for a head to appear.

9. Calculate the mean, median, mode, variance and standard deviation of the following discrete random variables:

i)

r	2	3	4	5	6
P(r)	0.1	0.2	0.3	0.2	0.2

ii)

r	100	200	250	300
P(r)	0.8	0.1	0.05	0.05

iii) $P(r = 1) = \frac{1}{2}$, $P(r = 2) = \frac{1}{4}$, $P(r = 3) = \frac{1}{8}$, $P(r = 4) = \frac{1}{8}$

iv)

	-3	-2	-1	0	1	2
P(x)	0.1	0.2	0.3	0.1	0.05	0.25

10. A game of chance costs 30p to play, and involves throwing a dart at a dartboard made up of 20 numbered squares in a rectangular shape:

    8 of the rectangles have the number 0 marked,

    5 of the rectangles have the number 20 marked,

    4 of the rectangles have the number 30 marked,

    3 of the rectangles have the number 100 marked.

    If the player has one dart to throw and it is equally likely to go in any particular square, calculate whether the game is fair or not.

11. Two players A and B, each put 5p in the kitty. Each player flips one coin in turn until it shows heads; the player who flips the head wins the 10p. Player A starts the game. The winner starts the next game and so on.

    a) What is the probability that A wins the first game.

    b) How much does A win in three games?

12. Suppose a 'shooting stall' at a fair offers money prizes for shooting accuracy; each player has 5 shots at a small circular target, and pays 50p to play. If the player scores a bull with the 1st shot, they receive £3.00; if it takes 2 shots, they receive £1.00; if they take 3 or 4 shots they receive 50p; if they take 5 shots, they receive 25p.

    A certain player, Bill Cody, usually hits a bull with probability 0.1; will he win on average?

13. A machine fills 1 kg bags of sugar, but overfills 3% of them. If a sample of 200 bags were taken, how many would you expect to be overfilled?

14. A birdwatcher observes the birds flying to a particular tree. If 20% of birds in the area are crows, how many birds will come to the tree before a crow arrives, on average?

15. How many tosses of two ordinary, fair, six-sided dice would be required, on average, for

    i) a total of 7 to appear,

    ii) a product of 12 to appear

16. For the distributions of question 9, calculate the mean and variance of:

    a) $2R + 1$,

    b) $-R$,

    c) $-5R - 2$,

    d) $3R^2$

17. A man offers to pay out the same number of pence as three times the uppermost face of an ordinary, fair, six-sided die plus 10p; what would be a fair price to pay to play the game?

18. If r is the total score obtained from throwing two fair, six-sided dice, calculate:

    i) E [R] , V [R]

    ii) E [3R + 2] , V [3R + 2]

    iii) E [-5R - 3] , v[5R - 3]

    iv) V $[R^2]$,

    v) E $[2R^2]$

19. It is estimated that 90% of the adults in Aiti are illiterate. Find the probability that in Aiti a random sample of five adults will include

    i) none,

    ii) one,

    iii) at least one, who is literate.

20. Three players A, B and C throw two dice each, in the order ABCABCA ... until one of them throws a double six. Each player puts one pound into the pool at the beginning of the game, and the first player to throw a double six takes the pool. Find the expected net gains of the three players.

21. A binomial distribution has probabilities of 0, 1, 2, ... n successes in n trial given by the coefficients of powers of t in the expansion of $(q + pt)^n$. Show that the number of successes has a mean np and variance npq.

    At a pottery, when running normally, 20% of the teacups made are defective.

    Find the probability that a sample of five will contain

    i) no defective,

    ii) exactly one defective,

    iii) at least two defectives.

22. A bag contains 4 black and 7 white balls. Random samples of 5 balls are drawn from the bag and replaced for subsequent sampling. If *x* denotes the number of black balls drawn per sample, find the probability distribution of *x* and hence find the expected value and the variance of *x*.

23. A shopkeeper found that five per cent of the eggs received from a central distributing agency were stale on delivery, and reduced his prices by five per cent. A housewife requiring ten fresh eggs was advised to purchase a dozen, the shopkeeper claiming that it was more likely than not that all the eggs in the dozen would be fresh, and furthermore that there was only a one in ten chance of two eggs in the dozen being stale.

    Are the shopkeeper's claims valid?

    What is the probability that, of the dozen eggs purchased,

    i) one egg is stale,

    ii) not more than two eggs are stale?

    If each customer accepted the shopkeeper's claims, and increased his or her egg order in the same ratio as the housewife, determine the net percentage change in the shopkeeper's daily receipts from egg sales.

24. A random sample of eight digits is taken from a set of random numbers. Find the mean number of the non-zero digits in the sample. Find also the most likely number of non-zero digits in the sample of eight.

25. An electrical circuit contains 5 components, one of which is faulty. To isolate the fault, the components are tested one by one until the faulty one is found. The random variable X denotes the number of tests required to locate the fault; the test of the faulty component itself is always included, so that X takes values from 1 to 5 inclusive. Given that any of the five components is equally likely to be the faulty one, so that X has a uniform probability distribution, find the expectation and variance of X.

    The cost C (in suitable units) of locating a fault depends in part on the number of tests required and is given by the formula $C = 5 + 2X$. Find the expectation and variance of C.

26. One turn of a game is as follows.

    Two coins are tossed. If the exposed faces of the two coins are the same as each other, then both are tossed for a second time and the turn then ends. Otherwise, the turn ends after the first toss of the coins. The score, X, obtained in the turn, is equal to the total number of heads exposed during that turn.

    i) Show that $P(X = 3) = \frac{1}{8}$, and that $P(X = 1) = \frac{5}{8}$

    ii) Find the expectation and variance of X.

    The scores obtained on two randomly chosen turns are X_1 and X_2.

    State the value of $E(X_1 - X_2)$. Find $P(X_1 = X_2)$.

27. i) A gambler pays £1 to draw six cards, one from each of six ordinary packs. If at least four of the cards are spades, he receives £26. Otherwise, he receives nothing. How much can be expect to lose in 100 tries.

ii) The game is now changed so that the gambler draws six cards from one pack, without replacement, and receives £30, if there are four or more spades among the six drawn.

What is a fair price for him to pay?

28. Choose a random sample of size 5 from the following distributions:

a)

x	2	4	6	8	10
$P(x)$	0.3	0.05	0.1	0.5	0.05

b)

r	100	200	300	400
P(r)	0.1	0.2	0.3	0.4

c) Binomial distribution with $n = 5$ and $\beta = 0.1$

d) Geometric distribution with $\beta = 0.4$

(*Hint*: calculate probabilities correct to 3 decimal places only)

e) Uniform distribution with $k = 6$.

7 JOINT PROBABILITY DISTRIBUTIONS

In the last chapter, we considered individual discrete random variables; we now consider more than one variable at a time.

Suppose we have two fair, six-sided dice, one of which is numbered 1 to 6, the other numbered 1, 1, 2, 3, 3, 3. Both are thrown together, the number of the first die, X, and on the second die, Y, being recorded. The individual probability distributions of X and Y are shown below.

X	1	2	3	4	5	6
P(X)	$\frac{1}{6}$	$\frac{1}{6}$	$\frac{1}{6}$	$\frac{1}{6}$	$\frac{1}{6}$	$\frac{1}{6}$

Y	1	2	3
P(Y)	$\frac{1}{3}$	$\frac{1}{6}$	$\frac{1}{2}$

Now, suppose we add together the numbers on the two dice, to obtain the total T = X + Y; we can then construct the probability distribution of T by considering each possible total in turn. For example, the total

T = 3 can occur in two ways, ie when X = 1 and Y = 2 or when X = 2 and Y = 1.

Now, $P(X = 1 \text{ and } Y = 2) = P(X = 1)\,P(Y = 2)$

$$= \frac{1}{6} \cdot \frac{1}{6} = \frac{1}{36}$$

(by the product rule, since the two dice are independent of each other)

and, $P(X = 2 \text{ and } Y = 1) = P(X = 2)\,P(Y = 1)$

$$= \frac{1}{6} \cdot \frac{1}{3} = \frac{1}{18}$$

Thus, $P(T = 3) = P(X = 1 \text{ and } Y = 2 \textit{ or } X = 2 \text{ and } Y = 1)$

$$= P(X = 1 \text{ and } Y = 2) + P(X = 2 \text{ and } Y = 1)$$

$$= \frac{1}{36} + \frac{1}{18} = \frac{1}{12}$$

(by the addition rule).

Clearly, to find the probability of each such total T occuring, we need to know each of the *joint probabilities* of X and Y occuring together, eg P(X = 1 and Y = 2). *In this example*, since the two dice (and, hence, the numbers X and Y) are independent of each other, we have:

$$P(X \text{ and } Y) = P(X)\,.\,P(Y)$$

Thus, we can construct the *joint probability distribution* of X and Y below, by simply multiplying the corresponding probabilities of X and Y together.

X \ Y	1	2	3	4	5	6
1	$\frac{1}{18}$	$\frac{1}{18}$	$\frac{1}{18}$	$\frac{1}{18}$	$\frac{1}{18}$	$\frac{1}{18}$
2	$\frac{1}{36}$	$\frac{1}{36}$	$\frac{1}{36}$	$\frac{1}{36}$	$\frac{1}{36}$	$\frac{1}{36}$
3	$\frac{1}{12}$	$\frac{1}{12}$	$\frac{1}{12}$	$\frac{1}{12}$	$\frac{1}{12}$	$\frac{1}{12}$

Using this table, we can now construct the probability distribution of the total T, shown below.

T = X + Y	2	3	4	5	6	7	8	9
P(T)	$\frac{1}{18}$	$\frac{1}{12}$	$\frac{1}{6}$	$\frac{1}{6}$	$\frac{1}{6}$	$\frac{1}{6}$	$\frac{1}{9}$	$\frac{1}{12}$

We can use the joint probability distribution to construct the probability distribution of other functions of X and Y; for example, if we let S be the *product* of the two numbers, then, S = XY. Now, we can obtain a product of S = 6, say, in three ways, viz

X = 2 and Y = 3

X = 3 and Y = 2

X = 6 and Y = 1

Thus, P(S = 6) = P(X = 2 amd Y = 3) + P(X = 3 and Y = 2) + P(X = 6 and Y = 1)

$$= \frac{1}{12} + \frac{1}{36} + \frac{1}{18} = \frac{1}{6}$$

(from the joint probability distribution of X and Y)

Repeating this process for all of the possible products of X and Y, we obtain the probability distribution of S below:

S=XY	1	2	3	4	5	6	8	9	10	12	15	18
P(S)	$\frac{1}{18}$	$\frac{1}{12}$	$\frac{5}{36}$	$\frac{1}{12}$	$\frac{1}{18}$	$\frac{1}{6}$	$\frac{1}{36}$	$\frac{1}{12}$	$\frac{1}{36}$	$\frac{1}{9}$	$\frac{1}{12}$	$\frac{1}{12}$

Returning to the joint probability distribution of X and Y above, remember that, since the values of X and Y are independent of each other (since the two dice are thrown independently of each other), we could calculate each joint probability by multiplying the two individual probabilities of X and Y together (because of the product rule for two independent events), eg

$$P(X = 1 \text{ and } Y = 2) = P(X = 1)P(Y = 2)$$

ie $\frac{1}{36} = \frac{1}{6} \cdot \frac{1}{6}$

Putting the individual or so-called marginal probabilities of X and Y on the above joint probability distribution table, each joint probability ('cell') can be obtained by multiplying the two corresponding marginal probabilities together.

Y \ X	1	2	3	4	5	6	Marginal Probabilities
1	$\frac{1}{18}$	$\frac{1}{18}$	$\frac{1}{18}$	$\frac{1}{18}$	$\frac{1}{18}$	$\frac{1}{18}$	$\frac{1}{3}$
2	$\frac{1}{36}$ (circled)	$\frac{1}{36}$	$\frac{1}{36}$	$\frac{1}{36}$	$\frac{1}{36}$	$\frac{1}{36}$	$\frac{1}{6}$ (circled)
3	$\frac{1}{12}$	$\frac{1}{12}$	$\frac{1}{12}$	$\frac{1}{12}$	$\frac{1}{12}$	$\frac{1}{12}$	$\frac{1}{2}$
Marginal Probabilities	$\frac{1}{6}$ (circled)	$\frac{1}{6}$	$\frac{1}{6}$	$\frac{1}{6}$	$\frac{1}{6}$	$\frac{1}{6}$	

[The probabilities used in the example above have been circled in the above diagram]

Notice that each marginal probability can be obtained by adding the joint probabilities of its corresponding row or column; for example, the marginal probability $P(X = 1) = \frac{1}{6}$ is obtained as follows:

$$\frac{1}{6} = P(X = 1) = P(X = 1 \text{ and } Y = 1 \textit{ or } X = 1 \text{ and } Y = 2 \textit{ or } X = 1 \text{ and } Y = 3)$$

$$= P(X = 1 \text{ and } Y = 1) + P(X = 1 \text{ and } Y = 2) + P(X = 1 \text{ and } Y = 3)$$

$$= \frac{1}{18} + \frac{1}{36} + \frac{1}{12}.$$

Thus, providing X and Y are independent, each and every joint probability P(X and Y) can be obtained by multiplying the two corresponding marginal probabilities together, ie P(X and Y) = P(X).P(Y).

We can now define the independence of two discrete random variables.

Definition:

The two discrete random variables X and Y are said to be *independent* if, for *every* value of x and y, we have:

$$P(X = x \text{ and } Y = y) = P(X = x)\,P(Y = y)$$

or more simply,

$$P(X \text{ and } Y) = P(X).P(Y)$$

An example of two discrete random variables that are *not* independent is as follows:

Example: Two ordinary, fair, six-sided dice are thrown together. Let their total be X and their product Y, then the distributions of X and Y be given below:

X	2	3	4	5	6	7	8	9	10	11	12
P(X)	$\frac{1}{36}$	$\frac{2}{36}$	$\frac{3}{36}$	$\frac{4}{36}$	$\frac{5}{36}$	$\frac{6}{36}$	$\frac{5}{36}$	$\frac{4}{36}$	$\frac{3}{36}$	$\frac{2}{36}$	$\frac{1}{36}$

Y	1	2	3	4	5	6	8	9	10	12
P(Y)	$\frac{1}{36}$	$\frac{2}{36}$	$\frac{2}{36}$	$\frac{3}{36}$	$\frac{2}{36}$	$\frac{4}{36}$	$\frac{2}{36}$	$\frac{1}{36}$	$\frac{2}{36}$	$\frac{4}{36}$

Y	15	16	18	20	24	25	30	36
P(Y)	$\frac{2}{36}$	$\frac{1}{36}$	$\frac{2}{36}$	$\frac{2}{36}$	$\frac{2}{36}$	$\frac{1}{36}$	$\frac{2}{36}$	$\frac{1}{6}$

Obviously, it would be laborious to construct the joint distribution of X and Y; however, providing we can show that, for one pair of values of X and Y,

$$P(X \text{ and } Y) \neq P(X)\ P(Y)$$

then, the definition for X and Y to be independent will *not* be satisfied.

Consider the pair of values X = 7 and Y = 12.

From the above distributions, we see that:

$$P(X = 7) = \frac{1}{6} \qquad \text{and } P(Y = 12) = \frac{4}{36}$$

Hence, $P(X = 7)\,P(Y = 12) = \frac{1}{6} \cdot \frac{4}{36} = \frac{1}{54}$

Now, the event 'X = 7 and Y = 12' is satisfied by the outcomes (3, 4) and (4, 3); thus, from the original definition of the probability of an event occuring, we have:

$$P(X = 7 \text{ and } Y = 12) = \frac{2}{36} = \frac{1}{18}.$$

Thus, we have demonstrated that:

$$P(X = 7 \text{ and } Y = 12) \neq P(X = 7)P(Y = 12)$$

and so conclude that X and Y are *not* independent discrete random variables.

We now state, without proof, the mean of the sum of two discrete random variables.

Theorem:

If X and Y are two discrete random variables then the expectation (mean) of their sum X + Y is given by:

$$E[X + Y] = E[X] + [Y]$$

This result is demonstrated in the following example.

Example: Suppose we have two fair, six-sided dice, one numbered 1, 1, 2, 2, 3, 3, the other numbered 1, 1, 1, 2, 2, 2. Let X and Y be the numbers occuring on the first and second die, respectively in a single throw of both, then X and Y are uniformly distributed, as below:

X	1	2	3
P(X)	$\frac{1}{3}$	$\frac{1}{3}$	$\frac{1}{3}$

Y	1	2
P(Y)	$\frac{1}{2}$	$\frac{1}{2}$

Then, $E[X] = \sum x.P(x) = 1 \times \frac{1}{3} + 2 \times \frac{1}{3} + 3 \times \frac{1}{3} = 2$

and $E[Y] = \sum y.P(y) = 1 \times \frac{1}{2} + 2 \times \frac{1}{2} = 1\frac{1}{2}$

Now, construct the distribution of their sum T = X + Y, as below:

T = X + Y	2	3	4	5
P(T)	$\frac{1}{6}$	$\frac{1}{3}$	$\frac{1}{3}$	$\frac{1}{6}$

Then, $E[X + Y] = E[T] = \sum t.P(t)$

$$= 2 \times \frac{1}{6} + 3 \times \frac{1}{3} + 4 \times \frac{1}{3} + 5 \times \frac{1}{6}$$

$$= 3\frac{1}{2}$$

$$= E[X] + E[Y]$$

From this theorem, we have the following corollary:

> Corollary:
>
> If X and Y are two discrete random variables, and a and b are two constants, then the expectation (mean) of the linear combination aX + bY is given by:
>
> $E[aX + bY] = a.E[X] + b.E[Y]$

Proof: Now, aX and bY are two discrete random variables themselves derived from X and Y; thus, from the above theorem, we have:

$$E[aX + bY] = E[aX] + E[bY]$$

$$= a E[X] + bE[Y]$$

(from result on page 132, chapter 6)

Example: If X and Y are the two discrete random variables in the above example, calculate

i) $E[3X + 2Y]$

ii) $E[X - 2Y]$

iii) $E[X - Y)$

iv) $E[\frac{1}{2}(x + y)]$

i) $E[3X + 2Y] = 3E[X] + 2E[Y]$

$= 3 \times 2 + 2 \times 1\frac{1}{2} = \underline{9}$

ii) $E[X - 2Y] = E[X] - 2E[Y]$

$= 2 - 2 \times 1\frac{1}{2} = \underline{-1}$

iii) $E[X - Y] = E[X] - E[Y] = 2 - 1\frac{1}{2} = \underline{\frac{1}{2}}$

iv) $E[\frac{1}{2}(X + Y)] = \frac{1}{2} E[X + Y] = \frac{1}{2} \times 3\frac{1}{2} = \underline{1\frac{3}{4}}$

A similar result for the *variance* of the sum of two discrete random variables is as follows:

Theorem:

If X and Y are two *independent* discrete random variables, then the variance of their sum X + Y is given by

$$V[X + Y] = V[X] + V[Y]$$

From which we have the immediate corollary:

Corollary:

If X and Y are two *independent* discrete random variables, a and b constants, then the variance of the linear combination aX + bY is given by:

$$V[aX + bY] = a^2V[X] + b^2V[Y]$$

Proof: Since aX and bY are both independent discrete random variables, then from the above theorem, we have:

$$V[aX + bY] = V[aX] + V[bY]$$
$$= a^2 V[X] + b^2 V[Y]$$

(from the result on page 138, chapter 6)

Example: If X and Y are the same two discrete variables as in the above examples, calculate

i) V [X + Y]

ii) V [2X + 3Y]

iii) V [2X - 3Y]

iv) V [X - Y]

Now, first, X and Y are independent, since they are the numbers appearing on two separate dice.

Also, $V[X] = E[X^2] - (E[X])^2$

$= E[X^2] - 2^2$

But, $E[X^2] = \Sigma x^2 P(x) = 1^2 \times \frac{1}{3} + 2^2 \times \frac{1}{3} + 3^2 \times \frac{1}{3} = \frac{14}{3}$

Hence, $V[X] = \frac{14}{3} - 2^2 = \frac{2}{3}$

Also, $V[Y] = E[Y^2] - (E[Y])^2$

$= E[Y^2] - \left(1\frac{1}{2}\right)^2$

But, $E[Y^2] = \Sigma y^2.P(y)$

$= 1^2 \times \frac{1}{2} + 2^2 \times \frac{1}{2} = \frac{5}{2}$

Hence, $V[Y] = \frac{5}{2} - \left(1\frac{1}{2}\right)^2$

$= \frac{5}{2} - \frac{9}{4} = \frac{1}{4}$

i) Hence, from the above theorem,

$V[X + Y] = V[X] + V[Y]$

$= \frac{2}{3} + \frac{1}{4} = \underline{\frac{11}{12}}$

ii) From the above corollary,

$V[2X + 3Y] = 2^2 V[X] + 3^2 V[Y]$

$= 4 \times \frac{2}{3} + 9 \times \frac{1}{4} = \underline{\frac{59}{12}}$

iii) $V[2X - 3Y] = 2^2 V[X] + (-3)^2 V[Y]$

$= 2^2 V[X] + 3^2 V[Y] = \underline{\frac{59}{12}}$

From ii)

iv) $V[X - Y] = V[X] + (-1)^2 V[Y]$

$= V[X] + V[Y]$

$= \frac{2}{3} + \frac{1}{4} = \underline{\frac{11}{12}}$

This last result illustrates that variances are always *added.*

Finally, we have the following result (without proof) for the expectation (mean) of the *product* of two discrete random variables.

<u>Theorem</u>:

If X and Y are two *independent* discrete random variables, then, the expectation of their product XY, is given by:

$$E[XY] = E[X] \cdot E[Y]$$

Example: If X and Y are the same independent discrete random variables in the above example, then, calculate:

i) $E[XY]$

ii) $E[(X+Y)^2]$

iii) $E[X(X-Y)]$

i) From the above theorem, we have:

$$E[XY] = E[X]\ E[Y] = 2 \times 1\tfrac{1}{2} = 3$$

ii)
$$E[(X+Y)^2] = E[X^2 + 2XY + Y^2]$$
$$= E[X^2] + 2E[XY] + E[Y^2]$$
$$= \frac{14}{3} + 2 \times 3 + \frac{5}{2} = 13\tfrac{1}{6}$$

(from the last example)

iii)
$$E[X(X-Y)] = E[X^2 - XY]$$
$$= E[X^2] - E[XY]$$
$$= \frac{14}{3} - 3 = 1\tfrac{2}{3}$$

There is no similar result for the *variance* of the product of two discrete random variables.

Exercise 7

1. Construct the joint probability distributions of the following pairs of discrete random variables.

   a) X = a digit chosen at random from the digits 0 to 5.

      Y = a digit chosen at random from the digits 3 to 7.

   b) X = number on a fair, six-sided die with faces numbered 1, 2, 3, 4, 4, 5.

      Y = number on a fair, six-sided die with faces numbered 2, 2, 4, 4, 6, 6.

   c) S = X + Y and T = XY where X and Y are as in b).

   d) X = number of red balls in a random sample of 3 balls chosen from a bag containing 4 red and 6 black balls.

      Y = number of green balls in a sample of size 4 chosen from a bag containing 4 green and 4 black balls.

   e) X = number of yellow balls,

      Y = number of red balls, in a sample of size 6 balls chosen from a bag containing 4 red, 4 yellow and 4 blue balls.

   f) X = the highest, Y = the lowest card in a hand of 5 cards dealt from an ordinary pack of 52 cards with all the picture cards and cards over 5 removed (Ace counts as 1).

   NB This question is lengthy.

2. State, with reasons, which of the above pairs of the discrete random variables are independent.

3. Calculate, where possible, using only the rules described above, the values of the following for each of the above pairs of discrete random variables:

   i) E [X] , V [X] , E [Y] , V [Y] ;

   ii) E [X + Y] , V [X + Y] ;

   iii) E [XY] ;

   iv) E [3X + 2y] , V [3X + 2Y] ;

   v) E [X - Y] , V [X - Y] ;

   vi) E $[(X + 2Y)^2]$.

4. The joint probability distribution of the two discrete random variables X and Y is as below. Are X and Y independent?

Y \ X	2	4	5
3	0.1	0.1	0.2
4	0.1	0.2	0.1
7	0.1	0.05	0.05

5. Three ordinary fair, six-sided dice are thrown together. If X is the highest, and Y the lowest number occuring, are X and Y independent?

8 CONTINUOUS RANDOM VARIABLES

8.1 Introduction

The last two chapters have considered *discrete* random variables, based upon *discrete* variates; we now develop an analagous theory associated with *continuous* variates.

Consider the continuous variate x taking any value in the range 0 - 5; we can represent this range by a straight line, as below:

Clearly, there are an infinite number of values x can take; indeed, between any two values of x, say 1 and 2, there is an infinity of values. This makes it impossible to specify a probability $P(x)$ for each value of x. How, then, can we define a *continuous* random variable? We will be able to answer this question after considering the following example.

Suppose that a machine is set to produce pieces of metal of length 50 cm, but that, in fact, the pieces of metal produced are between 0 and 5 mm longer than this. Furthermore, suppose that the errors x in the lengths of pieces of metal are equally uniformly distributed over the ranges 0 to 5 mm; thus, if we took a random sample of 100 pieces of metal from the production line and grouped the errors into say, 5 equal classes of width 1mm, we would expect that each class would contain 20 pieces of metal, as in the grouped frequency distribution below:

Error (x)	Frequency
0 - 1	20
1 - 2	20
2 - 3	20
3 - 4	20
4 - 5	20
Total	100

This is illustrated in the histogram below.

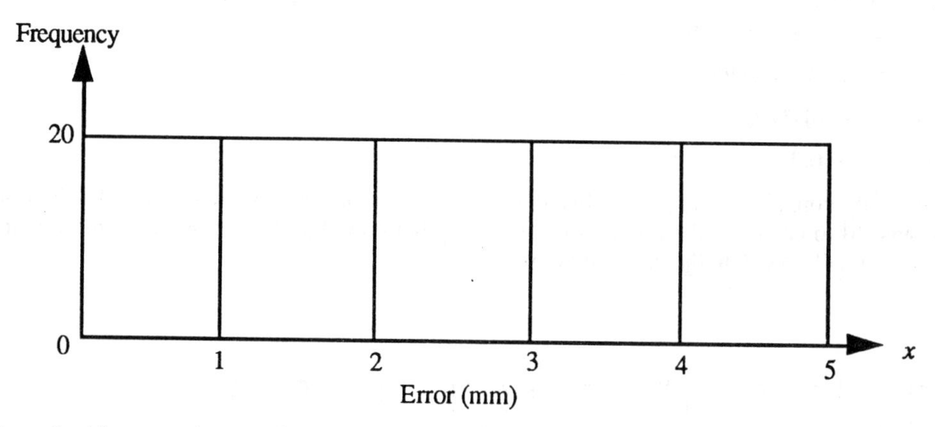

[*Note*: In this case, since we have chosen equal classes, we can make the *height* of the bars proportional to the frequency of the corresponding class; however, remember that it is the area of the bar that represents the frequency of the class.]

Now, returning to the grouped frequency distribution, we can *estimate* the probability that a piece of metal has an error lying in any of the given classes by calculating the relative frequency of that class (as for discrete random variables). For example, consider the class 3-4 mm; then,

P(error is in the class 3-4 mm)

$$= \quad P(3 \leq x < 4)$$

$$\simeq \quad \frac{\text{number of pieces of metal whose error is in class 3-4 mm}}{\text{total number of pieces of metal}}$$

$$= \quad \frac{20}{100} = 0.2$$

In this case, we see that each class has the same (uniform) corresponding probability of 0.2; we can depict this by replacing the frequency 20 in the above histogram by the probability 0.2, to obtain the diagram below.

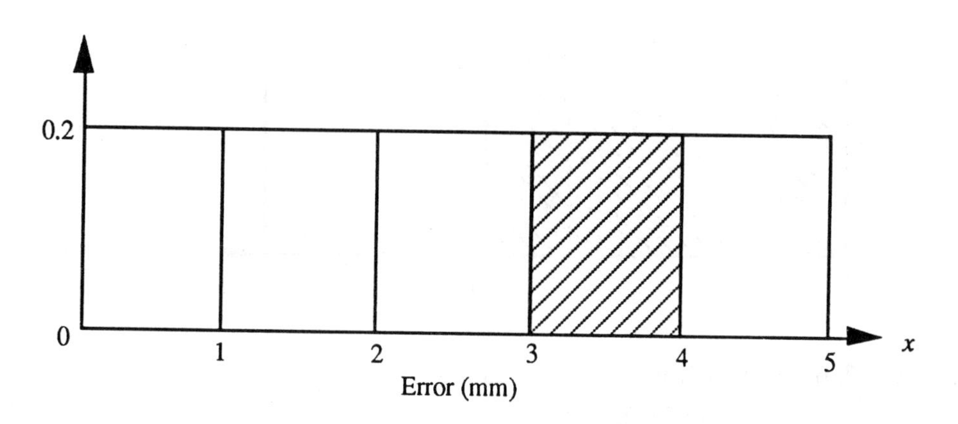

Just as the area of each bar of the *histogram* represents the *frequency* of the class, the area of each bar in *this diagram* actually equals the *probability* of x lying in the corresponding class, eg

$$\begin{aligned} P(3 \le x < 4) &= \text{area of bar (shaded)} \\ &= \text{class width x height} \\ &= (4-3)\ 0.2 \\ &= 0.2 \end{aligned}$$

So far we have only considered the probability that x lies in one of the above five classes. Suppose, however, we wished to calculate the probability of x lying between 3 and 3.5 mm; we would simply divide each class in half, to obtain the diagram below:

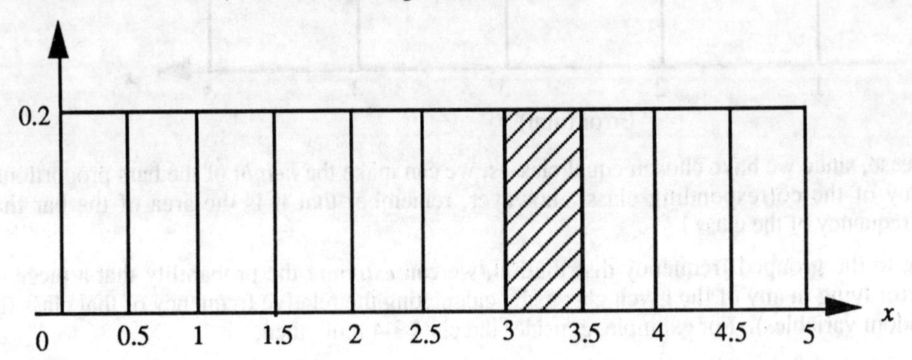

We then have:

$$\begin{aligned} P(3 \le x < 3.5) &= \text{area of bar (shaded)} \\ &= \text{class width x height} \\ &= (3.5 - 3)\ 0.2 \\ &= 0.1 \end{aligned}$$

In a similar way, we could find the probability of x lying between any two values. For example, the probability of x lying between the two values 1.2 and 3.5 would be the area of the bar shown below:

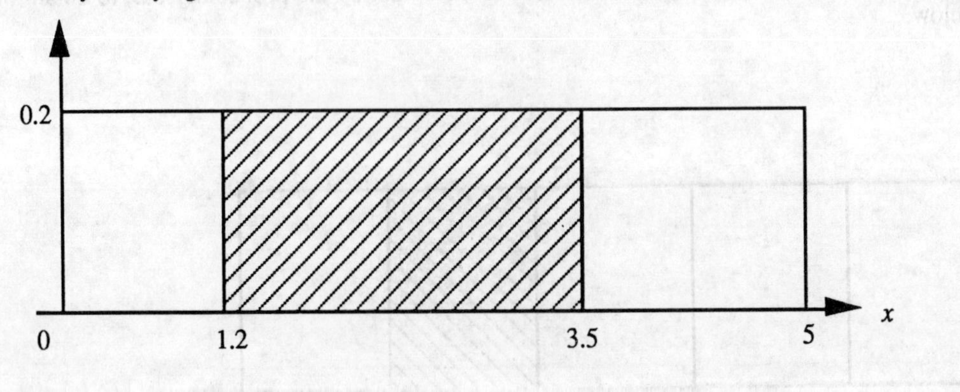

ie

$$\begin{aligned} P(1.2 \le x < 3.5) &= \text{area of bar} \\ &= \text{width x height} \\ &= (3.5 - 1.2)\ 0.2 \\ &= 0.46 \end{aligned}$$

Thus, we can use the diagram below to estimate the probability of x lying between any two values, by simply calculating the area under the horizontal line $f(x) = 0.2$ between those two values.

Notice that the total area under the line is 5 x 0.2 = 1, corresponding to the total probability.

In this sense, the above diagram is a *model* of the distribution of errors in lengths of metal produced by the machine.

We can define this model entirely by simply specifying the equation of the horizontal line in the above diagram, viz

$$f(x) = 0.2 , \quad 0 \leq x \leq 5$$
$$= 0 , \quad \text{otherwise*}$$

This function is called a probability density function (pdf), and is defined over the range $0 \leq x < 5$; outside of this range, the function is assumed to be 0, ie the x-axis. (See * above.)

Now, suppose that small errors were more likely to occur than large errors, as might more realistically be expected; repeating the above process would *not* yield the horizontal line in the above diagram, but rather a sloping line, perhaps as in the diagram below. (*Note*: these lines do not necessarily have to be straight.)

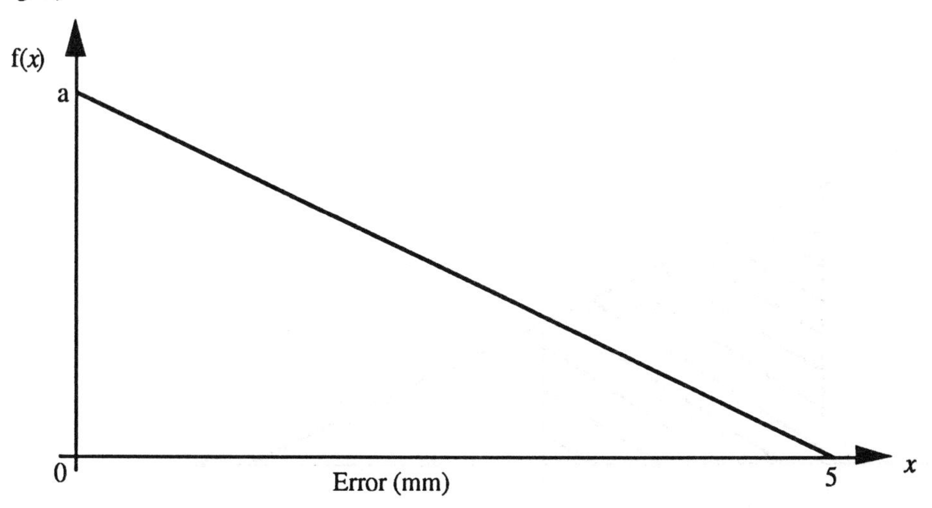

As in the above case, the probability of x lying between any two values is the area below the line between these two values; thus we can see that there is a greater probability (area) of x lying between O and 1 than between 4 and 5. Furthermore, again the total area under the line must equal 1. This enables us to find the value of height a since:

Total area $= \frac{1}{2}$ base x height

$= \frac{1}{2}(5 - 0)\ a = 1$

Thus, $a = \frac{2}{5} = 0.4$

We can now state the equation of the above straight line ie the probability density function,

$$f(x) = 0.4(1 - 0.2x), \quad 0 \le x \le 5.$$
$$= 0, \quad \text{otherwise}$$

Using this equation we can calculate the area under the curve between two given values by integration. Thus, we have for example:

$P(1 \le x < 3)$ = area under curve between $x = 1$ and $x = 3$

$$= \int_1^3 f(x).dx$$

$$= \int_1^3 0.4(1 - 0.2x)dx$$

$$= [0.4(x - 0.1x^2)]_1^3$$

$$= [0.4(3 - 0.9) - 0.4(1 - 0.1)]$$

$$= 0.48$$

This is illustrated on the diagram below:

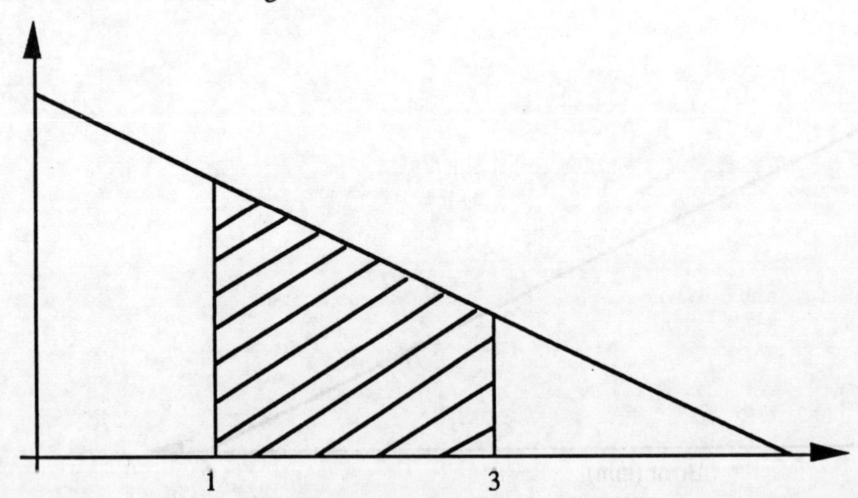

So far, we have modelled two situations involving continuous variates by specifying their probability density functions f(x), which can then be used to calculate probabilities associated with the situation.

We are now in a position to define a *continuous random variable.*

<u>Definition</u>:

A continuous random variable X, say, comprises a continuous variate x and a probability density function f(x) such that, if a and b are any two values of x, the probability P(a < x < b) of x lying between these values is given by the integral:

$$P(a < x < b) = \int_a^b f(x).dx$$

Notice that not *any* function f(x) can be a probability density function. Since the area under the curve must be 1, we must have that:

$$\int_{-\infty}^{\infty} f(x)\ dx) = 1$$

Thus, in order for a function f(x) to be a possible p.d.f., ie in order for it to be a *well-defined* p.d.f., it must satisfy the equation:

$$\int_{-\infty}^{\infty} f(x).dx = 1$$

Example: Which of the following functions are well-defined p.d.f.

a) $f(x) = x; \quad 0 \leq x \leq \sqrt{2}$

$\quad = 0, \qquad$ otherwise

Here, we have:

$$\int_{-\infty}^{\infty} f(x)\ dx = \int_0^{\sqrt{2}} x.dx$$

[since the function f(x) is 0 outside of the range $0 \leq x \leq \sqrt{2})$]

$$= \left[\frac{x^2}{2}\right]_0^{\sqrt{2}} = \left(\frac{2}{2} - 0\right) = 1$$

Thus, f(x) is well-defined.

b) $f(x) = x^2, \ -1 \leq x \leq 1$

$\quad = 0, \qquad$ otherwise

Here, we have:

$$\int_{\infty}^{\infty} f(x)dx = \int_{-1}^{+1} x^2.dx = \left[\frac{x^3}{3}\right]_{-1}^{+1}$$

$$= \frac{1}{3} + \frac{1}{3} = \frac{2}{3}$$

Thus, $f(x)$ is *not* well-defined.

c) $f(x) = \text{Sin}(x), \qquad 0 \leq x \leq \frac{\pi}{2}$

Here, we have:

$$\int_{-\infty}^{\infty} f(x)dx = \int_{0}^{\frac{\pi}{2}} \text{Sin}(x)dx \qquad = [-\text{Cos}(x)]_{0}^{\frac{\pi}{2}}$$

$$= 0 - (-1) = 1$$

Thus, $f(x)$ is well-defined.

Now, with a *discrete* random variable, X, say, it was possible to specify the probability of an individual value of x occuring, ie $P(x)$; however, for a *continuous* random variable X we have the following result.

<u>Theorem</u>:

If X is a continuous random variable, and b is any value, then:

$P(x = b) = 0$

Proof: Suppose we construct a class of width w, say, which contains the single value b, as shown in the diagram below.

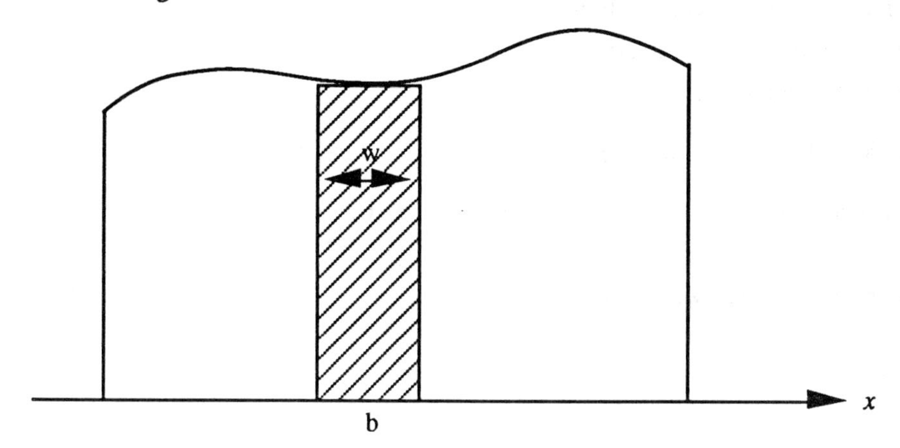

Then, certainly $P(x = b) \leq P(x \text{ is in the class containing b})$

But, $P(x$ is in the class$)$ = area under curve (shaded on diagram).

Now, if we gradually *reduce* the class width w, the area under the curve correspondingly decreases, and, hence $P(x$ is in the class$)$ decreases. Eventually, the class will *only contain the single value b*, and the area under the curve will have become O (the area of a vertical line = 0).

Hence, $P(x = b) = 0$.

Note: Some questions ask you to calculate $P(x = b)$ for some value of b, simply to test your knowledge of this theorem.

Corollary:

If X is a continuous random variable then:

$P(a \leq x \leq b) = P(a < x \leq b) = P(a \leq x < b) = P(a < x < b)$

This corollary says that the kind of inequality sign does not matter with a *continuous* random variable; we prove just one of the above results.

$$P(a \leq x < b) = P(x = a \text{ or } a < x < b)$$

$$= P(x = a) + P(a < x < b)$$

$$= O + P(a < x < b)$$

(from the above Theorem)

$$= P(a < x < b)$$

Thus, for example,

$$P(2 \leq x \leq 3) = P(2 < x \leq 3) = P(2 \leq x < 3) = P(2 < x < 3)$$

We can immediately solve a common type of problem appearing on the A level paper, as follows.

Example: The continuous random variable X has the pdf given by:

$$f(x) = Ax^2 (1 - x), \quad 0 \le x \le 1.$$
$$= 0, \quad \text{otherwise.}$$

a) Find the value of A.

b) Calculate $P\left(0 \le x \le \frac{1}{2}\right)$

c) Sketch the function f(x) over the range $0 \le x \le 1$.

a) First, for f(x) to be well-defined, we must have:

$$\int_{-\infty}^{\infty} f(x)\, dx = 1$$

$$\text{Now, } \int_{-\infty}^{\infty} f(x)\, dx = \int_0^1 Ax^2 (1 - x) dx = A\left[\frac{x^3}{3} - \frac{x^4}{4}\right]_0^1$$

$$= A\left[\frac{1}{3} - \frac{1}{4}\right]$$

$$= \frac{A}{12}$$

Hence, $\frac{A}{12} = 1$, ie $\underline{A = 12}$

b) Thus, we have:

$$f(x) = 12x^2 (1 - x), \quad 0 \le x \le 1.$$
$$= 0, \quad \text{otherwise}$$

$$\text{So, } P\left(0 \le x \le \frac{1}{2}\right) = \int_0^{\frac{1}{2}} f(x)\, dx$$

$$= \int_0^{\frac{1}{2}} 12x^2 (1 - x) dx$$

$$= 12\left[\frac{x^3}{3} - \frac{x^4}{4}\right]_0^{\frac{1}{2}}$$

$$= 12\left[\frac{1}{24} - \frac{1}{64}\right]$$

$$\text{ie} \quad P\left(0 \le x \le \frac{1}{2}\right) = \underline{\frac{5}{16}}$$

c) It is only necessary here to show the general shape of the function $f(x)$; hence, only five points will be calculated.

When $x = 0$, $\quad f(x) = 12(0)^2\ (1 - 0) \quad = 0$

When $x = \frac{1}{4}$ $\quad f(x) = 12\left(\frac{1}{4}\right)^2 \left(1 - \frac{1}{4}\right) \quad = 0.6$

When $x = \frac{1}{2}$ $\quad f(x) = 12\left(\frac{1}{2}\right)^2 \left(1 - \frac{1}{2}\right) \quad = 1.5$

When $x = \frac{3}{4}$ $\quad f(x) = 12\left(\frac{3}{4}\right)^2 \left(1 - \frac{3}{4}\right) \quad = 1.7$

When $x = 1$ $\quad f(x) = 12(1)^2 \quad (1 - 1) \quad = 0$

Thus, the function $f(x)$ looks like:

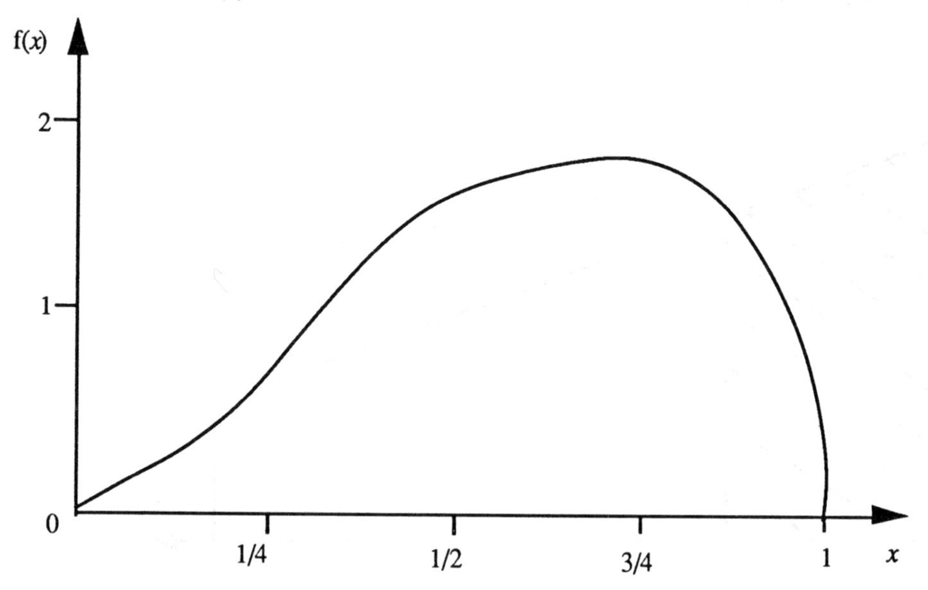

A different type of problem is as follows:

Example: If X is the continuous random variable with pdf $f(x)$ given by:

$$f(x) = \frac{1}{2}(3 - 2x), \quad 0 \le x \le 1$$
$$= 0, \quad \text{otherwise}$$

find a) $P(0.6 \le x \le 0.8)$

b) the value of b such that:

$$P(0 \le x \le b) = \frac{1}{2}$$

a) $P(0.6 \leq x \leq 0.8) = \int_{0.6}^{0.8} f(x).dx$

$= \int_{0.6}^{0.8} \frac{1}{2}(3 - 2x)\, dx$

$= \frac{1}{2}\,[3x - x^2]_{0.6}^{0.8}$

$= \frac{1}{2}\,[(2.4 - 0.64) - (1.8 - 0.36)]$

$= \underline{0.16}$

b) A sketch of the pdf is shown below.

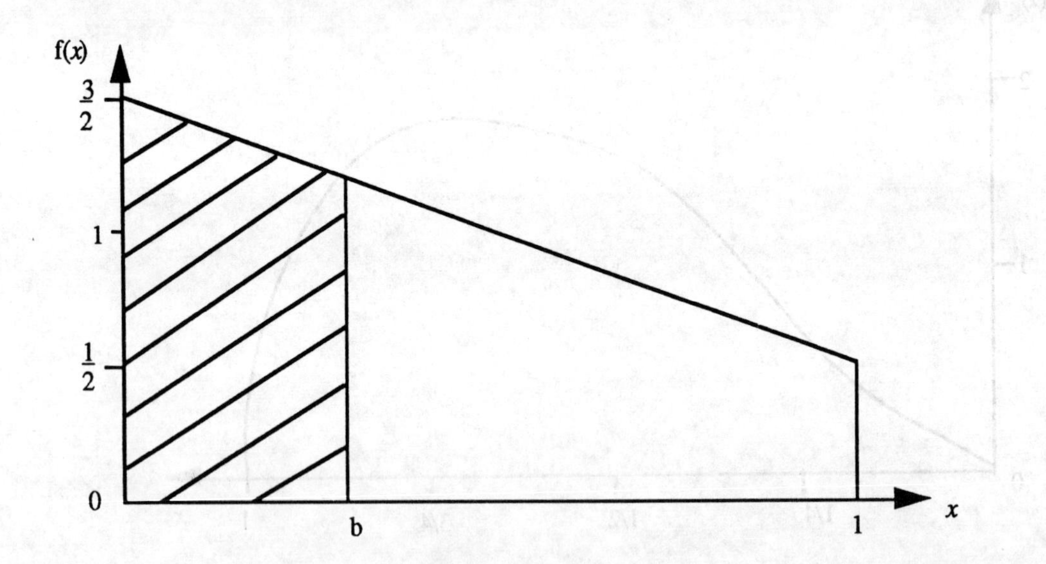

The shaded area below the value $x = b$ equals $\frac{1}{2}$; this area is a trapesium, and we therefore have:

$\frac{1}{2}$(sum of parallel sides) x (distance between them) $= \frac{1}{2}$

ie $\frac{1}{2}[f(0) - f(b)]\, b = \frac{1}{2}$

ie $\frac{1}{2}\left[\frac{3}{2} - \frac{1}{2}(3 - 2b)\right] b = \frac{1}{2}$

ie $2b^2 = 1$ ie $b = \frac{1}{2}\sqrt{2}$

[Notice that it is not always necessary to integrate $f(x)$ to calculate probabilities, as in the above example.]

8.2 Expectation (mean) of a continuous random variable

Recall that, for a discrete random variable, X, the expectation E[X] is given by:

$$E[X] = \sum xP(x)$$

The analagous formula for the continuous random variable is found by replacing the Σ sign by the $\int$ sign, and $P(x)$ by $f(x)$, to obtain:

$$E[X] = \int_{-\infty}^{\infty} x.f(x).dx$$

Example: Calculate the expectation of the continuous random variable with pdf given by

$$f(x) = \frac{1}{2}(3 - 2x), \quad 0 \leq x \leq 1$$
$$= 0, \quad \text{otherwise}$$

From the definitions, we have:

$$E[X] = \int_{-\infty}^{\infty} x.f(x).dx$$

$$= \int_0^1 x.\frac{1}{2}(3 - 2x)dx$$

$$= \frac{1}{2}\left[\frac{3x^2}{2} - \frac{2x^3}{3}\right]_0^1$$

$$= \frac{1}{2}\left(\frac{3}{2} - \frac{2}{3}\right) = \frac{5}{12}$$

8.3 Variance of a continuous random variable

Recall that, for a discrete random variable, X, the variance V [X] is given by the formula:

$$V[X] = E[X^2] - (E[X])^2$$

where $E[X^2] = \Sigma x^2.P(x)$

> Similarly, if X is a continuous random variable with pdf f(x), we have:
>
> $$V[X] = E[X^2] - (E[X])^2$$
>
> where $E[X^2] = \int_{-\infty}^{\infty} x^2.f(x).dx$

Example: If X is the continuous random variable with the pdf given by:

$$f(x) = \frac{1}{2}(3 - 2x), \quad 0 \le x \le 1$$
$$= 0, \quad \text{otherwise}$$

Calculate a) $E[X^2]$,

b) $V[X]$

a) Firstly, from the above formula, we have:

$$E[X^2] = \int_{-\infty}^{\infty} x^2.f(x)dx$$

$$= \int_0^1 x^2 . \frac{1}{2}(3 - 2x)dx$$

$$= \frac{1}{2}\left[x^3 - \frac{x^4}{2}\right]_0^1$$

$$= \frac{1}{2}\left(1 - \frac{1}{2}\right) = \frac{1}{4}$$

b) From the last example, we saw that

$$E[X] = \frac{5}{12}$$

Thus, from the above formula, we have:

$$V[X] = E[X^2] - (E[X])^2$$

$$= \frac{1}{4} - \left(\frac{5}{12}\right)^2$$

$$= \frac{1}{4} - \frac{25}{144} = \frac{11}{144}$$

8.4 Median of a continuous random variable

Now, for a *frequency* distribution of the variate x, the median, M, is the value of x below which 0.5 of the values lie, ie the middle largest value. Since the area under a *histogram* represents frequency, the value M divides the histogram into two equal halves, as illustrated in chapter 4 (page 55).

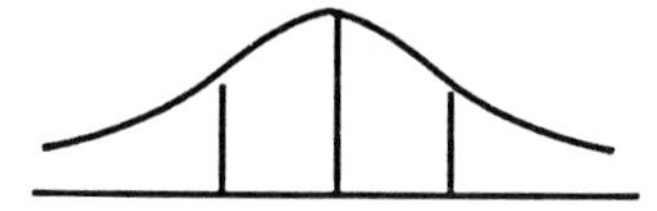

Similarly, the area under the graph of the pdf of a continuous random variable, X, represents probability, and so the value M divides this area into two equal halves, making $P(x \leq M) = \frac{1}{2}$; thus, we have the following definition.

> Definition:
>
> The median M of the continuous random variable, X, with pdf f(x) is the value of x such that:
>
> $$P(x \leq M) = \int_{-\infty}^{M} f(x)\,dx = \frac{1}{2}$$

Example: For the continuous random variable, X, with pdf given by

$$f(x) = \frac{1}{2}(3 - 2x), \quad 0 \leq x \leq 1$$

$$= 0, \quad \text{otherwise}$$

we showed in the sample on page 168 above that the value of b for which

$P(0 \leq x \leq b) = \frac{1}{2}$ was $\frac{\sqrt{2}}{2}$; thus, $M = \frac{5}{12}$

8.5 Mode of a continuous random variable

Recall that, for a grouped frequency distribution, the *modal class* is the class with highest frequency; on the histogram of the frequency distribution, it is the class corresponding to the highest bar.

> In an analagous way, the mode(s) of the continuous random variable X with pdf f(x) is defined as the value(s) of x at which f(x) has a local maximum.

This is illustrated in the diagram below.

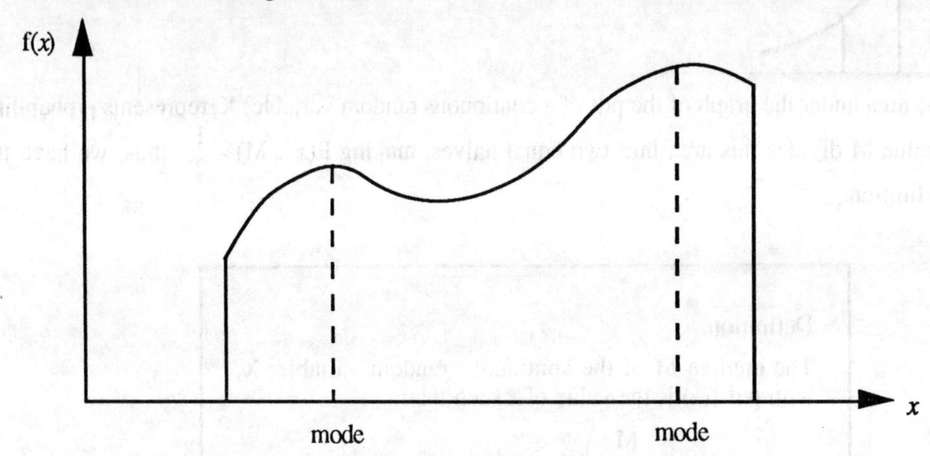

Example: For the continuous random variable X with pdf given by:

$$f(x) = \frac{1}{2}(3 - 2x), \quad 0 \leq x \leq 1$$
$$= 0, \quad \text{otherwise}$$

The mode can be seen to be $x = 0$. (See diagram on page 168)

Example: Find the mode of the continuous random variable with pdf

$$f(x) = 12x^2(1 - x), \quad 0 \leq x \leq 1$$
$$= 0, \quad \text{otherwise.}$$

Now, differentiating f(x), we have:

$$f'(x) = 12(2x - 3x^2)$$

Thus, $f'(x) = 0$ when $12(2x - 3x^2) = 0$

ie when $x(2 - 3x) = 0$

ie when $x = 0$ or $\frac{2}{3}$

Now, when $x = 0$, the function is a minimum, as can be seen on the sketch of f(x) on page 167.

However, the sketch shows a maximum between 0 and 1, which must, therefore be at $x = \frac{2}{3}$.

Hence, the mode is $\frac{2}{3}$.

The following results, proved for discrete random variables, also hold for continuous random variables, and are stated without proof.

<u>Theorem</u>:

If X and Y are continuous random variables, a and b constants, then:

i) E [a X + b] = a.E [X] + b

ii) V [a X + b] = a2.V [X]

iii) E [a X + b Y] = a.E [X] + b.E [Y]

iv) V [a X + b Y] = a2.V [X] + b2.V [X]

Note, however, results i) and iv) are true only if X and Y are also independent, as before.

We now consider two special continuous random variables, viz the uniform and exponential; the proofs of the mean and variance of these distributions are required for the examination, and it is strongly advised that they are learned.

8.6 The continuous uniform distribution

The first example considered in this chapter was of the continuous uniform distribution, wherein the errors, x, of lengths of metal were equally or uniformly distributed throughout the range 0 - 5.

We now make a formal definition of the general uniform distribution, and derive its mean and variance.

<u>Definition</u>:

The continuous random variable X is said to be *uniformly distributed* over the range $a \le x \le b$ if the pdf f(x) is defined as:

$$f(x) = \frac{1}{b - a}, \quad a \le x \le b$$
$$= 0, \quad \text{otherwise}$$

The graph of f(x) is shown below:

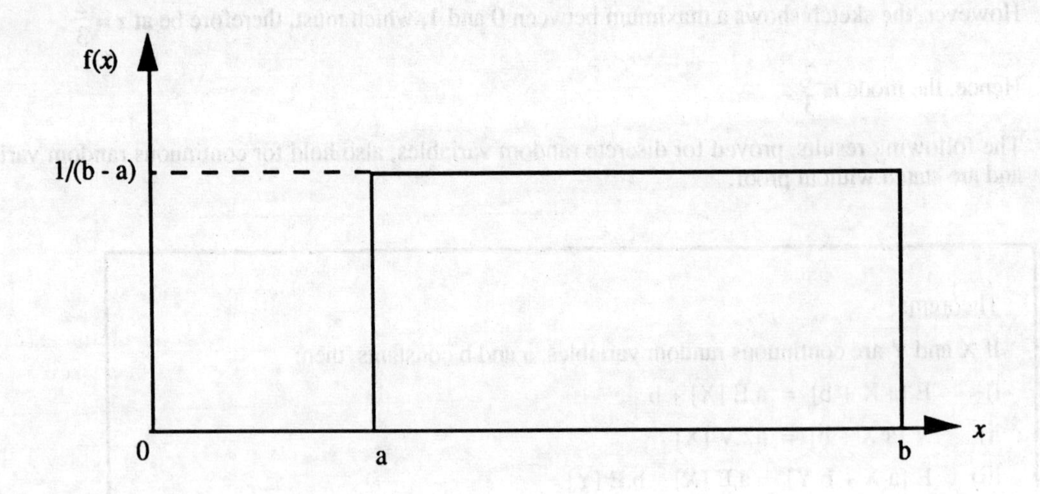

In the example referred to above, since x took any value between 0 and 5, we have:

$$a = 0 \quad \text{and } b = 5$$

giving the pdf:

$$f(x) = \frac{1}{5 - 0}, \qquad 0 \le x \le 5$$

$$\text{ie } f(x) = 0.2, \; 0 \le x \le 5.$$

Mean

From the definition of the mean, we have:

$$E[X] = \int_{\infty}^{\infty} x.f(x)\, dx$$

$$= \int_{a}^{b} x.\left(\frac{1}{b - a}\right) dx$$

$$= \frac{1}{b - a}\left[\frac{x^2}{2}\right]_a^b$$

$$= \frac{a}{b - a} \bullet \frac{b^2 - a^2}{2} = \frac{a + b}{2}$$

Thus, the mean of the uniform distribution is the mid-point of the range a $\le x \le$ b, and is depicted below. (Notice that this is also the same value as the median, since it divides the area under f(x) in half.)

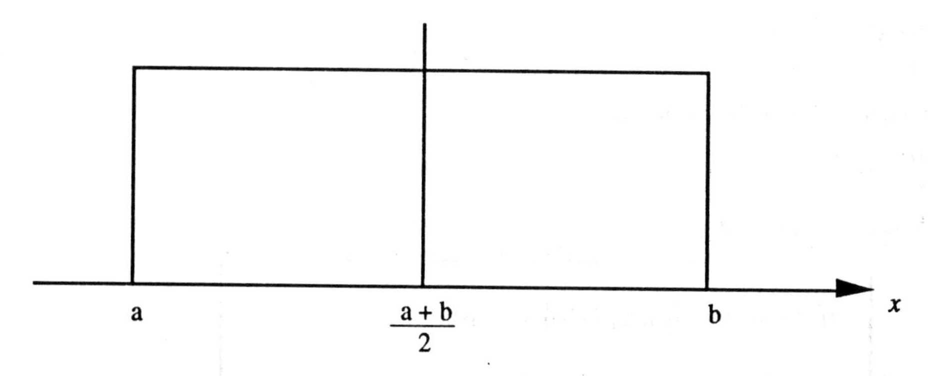

In the first example considered a = 0 and b = 5, thus, $E[X] = \frac{1}{2}(0 + 5) = 2\frac{1}{2}$; hence, the mean error is 2.5 mm.

Variance

From the definition, we have

$$V[X] = E[X^2] - E[X]^2$$
$$= E[X^2] - \left(\frac{a+b}{2}\right)^2$$

Now, $$E[X^2] = \int_{-\infty}^{\infty} x^2.f(x).dx$$
$$= \int_a^b x^2 \left(\frac{1}{b-a}\right) dx$$
$$= \frac{1}{b-a}\left[\frac{x^3}{3}\right]_a^b$$
$$= \frac{b^3 - a^3}{3(b-a)}$$
$$= \frac{b^2 + ab + a^2}{3}$$

(since b - a is a factor of $b^3 - a^3$)

Thus, $$V[X] = \frac{b^2 + ab + a^2}{3} - \left(\frac{a+b}{2}\right)^2$$
$$= \frac{b^2 + ab + a^2}{3} - \frac{a^2 + 2ab + b^2}{4}$$
$$= \frac{4b^2 + 4ab + 4a^2 - 3a^2 - 6ab - 3b^2}{12}$$
$$= \frac{b^2 - 2ab + a^2}{12}$$

$$= \frac{(b - a)^2}{12}$$

In the above example, a = 0 and b = 5, and so,

$$V[X] = \frac{(5 - 0)^2}{12} = \frac{25}{12}.$$

Summarising these two results, we have:

> If X is uniformly distributed, with pdf
>
> $$f(x) = \frac{1}{b - a}, \quad a \leq x \leq b$$
> $$= 0, \quad \text{otherwise}$$
>
> $$\text{then } E[X] = \frac{a + b}{2} \text{ and } V[X] = \frac{(b - a)^2}{12}$$

Example: A number x is chosen at random between 5 and 10.
Find the mean and variance of such numbers.
A rectangle of sides x and $(x + 2)$ cm is constructed.
Calculate

a) its mean perimeter

b) its mean area

c) the probability that its perimeter exceeds 35 cm.

d) the probability that its area is less than 48 cm^2.

Since x is evenly distributed throughout the range 5 to 10, it is uniformly distributed with pdf.

$$f(x) = \frac{1}{10 - 5}, \quad 5 \leq x \leq 10$$

ie $$f(x) = \frac{1}{5}, \quad 5 \leq x \leq 10$$

From the above results, we have:

$$E[X] = \frac{5 + 10}{2} = \underline{7.5}$$

and $$V[X] = \frac{(10 - 5)^2}{12} = \underline{\frac{25}{12}}$$

a) Now, the perimeter, P say, of the rectangle in terms of x is:

$P = x + x + (x + 2) + (x + 2) = 4x + 4$

Hence, $E[P] = E[4X + 4] = 4E[X] + 4$

(using result E[X] = 7.5 above) $= 4 \times 7.5 + 4 = \underline{34 \text{ cm}}$

b) Furthermore, if A is the area of the rectangle, then:

$A = x(x + 2) = x^2 + 2x$

Thus, $E[A] = E[X^2 + 2X] = E[X^2] + 2E[X]$

(using result $E[X] = 7.5$ above) $= E[X^2] + 2 \times 7.5$

$= E[X^2] + 15$

Now, from the proof of the variance on page 175, we see that:

$$E[X^2] = \frac{b^2 + ab + a^2}{3}$$

$$= \frac{10^2 + 5 \times 10 + 5^2}{3}$$

$$= \frac{175}{3}$$

Thus, $E[A] = \frac{175}{3} + 15 = 73\frac{1}{3}\ \text{cm}^2$

c) Here, we require $P(P \geq 35\ \text{cm})$

ie $P(4x + 4 > 35)$

ie $P(x > 7.75)$

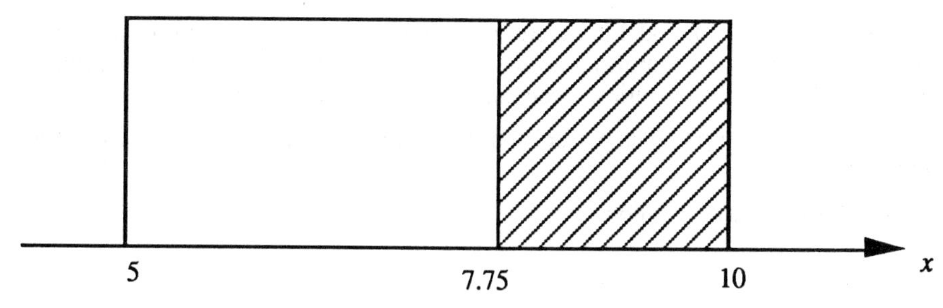

From the diagram, this is the area of the shaded part,

ie $P(x > 7.75) = (10 - 7.75)\ \frac{1}{10 - 5}$

$= 0.45$

d) Here, we require $P(A < 48)$

Now, $P(A < 48) = P(x^2 + 2x < 48)$

$= P(x^2 + 2x - 48 < 0)$

$= P[(x + 8)(x - 6) < 0]$

$= P(x - 6 < 0)$

$= P(x < 6)$

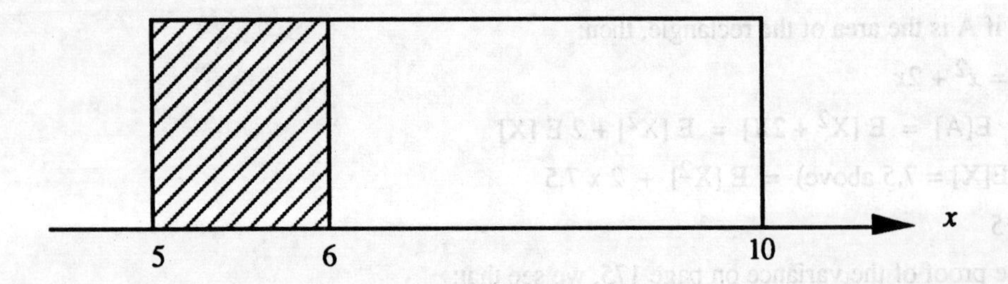

From the diagram, this is the area of the shaded part,

ie $P(x < 6) = (6 - 5)\left(\frac{1}{10 - 5}\right)$

$= 0.2$

[**Note*: The inequality $(x + 8)(x - 6) < 0$ is satisfied by the values $-8 < x < 6$; and so, the probability becomes:

$P(-8 < x < 6) = P(5 < x < 6)$,

since $f(x) = 0$ when $x < 5$.]

8.7 The exponential distribution

This distribution can often be used as a model in several real life situations, including life expectancy of products, and radioactive decay. The examples considered in this section will involve the former; for instance, the life of electric light bulbs, batteries, electrical circuit components, etc.

In such cases, the probability of such products performing satisfactorily diminishes as time increases - in fact, the decrease is exponential.

We define the exponential distribution as follows:

Definition:

The continuous random variable X, is said to be *exponentially distributed* if, for some positive constant μ, its pdf is:

$f(x) = \mu.e^{-\mu x}, \quad x \geq 0$

$= 0$, otherwise

The graph f(x) looks like:

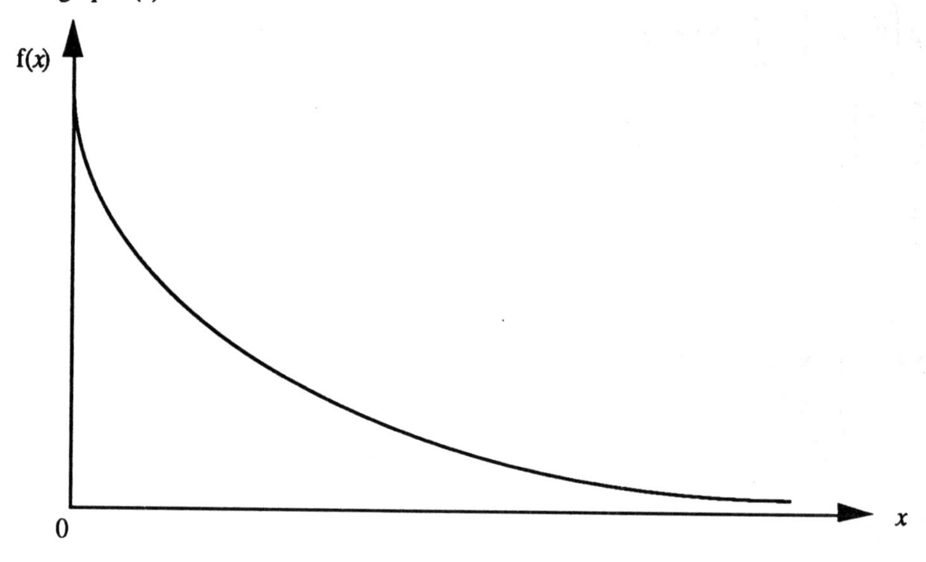

As can be seen, as x increases, probability decreases. Now, we must first show that f(x) is a well-defined pdf.

$$\int_{-\infty}^{\infty} f(x).dx = \int_{0}^{\infty} \mu e^{-\mu x}.dx = \mu\left[\frac{e^{-\mu x}}{-\mu}\right]_0^{\infty}$$

$$= \left[-e^{-\mu x}\right]_0^{\infty}$$

$$= 0 - (-1) = 1$$

Hence, f(x) is well-defined.

Mean

From the definition:

$$E[X] = \int_{-\infty}^{\infty} x.f(x)\, dx$$

$$= \int_{0}^{\infty} x.\mu e^{-\mu x}\, dx$$

$$= \mu \int_{0}^{\infty} x.e^{-\mu x}\, dx$$

(intregrating by parts) $= \mu\left(\left[x \bullet \frac{e^{-\mu x}}{-\mu}\right]_0^{\infty} - \int_0^{\infty} 1 \bullet \frac{e^{-\mu x}}{-\mu}\, dx\right)$

$$= \mu\left(0 + \frac{1}{\mu}\int_0^\infty e^{-\mu x} \bullet dx\right)$$

$$= \left[\frac{e^{-\mu x}}{-\mu}\right]_0^\infty \qquad = -\frac{1}{\mu}(0 - 1)$$

$$= \frac{1}{\mu}$$

Variance

From the definition:

$$V[X] = E[X^2] - (E[X])^2$$

$$= E[X^2] - \left(\frac{1}{\mu}\right)^2$$

But, $\quad E[X^2] = \int_{-\infty}^{\infty} x^2.f(x).dx$

$$= \int_0^\infty x^2.\mu e^{-\mu x}\,dx$$

$$= \mu\int_0^\infty x^2.e^{-\mu x}\,dx$$

$$(\text{By parts}) = \left(\left[x^2 . \frac{e^{-\mu x}}{-\mu}\right]_0^\infty - \int_0^\infty 2x. \frac{e^{-\mu x}}{-\mu}\,dx\right)$$

$$= \mu\left(0 + \frac{2}{\mu}\int_0^\infty x.e^{-\mu x}.dx\right)$$

$$= 2\int_0^\infty x.e^{-\mu x}\,dx$$

But, from the proof of the mean, we showed

$$\mu\int_0^\infty x.e^{-\mu x}\,dx = \frac{1}{\mu}$$

Hence, $\quad \int_0^\infty x.e^{-\mu x}.dx = \frac{1}{\mu^2}$

Thus, $\quad E[X^2] = 2\left(\frac{1}{\mu^2}\right) = \frac{2}{\mu^2}$

So, $\quad V[X] = \frac{2}{\mu^2} - \left(\frac{1}{\mu}\right)^2 = \frac{1}{\mu^2}$

Summarising these two results, we have:

> If X is exponentially distributed with pdf
>
> $f(x) = \mu.e^{-\mu x}, \quad x \geq 0$
>
> $\quad\;\; = 0, \quad$ otherwise
>
> then, $E[X] = \dfrac{1}{\mu}$ and $V[X] = \dfrac{1}{\mu^2}$

Example: The working life, t, measured in hours, of a certain brand of battery is exponentially distributed, with mean life 120 hrs. State the pdf of the continuous random variable T, and the variance of battery lives. Furthermore, calculate:

a) the probability of a battery lasting longer than 150 hours,

b) the life expectancy below which 80% of the batteries last.

If the manufacturer offers a refund if the batteries expire within 10 hours, how many would you expect to be refunded out of a batch of 1,000?

First, since T is exponentially distributed, we know that, for some μ, the pdf is:

$$f(t) = \mu e^{-\mu t}, \quad t \geq 0$$
$$= 0 \quad, \text{ otherwise.}$$

Now, we know that the mean of the exponential distribution is $\dfrac{1}{\mu}$.

Hence, since we are given that this is 120, we have:

$$\frac{1}{\mu} = 120$$

$$\therefore \quad \mu = \frac{1}{120}$$

Thus, the required pdf is

$$f(t) = \frac{1}{120} e^{-t/120}, \quad t \geq 0$$
$$= 0, \quad \text{otherwise}$$

The sketch of the distribution looks like:

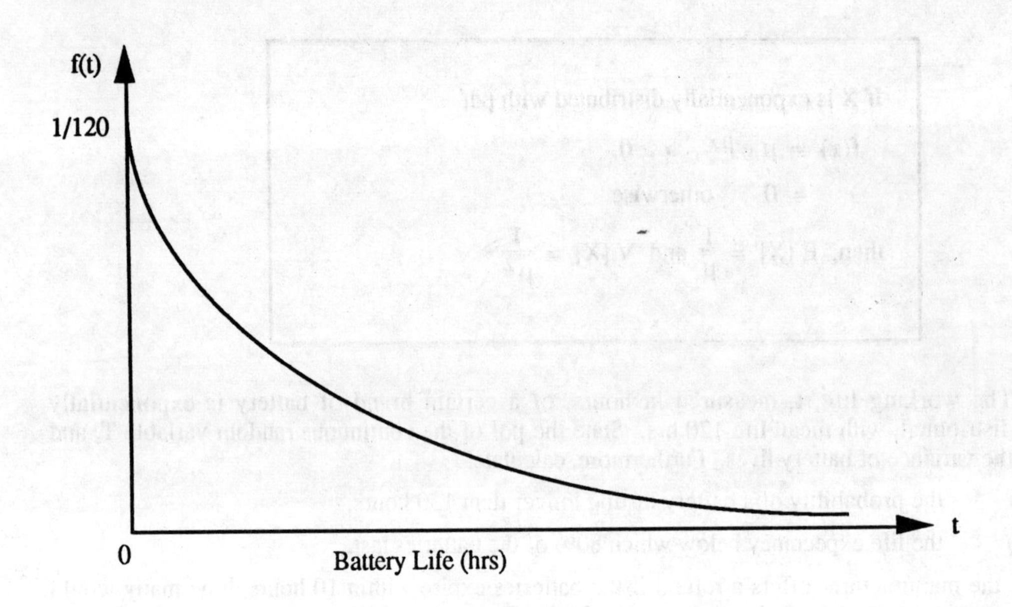

From the above results, we have:

$$V[T] = \frac{1}{\mu^2} = \frac{1}{(1/120)^2} = \underline{14{,}400}\ \text{hr}^2$$

a) We require P(t > 150), illustrated below.

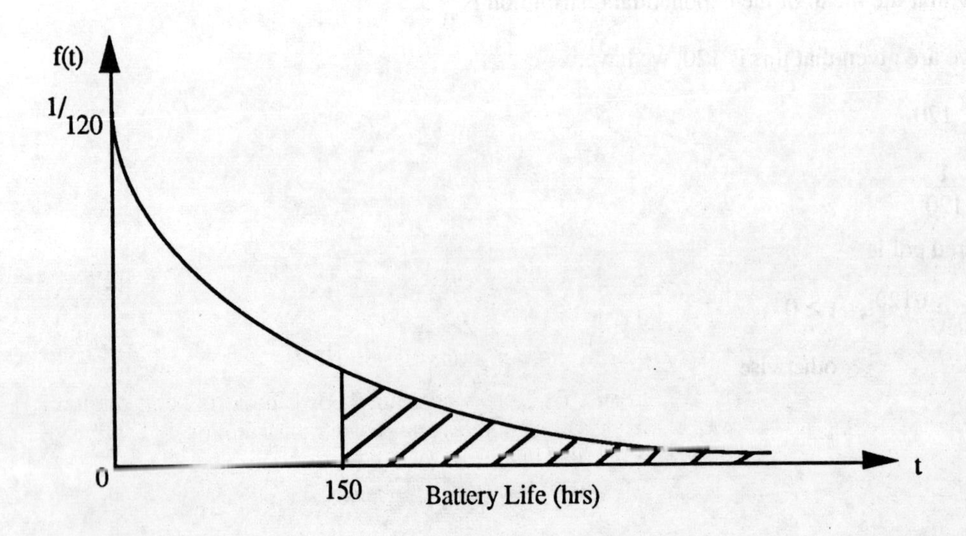

$$\text{Now, } P(t > 150) = \int_{150}^{\infty} \frac{1}{120} e^{-t/120}\, dt$$

$$= \frac{1}{120}\left[\frac{e^{-t/120}}{-1/120}\right]_{150}^{\infty}$$

$$= -0 - e^{-150/120}$$

$$= e^{-150/120} = 0.29$$

b) Let t_o be the required life expectancy below which 80% of batteries last, then:

$$P(t \le t_o) = 0.8$$

$$\text{ie} \quad \int_0^{t_o} \frac{1}{120} . e^{-t/120} . dt = 0.8$$

$$\text{ie} \quad \frac{1}{120}\left[\frac{e^{-t/120}}{-1/120}\right]_0^{t_o} = 0.8$$

$$\text{ie} \quad 1 - e^{\frac{-to}{120}} = 0.8$$

$$\text{ie} \quad e^{\frac{-to}{120}} = 0.2$$

$$\text{ie} \quad -t_o/120 = \log_e(0.2)$$

$$\text{ie} \quad t_o = -120 \log_e(0.2)$$

$$= \underline{193 \text{ hrs}}$$

Now, the probability that a battery lasts less than 10 hrs is $P(t < 10)$,

$$\text{ie} \quad P(t < 10) = \int_0^{10} \frac{1}{120} e^{-t/120}\, dt$$

$$= 1 - e^{-10/120} = 0.07996$$

Thus, we would expect 0.07996 x 1000 = 80 batteries to be refunded.

8.8 Choosing a random sample from a continuous distribution

Finally, we consider how to choose a random sample from a continuous distribution, using random number tables (recall the method shown in chapter 6 page 138 for a discrete distribution). The basic method is shown using the continuous uniform distribution in the example below.

Example: Choose a random sample of size 5 from the continuous uniform distribution defined over the interval $x = 2$ to $x = 7$.

The pdf for this distribution is

$$f(x) = \frac{1}{5} \text{ for } 2 \le x \le 7$$

$$= 0, \qquad \text{otherwise.}$$

Furthermore, for any value a of the distribution we have:

$$P(x \le a) = \int_{\infty}^{a} f(x)\,dx = \int_{2}^{a} \frac{1}{5}\,dx = \frac{a-2}{5}$$

As before, we choose 5 such probabilities from the random number tables, thus yielding 5 values of a, making up our random sample. Working correct to 3 decimal places, and starting from the top left digit of table 8 on page 12, we have the 5 probabilities:

0.210 0.742 0.282 0.317 0.596

From the first of these, $\frac{a-2}{5} = 0.201$

giving $a = 3.005$

Repeating this process, we obtain the required sample:

3.005 5.71 3.41 3.585 4.98

The same method can be used with more complicated distributions, as illustrated in the following example.

Example: Choose a random sample of size 4 from the exponential distribution whose mean is 50.

Here $\frac{1}{\mu} = 50$ and so $\mu = \frac{1}{50}$

Thus, the pdf is: $f(x) = \frac{1}{50} e^{\frac{-x}{50}}, \; x \ge 0$

$$= 0, \qquad \text{otherwise}$$

Furthermore, for any value $a \ge 0$ we have,

$$P(x < a) \int_{0}^{a} f(x)\,dx = \int_{0}^{a} \frac{1}{50} e^{\frac{-x}{50}}.dx$$

$$= 1 - e^{\frac{-a}{50}}$$

From the random number tables, choose four random probabilities, say:
0.895, 0.569, 0.968, 0.721. For the first of these, we have:

$$1 - e^{\frac{-a}{50}} = 0.895$$

$$\therefore \quad e^{\frac{-a}{50}} = 1 - 0.895 = 0.105$$

$$\therefore \quad -\frac{a}{50} = \log_e(0.105) = -2.254$$

$$\therefore \quad a = 112.7$$

Repeating this process, we obtain the sample:

112.7 42.1 172.1 63.6

The next chapter considers the most important continuous probability distribution viz the normal distribution.

Exercise 8

1. Show that the following functions are well-defined pdf

a) $f(x) = \frac{2}{9}(3 - x), \quad 0 \leq x \leq 3$
$\quad = 0, \quad$ otherwise

b) $f(x) = \frac{3}{2}x^2, \quad -1 \leq x \leq 1$
$\quad = 0, \quad$ otherwise

c) $f(x) = \text{Sec}^2(x), \quad 0 \leq x \leq \frac{\pi}{4}$
$\quad = 0, \quad$ otherwise

d) $f(x) = -6(x - 3)(x - 2), \quad 2 \leq x \leq 3)$
$\quad = 0, \quad$ otherwise

e) $f(x) = x, \quad 0 \leq x \leq 1$
$\quad = 2 - x, \quad 1 \leq x \leq 2$
$\quad = 0, \quad$ otherwise.

Sketch each of the above pdf.

2. Find the value of A which makes the following functions well-defined.

a) $f(x) = A(2 - x)$, $\quad 0 \le x \le 2$
$= 0$, otherwise

b) $f(x) = A(x^2 - 4)$, $\quad 2 \le x \le 3$
$= 0$, otherwise

c) $f(x) = A.\text{Cos}(x)$, $\quad 0 \le x \le \frac{\pi}{2}$
$= 0$, otherwise

d) $f(x) = A.e^x$, $\quad 0 \le x \le 1$
$= 0$, otherwise

e) $f(x) = \frac{x}{3}$, $\quad 0 \le x \le A$
$= \frac{A(x - 3)}{3(A - 3)}$ $\quad A \le x \le 3$

(*Hint*: Draw a sketch first)

Sketch the above pdf

f) $f(x) = \frac{1}{27}x^2$ $\quad 0 \le x \le 3$
$= \frac{1}{3}$ $\quad 3 \le x \le 5$
$= 0$ otherwise

Sketch f(x) and calculate mean and variance.

g) $f(x) = \frac{1}{2}(3 - 2x)$ $\quad 0 \le x \le 1$
0 otherwise

i) Sketch 4(x)

ii) Find $P(.6 \le x \le .8)$

h) $f(x) = K\, x^2(2 - x)$ $\quad 0 \le x \le 2$
0 otherwise

i) Find K

ii) Sketch f(x)

iii) Find mean and variance

3. Calculate the means for the continuous random variables with pdfs in questions 1 and 2.

4. Calculate the medians for the continuous random variables with pdfs in question 1 except d), and 2 except b).

5. Calculate the modes for the continuous random variables with pdfs in questions 1 and 2.

6. Calculate the variances for the continuous random variables with pdfs in question 1 except c), and 2.

7. Calculate the required probabilities associated with continuous random variable whose pdfs are as in question 1 and 2 above.
   i) Question 1 a), $P(1 < x \leq 2)$
   ii) Question 1 b), $P(x \leq 0)$
   iii) Question 1 c), $P(x > \frac{\pi}{8})$
   iv) Question 1 d), $P(x = 2.5)$
   v) Question 1 e), (i) $P(x < 0.5)$ (ii) $P(x \leq 1.5)$
   vi) Question 2 a), $P(0 \leq x < 1)$
   vii) Question 2 b), $P(x > 2.5)$
   viii) Question 2 c), $P\left[\frac{\pi}{5} < x < \frac{\pi}{3}\right]$
   ix) Question 2 d), $P(x > 0.5)$
   x) Question 2 e), (i) $P(x < 1)$ (ii) $P(x \geq 2.5)$

8. The continuous random variable X has the pdf:

$$f(x) = \frac{3}{4}x^2(2 - x), \qquad 0 \leq x \leq 2$$
$$= 0, \qquad \text{otherwise}$$

   By sketching f(x), show that the distribution is skewed. State the relationship between the mean, median and mode; hence, by calculating the mean and mode, state two values that the median must lie between.

9. Derive the mean and variance of the uniform distribution.

   Random number tables are used to choose numbers, X, at random from the range 10 to 100, correct to three decimal places. Assuming that the uniform distribution is a good model of the variable X, state the mean and variance of X. Calculate the following probabilities.

   i) $P(x > 70)$;
   ii) $P(71 \leq x \leq 91.2)$

   A circle is drawn whose radius is x cm.

   Calculate:

   a) mean and variance of its circumference, C;
   b) mean of its area, A;
   c) the probability that its circumference exceeds 370 cm;
   d) the probability that its area is less than 10,000 cm^2.

10. Assuming that the ages X (to the nearest day) from a large school A are uniformly distributed over the range 3 to 8 years, inclusive, and the ages Y from a large school B are uniformly distributed over the range 5 to 11 years, write down the mean and variance of ages for each school.

    Write down the mean and variance of the following derived variables:

    a) $3X + 4$;

    b) $5 - Y$;

    c) $2X + 3Y$;

    d) $X - Y$

    Write down the mean and variance of the total age, T, of 3 pupils chosen from school A and 2 pupils chosen from School B.

    What assumption will you need to make?

11. Derive the mean and variance of the exponential distribution.

    The mean life of an electric light bulb is 500 hrs. Assuming bulb life, t, to be exponentially distributed, state and sketch the pdf of the random variable T.

    What is the variance of T?

    Calculate:

    a) the probability a light bulb lasts longer than 700 hrs;

    b) the percentage of light bulbs lasting exactly 500 hrs;

    c) the life below which 95% of bulbs last.

    What shape is the distribution? Demonstrate the following relationship:

    mode $\geq$ median $\geq$ mean.

    What is the probability that a light bulb has a life expectancy between the mean and mode?

12. The number of d km that a certain type of tyre will last is exponentially distributed with the pdf given below:

    $$P(d) = a.e^{-d/5000}, \quad d \geq 0.$$
    $$= 0, \quad \text{otherwise.}$$

    State the value of a.

    State the mean and variance.

    Calculate:

    a) the proportion of tyres lasting longer than 9,000 km;

    b) the probability that at least 2 out of 3 randomly chosen tyres last longer than 9,000 kg;

    c) the distance below which 20% of the tyres last.

13. Choose random samples of size 6 from the following distributions:

a) Uniform distribution defined on $-2 < x < 2$;

b) Exponential distribution with mean 4;

c) Continuous distribution with pdf:

i) $f(x) = \mathrm{Cos}(x)$, $\quad 0 \le x \le \frac{\pi}{2}$

ii) $f(x) = \frac{2}{9}(3 - x)$; $\quad 0 \le x \le 3$

14. The probability density function, $\phi(x)$, of a random variable x, is given by

$\phi(x) = \alpha x(4 - x) \qquad$ if $0 < x < 4$) $\qquad$ (= 0 otherwise).

Find the value of α, and hence find the mean and variance of x. What is the probability that x lies between 0 and 1?

15. The exponential distribution is a continuous distribution which has many applications in statistics. It is a probability distribution with probability density given by

$\phi(x) = a\,c^{-ax}$.

Find the mean and variance of this distribution.

The interval between successive reports of a certain rare disease is found to be distributed exponentially with a mean of 80 days. Find the probability of there being an interval of at least 120 days between one report of the disease and the next.

16. A random variable x, whose probability density funciton is $f(x)$ has a mean μ and μ' is the expected value of x^2. Show that the standard deviation σ is given by

$\sigma^2 = \mu' - \mu^2$

Determine c in order that

$f(x) = c \exp(-x/\sigma), \ (x \ge 0)$

$\quad = 0, \ (x < 0)$

may represent the probability density function of a random variable x.

17. The probability density function $f(x)$ of a variable x is given by

$f(x) = kx \sin \pi x \qquad (0 \le x \le 1)$

$\quad = 0$ for all other values of x.

Show that $k = \pi$ and deduce that the mean and the variance of the distribution are

$\left(1 - \frac{4}{\pi^2}\right)$ and $\frac{2}{\pi^2}\left(1 - \frac{8}{\pi^2}\right)$ respectively.

18. A random variable x has a probability density function

$$f(x) = Ax(6-x)^2, \quad 0 \leq x \leq 6$$
$$= 0 \text{ elsewhere.}$$

Find the value of the constant A.

Calculate the arithmetic mean, mode, variance and standard deviation of x.

19. The quality of an animal feedstuff depends both upon the raw materials used and the production process. One measure of this quality is the Nutritional Index, which varies between 0 and 1. A particular mill produces an animal feedstuff in batches of constant size, and the Nutritional Index of any batch may be considered to be an observation of a random variable X having probability density function

$$f(x) = kx(1-x^2) \quad (0 < x < 1)$$
$$= 0 \text{ otherwise}$$

a) Show that $k = 12$, and sketch $f(x)$.

b) Calculate $P(X < 0.25)$.

c) Determine the mean of X.

d) Batches of this mill's feedstuff may be sold for £500 each if the index is 0.8 or more, and £350 otherwise. The cost of producing a batch is £30. What is the expected profit per batch?

9 THE NORMAL DISTRIBUTION

9.1 Introduction

The normal distribution arises from the binomial distribution as n becomes large. The diagram below illustrates binomial distributions for increasingly large values of n ($\beta = \frac{1}{2}$ in these cases, though it could be any value); as n increases, so the polygonal shape of the distribution approximates a symmetrical, bell-shaped curve - this is the shape of the normal distribution.

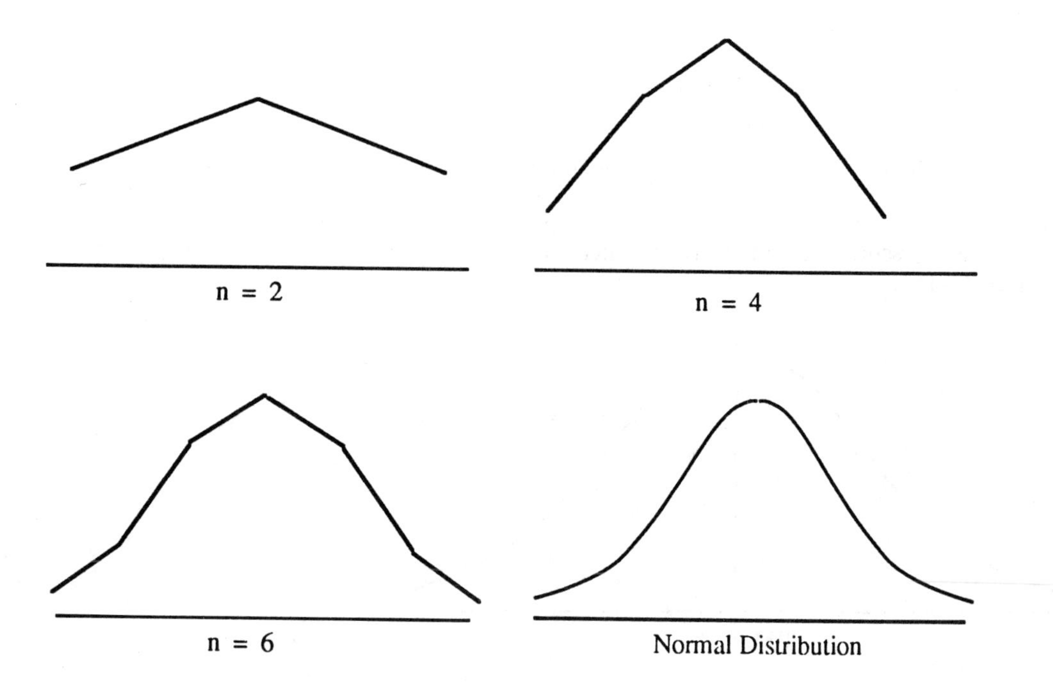

In order to specify the normal distribution, we need to state its pdf, ie the equation of the bell-shaped curve; this is a complicated function, and is as follows:

$$f(x) = \frac{1}{\sigma\sqrt{2\pi}} e^{-\frac{1}{2}\left[\frac{x-\mu}{\sigma}\right]^2} \qquad -\infty < x < \infty$$

As can be seen, the pdf is defined for all value of x. Furthermore, it is also well-defined, since:

$$\int_{-\infty}^{\infty} \frac{1}{\sigma\sqrt{2\pi}} e^{-\frac{1}{2}\left[\frac{x-\mu}{\sigma}\right]^2} dx = 1$$

However, the proof of this result is not required.

We will show later that the mean and standard deviation of the above normal distribution are μ and σ, respectively. This can be illustrated using the diagram below. We denote this normal distribution by $N(\mu, \sigma^2)$, the variance σ^2 being used in preference to the standard deviation σ.

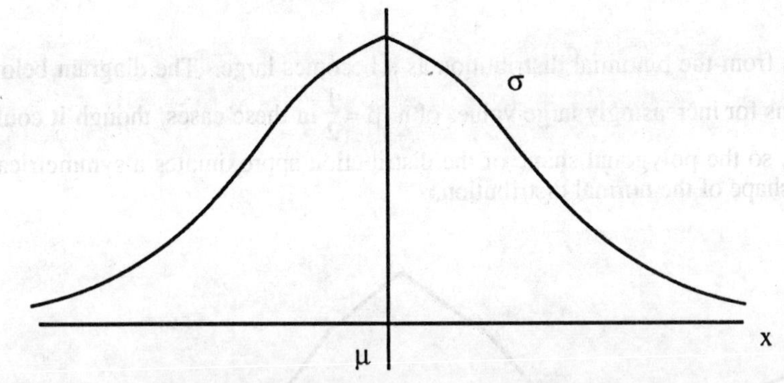

The reason that the normal distribution is so important is due to the fact that many populations in everyday life are approximately normally distributed, eg weight, height, IQ, masses of bags of sugar, errors of measurement. Each of these different normal distributions will have different values of μ and σ; for example, IQ scores are normally distributed with mean $\mu = 100$ and standard deviation $\sigma = 15$, and is depicted below.

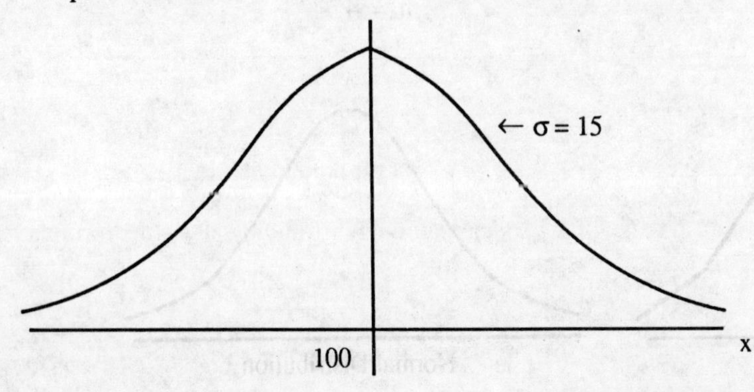

Suppose now we wish to calculate the probability that a randomly chosen adult has an IQ of below 120, say; as for other continuous distributions, we would have to calculate the shaded area in the diagram below using integration.

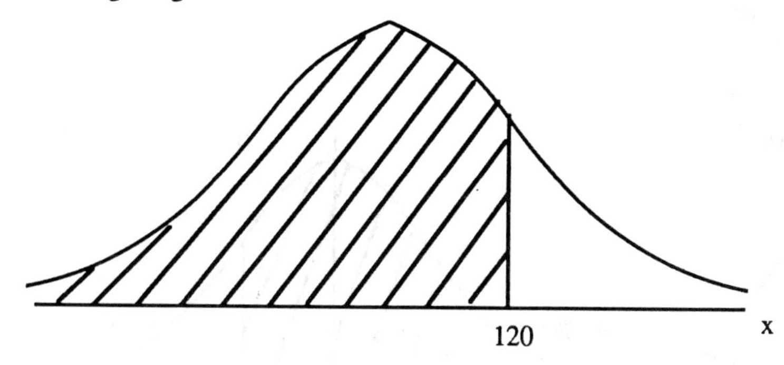

However, the pdf for this normal distribution is

$$f(x) = \frac{1}{15\sqrt{2\pi}} e^{-\frac{1}{2}\left[\frac{x - 100}{15}\right]^2} \qquad (-\infty < x < \infty)$$

which is extremely difficult to integrate.

How then are we to calculate the required probability $P(x < 120)$?

The answer is by using the normal distribution tables (Table 1, page 4 of the Cambridge Elementary Statistical Tables). These give the areas under the curve for different values of x. Obviously, however, as there are many different normal distributions, these tables cannot be used *directly* since the scale of x will be different in each case (notice that the values of x range from 0 to 4 in the tables, whereas we are interested in calculating the area under the curve and to the left of $x = 120$).

We must first transform our normal distribution of IQ scores into the so-called standard normal distribution that the tables correspond to; this is a unique normal distribution with mean $\mu = 0$ and standard deviation $\sigma = 1$. We can depict this transformation using the diagram below.

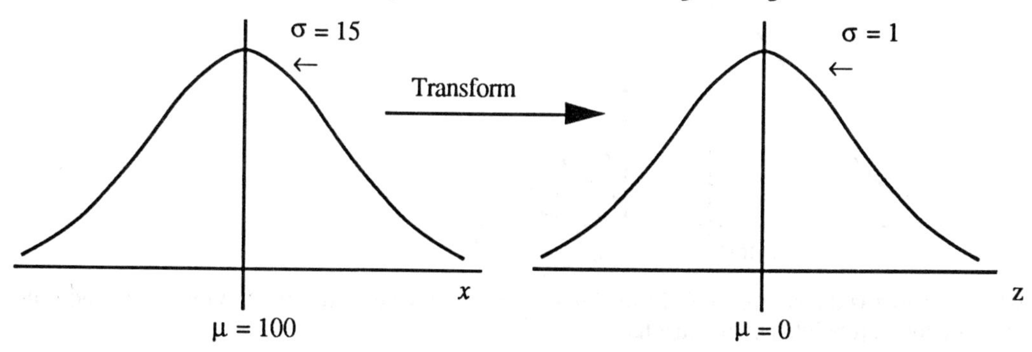

The nature of this transformation is to convert the scale of a *general* normal distribution (eg IQ scores) to that of the standard normal distribution (the one used in the tables); this is achieved by transforming each value of x into its corresponding Z-value of the standard normal distribution, using the formula:

$$Z = \frac{x - \mu}{\sigma}$$

Thus, we can transform our IQ score $x = 120$ into its corresponding Z-value (denoted Z_{120}):

$$Z_{120} = \frac{120 - 100}{15} = 1.33$$

This is depicted in the diagram below.

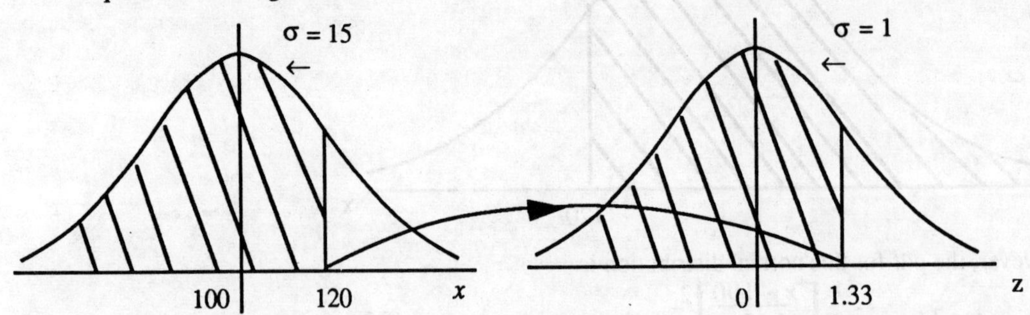

Having transformed our x-value into a Z-value, we can now look up the corresponding area in the tables (notice that the Cambridge Elementary Statistical Tables nevertheless use x as the variable, instead of Z; this is somewhat confusing). Reading down the third vertical column headed $\phi\ (x)$ we find the area corresponding to 1.33 is 0.9082. Thus, since this is the area to the left of Z = 1.33, we can write:

$$P(Z < 1.33) = 0.9082.$$

But, as the diagram above indicates, this is the *same* area to the left of $x = 120$; thus, we have finally found the probability of an adult having an IQ of less than 120, and can write:

$$P(x < 120) = P(Z < 1.33) = \underline{0.9082}$$

Suppose now we wish to calculate the probability of a randomly chosen adult having an IQ *exceeding* 120,

ie $P(x > 120)$; this is the shaded area in the diagram below.

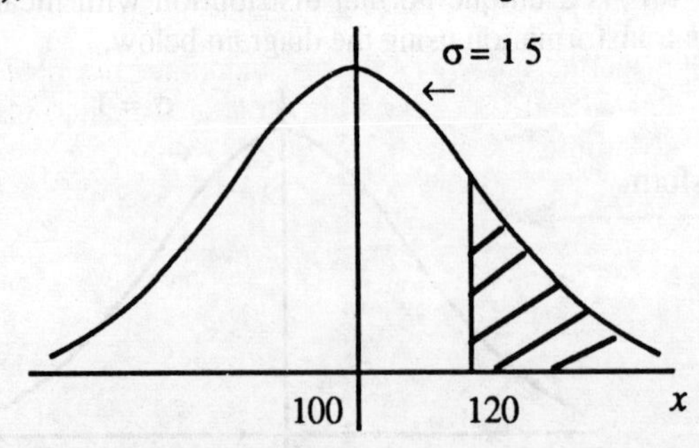

Now, since this area is the complement of the area calculated above, and since the total area under the curve is 1 (the total probability), we must have:

$$\begin{aligned} P(x > 120) &= \text{area to the right of } 120 \\ &= 1 - \text{area to the left of } 120 \\ &= 1 - P(x < 120) \\ &= 1 - 0.9082 \\ &= \underline{0.0918} \end{aligned}$$

If we now wish to calculate the probability of a randomly chosen adult having an IQ of above 90, say, the symmetry of the normal distribution will have to be used. First, we must calculate Z_{90}, the corresponding Z-score:

$$Z_{90} = \frac{90 - 100}{15} = -0.67$$

Thus, $P(x > 90) = P(Z > -0.67)$

As can be seen, Z_{90} is *negative*, whereas the normal distribution tables range only from $Z = 0$ to $Z = 4$; however, the diagram below shows that the area *above* $Z = -0.67$ is the same as the area *below* $Z = 0.67$.

Thus we can write:

$$P(x > 90) = P(Z > -0.67)$$
$$= P(Z < 0.67) = \underline{0.7486}\text{ , from the tables.}$$

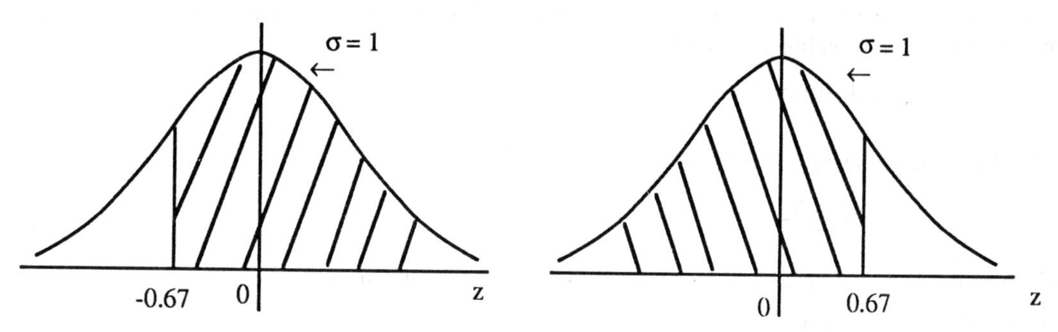

Finally, we can now find $P(x < 90)$ by again using the fact that the total area under a pdf curve is 1. Thus,

$$P(x < 90) = 1 - P(x > 90)$$
$$= 1 - 0.7486 = 0.2514$$

Now, by combining several of the previous results together, we can calculate the probability of a randomly chosen adult having an IQ of between 90 and 120, say. This is the area shaded in the diagram below, and can be seen to be the area below 120 minus the area below 90.

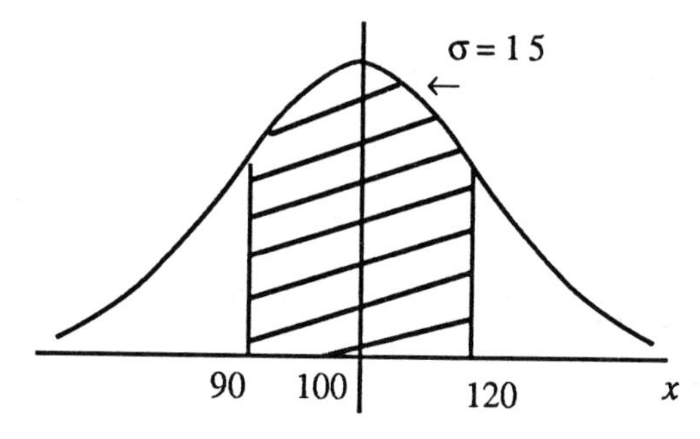

Thus, $P(90 < x < 120) = P(x < 120) - P(x < 90)$

$$= 0.9082 - 0.2514$$
$$= \underline{0.6568}$$

In actual fact, a Z-value is the number of standard deviations that its corresponding x value differs from the mean μ by; this is illustrated in the diagram below using the distribution of IQ scores. Here, each x-value marked is a whole number of standard deviations (15) from the mean (100).

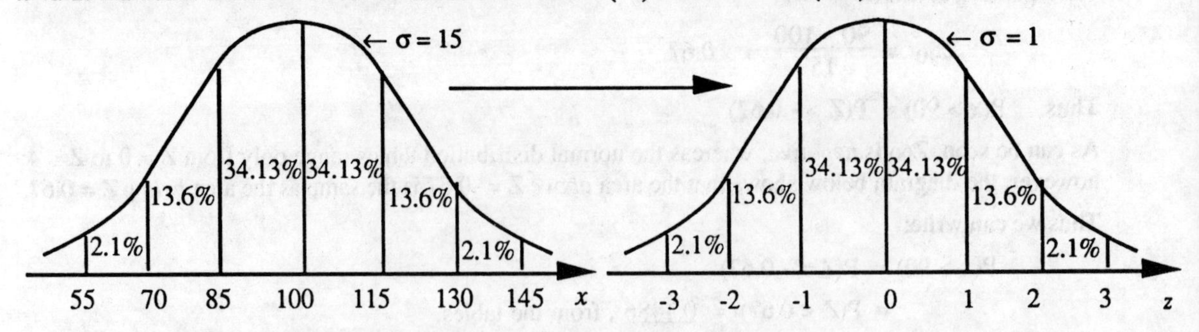

Thus, for example, the value x = 130 is 2 standard deviations from the mean, ie 130 = 100 + 2 x 15. Furthermore, the corresponding Z-value is:

$$Z_{130} = \frac{130 - 100}{15} = 2$$

Again, for x = 85, we have:

$$Z_{85} = \frac{85 - 100}{15} = -1\,,$$

which indicates that x = 85 is one standard deviation to the *left* of the mean.

Using the tables, we see that:

$$P(Z < 1) = 0.8413$$

$$P(Z < 2) = 0.97725$$

$$P(Z < 3) = 0.99865.$$

Thus, over 99% of all values of the normal distribution lie within 3 standard deviations of the mean; other percentages are shown on the diagram above.

We now consider the reverse of the kind of problem solved above; for example, below which IQ score does 91% of the adult population lie?

Written algebraically, the problem is to find the IQ score x_0 such that:

$$P(x < x_0) = 0.91$$

This is illustrated in the diagram below.

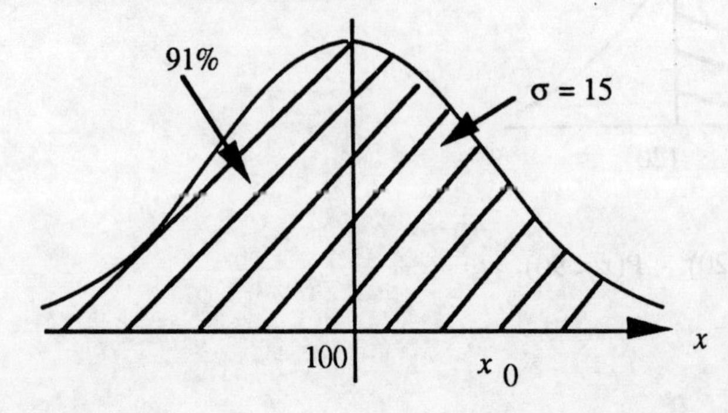

Now the IQ score x_0 corresponds to a certain z-value, say z_0, which is related to it by the formula:

$$z_0 = \frac{x_0 - 100}{15}$$

This is illustrated by the diagram below.

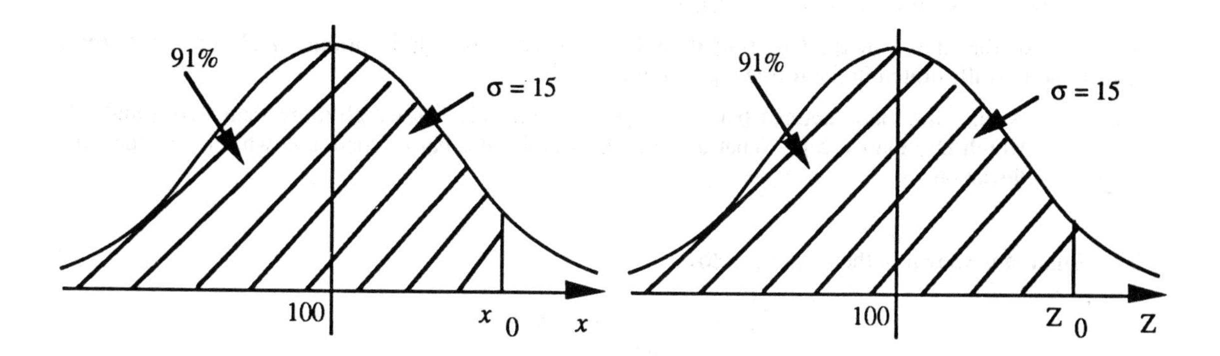

Furthermore, since we know that the area below Z_0 is 0.91, we can use the tables *in reverse* to find it; in fact,

$$Z_0 = 1.34$$

Thus, we can write:

$$1.34 = \frac{x_0 - 100}{15}$$

ie $\qquad x_0 = 100 + 1.34 \times 15 = \underline{120.1}$

[NB Here we have used the following transposition of the transformation formula: $x = \mu + z \cdot \sigma$]

Again, we can use the properties of the normal distribution (total probability = 1 and symmetry) to calculate other IQ scores. For example, suppose we wish to calculate the IQ score x_0 below which 28% of all adults lie (as depicted in the diagram below).

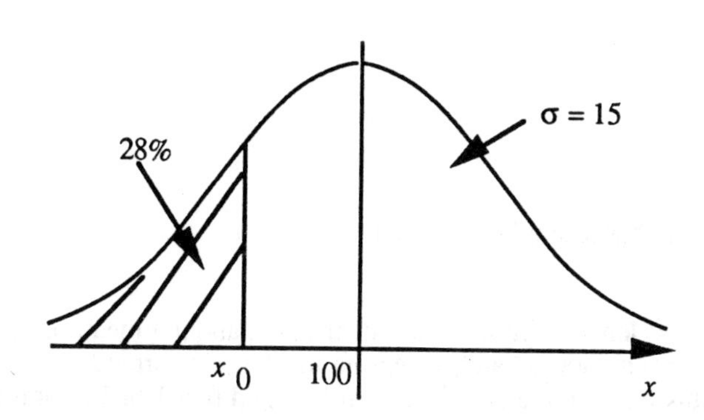

If the Z-score corresponding to x_0 is z_0, then z_0 is negative; so $-z_0$ is positive and has an area of 72% below it. From the tables then,

$$-Z_0 = 0.58$$

and so $$Z_0 = -0.58$$

$$\therefore \quad x_0 = 100 - 0.58 \times 15 = \underline{91.3}$$

We can use the above method to find μ and/or σ given enough information about the normal distribution, as illustrated in the following example.

Example: A machine manufactures bars of soap such that only 5% weigh more than 210 g and 1% weigh less than 185 g. What is the mean weight of a bar of soap, and what is the standard deviation?

The situation is shown in the diagram below.

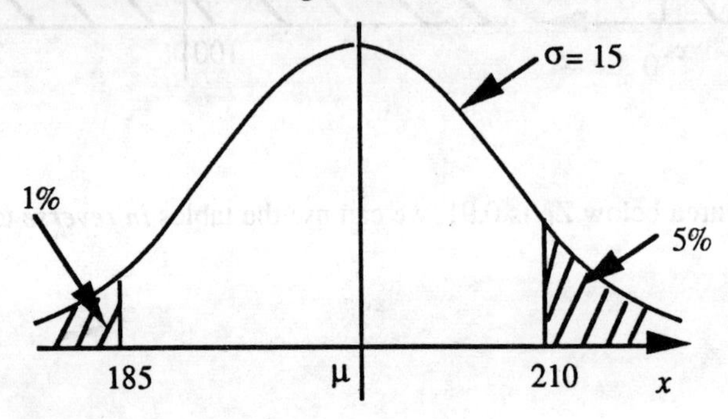

From the tables,

$$Z_{210} = 1.64 \text{ and } Z_{185} = -2.33$$

Hence, using the formula: $x = \mu + z \cdot \sigma$, we obtain the simultaneous equations:

$$210 = \mu + 1.64\,\sigma \quad (1)$$

$$185 = \mu - 2.33\,\sigma \quad (2)$$

(1) - (2): $$25 = 3.97\,\sigma$$

$$\therefore \quad \sigma = \underline{6.3 \text{ g}}$$

Substituting for σ into (1), we have:

$$210 = \mu + 1.64 \times 3.97$$

$$\therefore \quad \mu = \underline{203.5 \text{ g}}$$

Before proceeding with the next section, another example is worked.

Example:

A machine is producing components whose lengths are normally distributed about a mean of 6.50 cm. An upper tolerance limit of 6.54 cm has been adopted and, when the machine is correctly set, 1 in 20 components are rejected as exceeding this limit. On a certain day, it is found that 1 in 15 components are rejected for exceeding this limit.

a) Assuming that the mean has not changed but that production has become more variable, estimate the new standard deviation.

b) Assuming that the standard deviation has not changed but that the mean has moved, estimate the new mean.

c) If 1,000 components are produced in a shift, how many of them may be expected to have lengths in the average 6.48 cm to 6.53 cm if the machine is set as in a)?

When the machine is correctly set, the situation is as depicted below.

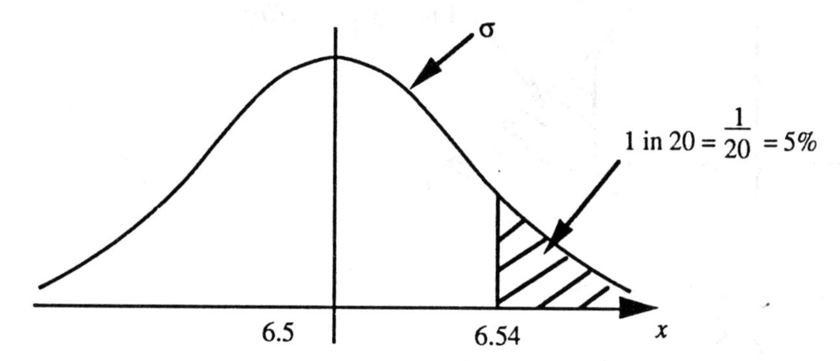

Here, $\mu = 6.5$, and we can find σ using the above method. From the tables, $Z_{6.54} = 1.64$, and so:

$$x = \mu + z \cdot \sigma$$

$$6.54 = 6.5 + 1.64\,\sigma$$

ie $\sigma = 0.024$ cm

When the machine is not correctly set, with 1 in 15 components being rejected, the situation changes.

a) Here, $\mu = 6.5$, but σ has changed to σ', say.

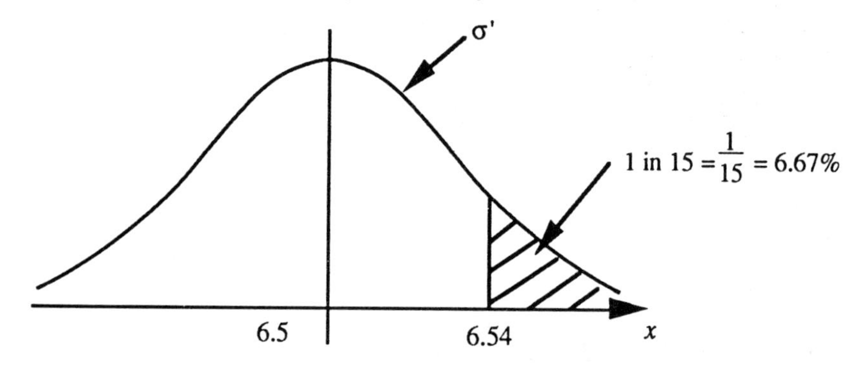

From the tables, $Z_{6.54} = 1.5$, and so:

$$6.54 = 6.5 + 1.5\sigma'$$

ie $\sigma' = 0.027$ cm

b) Here, $\sigma = 0.024$, but μ has changed to μ', say.

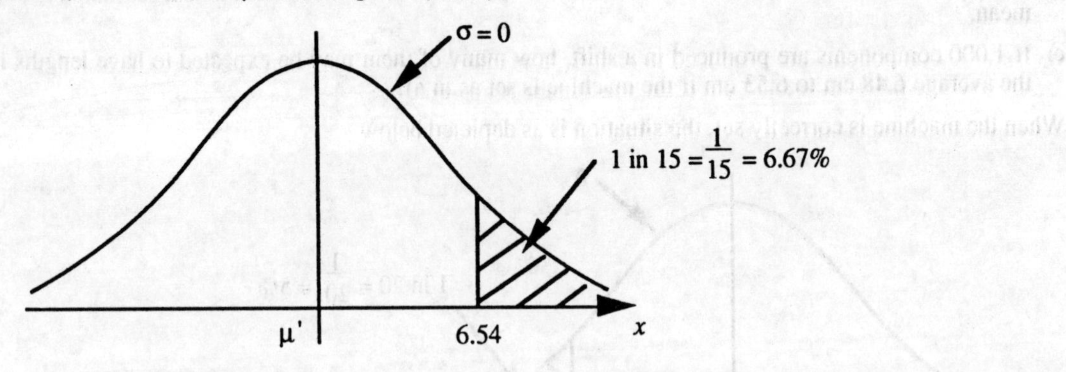

From the tables, $Z_{6.54} = 1.5$, as above, and so:

$$6.54 = \mu' + 1.5 \times 0.024$$

ie $\underline{\mu' = 6.504 \text{ cm}}$

c) Here, $\mu = 6.5$ and $\sigma = 0.027$, and the situation is as below.

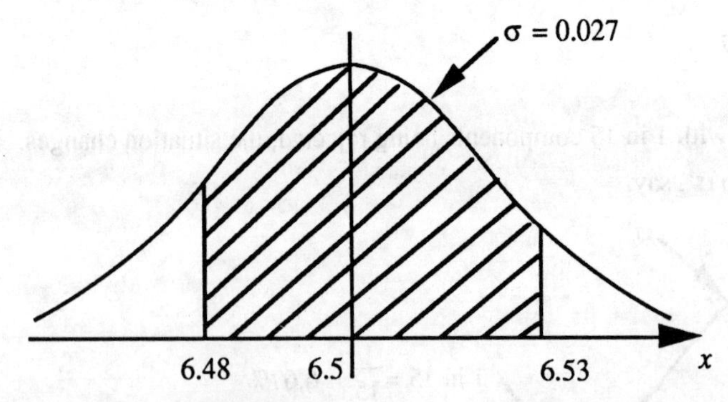

Now, $Z_{6.48} = \dfrac{6.48 - 6.5}{0.027} = -0.74$

and $Z_{6.53} = \dfrac{6.53 - 6.5}{0.027} = 1.11$

Hence, $P(6.48 < x < 6.53) = P(x < 6.53) - P(x < 6.48)$

(from the tables) $= 0.8665 - 0.2296$

$= 0.6369$

Thus, 63.69% of the 1,000 components lie between 6.48 cm and 6.53 cm ie <u>636.9 (= 637) components</u>.

9.2 Mean of the normal distribution

We now show that the mean of the normal distribution with pdf

$$f(x) = \frac{1}{\sigma\sqrt{2\pi}} e^{-\frac{1}{2}\left[\frac{x - \mu}{\sigma}\right]^2}, \quad -\infty < x < \infty$$

is μ, ie we will show that $E[X] = \mu$

Now, $E[X] = \int_{-\infty}^{\infty} x.f(x).dx$

$$= \int_{-\infty}^{\infty} x.\frac{1}{\sigma\sqrt{2\pi}} e^{-\frac{1}{2}\left[\frac{x-\mu}{\sigma}\right]^2} dx$$

$$= \frac{1}{\sigma\sqrt{2\pi}} \int_{-\infty}^{\infty} x.e^{-\frac{1}{2}\left[\frac{x-\mu}{\sigma}\right]^2} .dx$$

In order to simplify, make the substitution

$$t = \frac{x-\mu}{\sigma}$$

Then, $dt = \frac{dx}{\sigma}$, ie $dx = \sigma.dt$

and $x = \mu + \sigma t$; the limits do not change.

Thus, $$E[X] = \frac{1}{\sigma\sqrt{2\pi}} \int_{-\infty}^{\infty} (\mu + \sigma t).e^{-\frac{t^2}{2}}.\sigma dt$$

$$= \frac{\mu}{\sqrt{2\pi}} \int_{-\infty}^{\infty} e^{-\frac{t^2}{2}} dt + \frac{\sigma}{\sqrt{2\pi}} \int_{-\infty}^{\infty} t.e^{-\frac{t^2}{2}} dt.$$

We consider each of these integrals separately.

The first of these is too complicated to be considered here, and its evaluation is not required by the AEB. In fact, all questions requiring this proof state that the following assumption can be made:

$$\frac{1}{\sqrt{2\pi}} \int_{-\infty}^{\infty} e^{-\frac{t^2}{2}} . dt = 1$$

The second integral can be evaluated as follows:

$$\int_{-\infty}^{\infty} t.e^{-\frac{t^2}{2}} dt = \int_{0}^{\infty} t.e^{-\frac{t^2}{2}} dt + \int_{-\infty}^{0} t.e^{-\frac{t^2}{2}} dt$$

If we replace t by -t in the second of these integrals, then dt becomes -dt, and the limits become ∞ and 0; thus we have:

$$\int_{-\infty}^{\infty} t.e^{-t^2/2} dt = \int_{0}^{\infty} t.e^{-\frac{t^2}{2}} dt + \int_{\infty}^{0} -t.e^{-\frac{t^2}{2}}.(-dt)$$

$$= \int_{0}^{\infty} t.e^{-\frac{t^2}{2}} dt - \int_{0}^{\infty} t.e^{-\frac{t^2}{2}}.dt$$

$$= 0$$

Thus, we have:

$$E[X] = \mu.1 + \frac{\sigma}{\sqrt{2\pi}}.0 = \mu$$

9.3 Variance of the normal distribution

Here, we prove that the variance of the above normal distribution is σ^2,

ie $\quad V[X] = \sigma^2$

Now, $\quad V[X] = E[X^2] - (E[X])^2$

$\quad\quad = E[X^2] - \mu^2$, from above.

Also, $\quad E[X^2] = \int_{-\infty}^{\infty} x^2.f(x)dx$

$$= \int_{-\infty}^{\infty} x^2 . \frac{1}{\sigma\sqrt{2\pi}} e^{-\frac{1}{2}\left[\frac{x-\mu}{\sigma}\right]^2} dx$$

$$= \frac{1}{\sigma\sqrt{2\pi}} \int_{-\infty}^{\infty} x^2.e^{\left[\frac{x-\mu}{\sigma}\right]^2} .dx$$

Again, make the substitution $t = \frac{x-\mu}{\sigma}$, then, $dx = \sigma.dt$, and $x = \mu + \sigma t$.

Thus, $\quad E[X^2] = \frac{1}{\sigma\sqrt{2\pi}} \int_{-\infty}^{\infty} (\mu + \sigma t)^2 .e^{-\frac{t^2}{2}} .\sigma.dt$

$$= \frac{1}{\sqrt{2\pi}} \int_{-\infty}^{\infty} (\mu^2 + 2\mu\sigma t + \sigma^2 t^2) e^{-\frac{t^2}{2}} .dt$$

$$= \frac{\mu^2}{\sqrt{2\pi}} \int_{-\infty}^{\infty} e^{-\frac{t^2}{2}} .dt + \frac{2\mu\sigma}{\sqrt{2\pi}} \int_{-\infty}^{\infty} t.e^{-\frac{t^2}{2}} dt + \frac{\sigma^2}{\sqrt{2\pi}} \int_{-\infty}^{\infty} t^2.e^{-\frac{t^2}{2}} dt$$

$$= \mu^2.1 + \frac{2\mu\sigma}{\sqrt{2\pi}}.0 + \frac{\sigma^2}{\sqrt{2\pi}} \int_{-\infty}^{\infty} t^2 e^{-t^2/2}.dt \quad \left\{ \text{remember } \frac{1}{\sqrt{2\pi}} \int_{-\infty}^{\infty} e^{-\frac{t^2}{2}} .dt = 1 \right\}$$

(using the above results).

Now, $\quad \frac{1}{\sqrt{2\pi}} \int_{-\infty}^{\infty} t^2.e^{-\frac{t^2}{2}} dt = \frac{1}{\sqrt{2\pi}} \int_{-\infty}^{\infty} t.\left[t\, e^{-\frac{t^2}{2}}\right] dt$

which we can integrate by parts.

Taking $u = t$ and $dv = t.e^{-\frac{t^2}{2}}$,

we have: $du = 1$ and $v = \int t.e^{-\frac{t^2}{2}} dt = -e^{-\frac{t^2}{2}}$

(the latter integration by substituting $s = t^2$, say)

Now, the rule for integration by parts is as follows:

$\int u.dv = uv - \int v.du$

$$\text{Thus, } \frac{1}{\sqrt{2\pi}} \int_{-\infty}^{\infty} t.\left[te^{-\frac{t^2}{2}}\right] dt = \frac{1}{\sqrt{2\pi}} \left[t.-e^{-\frac{t^2}{2}}\right]_{-\infty}^{\infty} - \frac{1}{\sqrt{2\pi}} \int_{-\infty}^{\infty} 1.-e^{-\frac{t^2}{2}} .dt$$

$$= 0 + 1 = 1$$

Hence, $E[X^2] = \mu^2 + \sigma^2$, and so:

$V[X^2] = (\mu^2 + \sigma^2) - \mu^2 = \sigma^2$

The above proofs are tedious, but it is advisable to learn them in view of the high marks they carry.

9.4 Linear combination of normal variables

A common question often involves two or more normal variables, say X and Y, *simultaneously*.

We can combine these two normal variables together, using constants a and b, to form a new normal variable, viz:

$$aX = bY$$

Being another normal variable, we can then proceed to apply the theory so far developed to this new variable; however, we firstly need to recall the results of chapter 7, page 134/135.

$$E[aX + bY] = a.E[X] + b.E[Y]$$

and $$V[aX + bY] = a^2V[X] + b^2V[Y]$$

the latter result only if X and Y are independent.

Thus, if X and Y have means μ_x, μ_y and variances $\sigma^2{}_x$, $\sigma^2{}_y$, respectively, we have:

$$E[aX + bY] = a.\mu_x + b.\mu_y$$

and $$V[aX + bY] = a^2.\sigma^2{}_x + b^2.\sigma^2{}_y$$

NB You are often required to prove this result.

Consider the following example.

Example: Cubic building blocks are made in two sizes, the sides of which are normally distributed with means 4 cm, 8 cm and with variances 0.5 cm^2, 0.6 cm^2.

a) What is the average height of a column made up of one of each type?

What is the probability that this height exceeds 12.2 cm?

b) If a new type of block is produced, with sides normally distributed of mean length 12 cm and variance 0.75 cm^2, what is the mean height of a column made up of 3 small, 2 medium and 1 large block? What percentage of these exceed a height of 43 cm? Given that 7% of these columns lie below a certain height, what is this height?

a) Let X and Y denote the lengths of sides of the cubic building blocks; then we have:

$\mu_x = 4$ cm, $\quad \sigma^2{}_x = 0.5$ cm^2

$\mu_y = 8$ cm, $\quad \sigma^2{}_y = 0.6$ cm^2

Now, the height of a column is X + Y, which is normally variable, say W; then

$W = X + Y$

Using the above results, with a = 1 and b = 1, we have:

$E[W] = E[X + Y] = \mu_x + \mu_y = 4 + 8 = 12$ cm

and $V[W] = V[X + Y] = \sigma^2{}_x + \sigma^2{}_y = 0.5 + 0.6 = 1.1$ cm

Hence, $\mu_w = 12$ cm and $\mu^2{}_w = 1.1$

We require $P(W > 12.2)$, illustrated in the diagram.

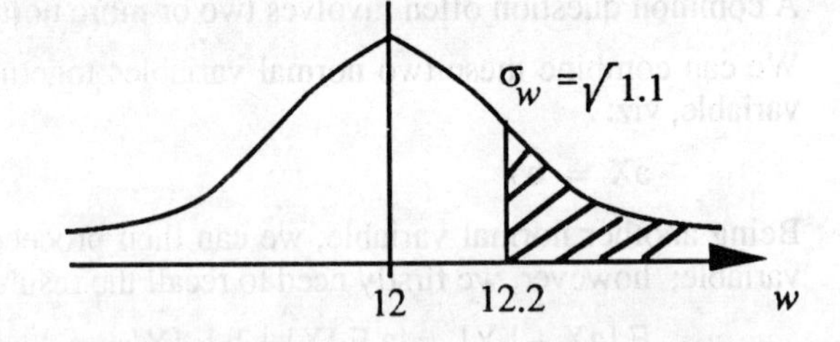

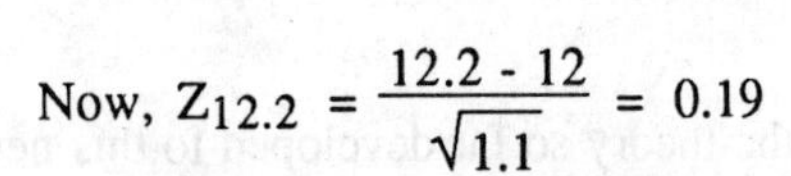

Now, $Z_{12.2} = \dfrac{12.2 - 12}{\sqrt{1.1}} = 0.19$

Hence, $P(W > 12.2) = P(Z > 0.19)$

$= 1 - P(Z < 0.19)$

$= 1 - 0.5753 = \underline{0.4247}$

b) Let V denote the third normal variable; then, $\mu_v = 12$ cm and $\sigma^2{}_v = 0.75$ cm^2.

If H is the height of a column, then:

$H = 3X + 2Y + V$

yet another normal variable.

Applying the above results, we have:

$E[H] = E[3X + 2Y + V] = 3\mu_x + 2\mu_y + \mu_v$

$= 3 \times 4 + 2 \times 8 + 12$

$= 40$ cm.

and $V[H] = V[3X + 2Y + V] = 9\sigma^2{}_x + 4\sigma^2{}_y + \sigma^2{}_v$

$= 9 \times 0.5 + 4 \times 0.6 + 0.75$

$= 7.65$ cm^2

Hence, $\mu_h = 40$ cm and $\sigma^2_h = 7.65$

We require P(h > 43)

Now, $Z_{43} = \frac{43 - 40}{\sqrt{7.65}} = 1.08$

Hence, $P(h > 43) = P(Z > 1.08)$

$= 1 - P(Z < 1.08)$

$= 1 - 0.8599$

$= 0.1401$

Thus, the percentage exceeding 43 cm is 14.01%

Furthermore, if h_o is the height below which 7% of all columns lie, then,

$P(h < h_o) = 0.07$

Thus, $Z_o = -1.475$ (from the tables)

Hence, ho $= 40 - 1.475\sqrt{7.65}$

$= 35.9$ cm

9.5 Normal approximation to the binomial distribution

As we saw in the beginning of this chapter, the binomial distribution gradually approaches a normal distribution as n increases; thus, for large values of n we can use the normal distribution (and its associated theory developed above) to approximate the binomial distribution.

Now, obviously, for a normal distribution to approximate a binomial distribution, their means and variances must be the same; recall from chapter 6 page 121 that the mean and variance of the binomial distribution are $n\beta$ and $n\beta(1 - \beta)$, respectively. Thus, for the normal distribution $N(\mu, \sigma^2)$ to approximate the binomial distribution with parameters n and β we must have:

$\mu = n\beta$ and $\sigma^2 = n\beta(1 - \beta)$

NB: Since the binomial variable R is *discrete*, with r = 0, 1, ..., n, and we are approximating it by the *continuous* normal variable, X, with $-\infty < x < \infty$, there are certain theoretical problems; however, since this is not on the AEB syllabus, they will be overlooked here.

Summarising this result, we have:

Providing n is large, the binomial distribution with parameters n and β can be approximated by $N(n\beta, n\beta(1 - \beta))$.

We can now solve problems involving binomial distributions with large values of n, eg n = 30, 50, 100, 200, which would involve tedious calculations otherwise.

Example: A fair, six-sided die is thrown 100 times; what is the probability of obtaining more than 20 sixes?

Here, the number of sixes is binomially distributed, with n = 100 and $\beta = \frac{1}{6}$. Since n is large, we can approximate using the normal distribution $N(\mu, \sigma^2)$, where

$$\mu = n\beta = 100 \times \frac{1}{6} = 16.67$$

$$\text{and } \sigma^2 = n\beta(1 - \beta) = 100 \times \frac{1}{6}\left(1 - \frac{1}{6}\right) = 13.89$$

We require the probability of more than 20 sixes, ie $P(x > 20)$.

$$\text{Now,} \quad Z_{20} = \frac{20 - 16.67}{\sqrt{13.89}} = 0.89$$

$$\begin{aligned} \text{Thus,} \quad P(x > 20) &= P(Z > 0.89) \\ &= 1 - P(Z < 0.89) \\ &= 1 - 0.8133 = 0.1867 \end{aligned}$$

9.6 The distribution of proportions

This is a continuous distribution arising from the binomial distribution when n becomes large. Recall that the binomial variable r is the number of 'successes' in n repeated trials; for example, if 200 people were asked whether they preferred margarine to butter, 70 might reply 'Yes' (a success). Here, then, r = 70 and n = 200, and we say that 70 out of 200 people prefer margarine to butter. However, we could equally well express 70 as a fraction (or *proportion*) of 200, and say that 70/200 or 35% or 0.35 of people prefer margarine to butter.

Thus, we have converted the value r = 70 into a proportion $p = \frac{70}{200}$, using the formula:

$$\boxed{p = \frac{r}{n}}$$

Clearly, all such proportions p lie between 0 and 1, and are continuous over this range.

Furthermore, if many such samples of size n are chosen and p calculated for each, we will end up with a continuous distribution of proportions p, whose mean and variance can be found as follows:

$$E[P] = E\left[\frac{R}{n}\right] = \frac{1}{n} \cdot E[R] = \frac{1}{n} \cdot n\beta = \beta$$

$$V[P] = V\left[\frac{R}{n}\right] = \frac{1}{n^2} \cdot V[R] = \frac{1}{n^2} \cdot n\beta(1-\beta) = \frac{\beta(1-\beta)}{n}$$

Summarising, we have:

$$\boxed{\begin{aligned} E[P] &= \beta \\ V[P] &= \frac{\beta(1-\beta)}{n} \end{aligned}}$$

Now, since this variable P is simply a multiple of the binomial variable R, then providing n is large, P will be approximately normal (since R is); thus, we can now solve problems involving proportions of samples using the theory of normal distributions.

Example: It is estimated that 23% of the population believe in ghosts. If a random sample of 150people were each asked 'do you believe in ghosts?' what is the probability that less than 30 people say 'yes'?

Here, n = 150, $\beta = 0.23$; since n is large we can assume that the distribution of proportions, P, of believers is normally distributed, with:

$$\mu = \beta = 0.23$$

and $$\sigma^2 = \frac{\beta(1-\beta)}{n} = \frac{0.23(1 - 0.23)}{150} = 0.00118$$

Now, we require the probability that less than 30 people out of a sample of 150 people are believers, ie a proportion of less $\frac{30}{150} = 0.2$ are believers, which can be written:

$$P(p < 0.2)$$

The Z-value for p = 0.2 is:

$$Z = \frac{0.2 - 0.23}{\sqrt{0.00118}} = -0.71 = -0.87$$

Thus, $P(p \leq 0.2) = P(Z < 0.71) = P(Z < -0.87)$

$= 1 - P(Z < 0.71) = 1 - P(Z < 0.87)$

$= 1 - 0.7611 = 0.2389 = 1 - 0.8078 = \underline{0.127}$

[Note: This problem could just as easily have been solved using a binomial approximation with n = 150 and $\beta = 0.23$.]

9.7 The sampling distribution of the mean

Suppose we draw many samples, size n, from the normal distribution $N(\mu, \sigma^2)$, and calculate the mean, $\bar{x}$ of each sample. This would give us a distribution of means, which we call the *sampling distribution of the mean.* We can easily find the mean and variance of this distribution as follows. Consider one such sample $x_1, x_2, ..., x_n$; then, the mean, $\bar{X}$, of this sample is given by:

$$X = \frac{1}{n}(x_1 + x_2 + ... x_n)$$

Thus, we have:

$$E[\bar{X}] = E\left[\frac{1}{n}(x_1 + x_2 + ... + x_n)\right]$$

$$= \frac{1}{n}(E[x_1] + E[x_2] + ... + E[x_n])$$

$$= \frac{1}{n}(\mu + \mu + ... + \mu) *$$

$$= \frac{1}{n}(n\mu) = \mu$$

[*This step follows since $E[x_1]$, for example, is the mean of all possible values x_1 could be, which is thus the mean of the underlying distribution $N(\mu, \sigma^2)$, and is thus μ; the same argument applies to every other member of the sample.]

$$V[\bar{X}] = V\left[\frac{1}{n}(x_1 + x_2 + \dots + x_n)\right]$$

$$= \frac{1}{n^2}(V[x_1] + V[x_2] + \dots + V[x_n])$$

$$= \frac{1}{n^2}(\sigma^2 + \sigma^2 + \dots + \sigma^2)$$

$$= \frac{1}{n^2}(n\sigma^2) = \frac{\sigma^2}{n}$$

Summarising, we have:

$$E[\bar{X}] = \mu$$

$$V[\bar{X}] = \frac{\sigma^2}{n}$$

We can illustrate the derivation of the sampling distribution of the mean using the following diagram. Here we have denoted

$E[\bar{X}]$ by $\mu^2_{\bar{x}}$ and $V[\bar{X}]$ by $\sigma^2_{\bar{x}}$.

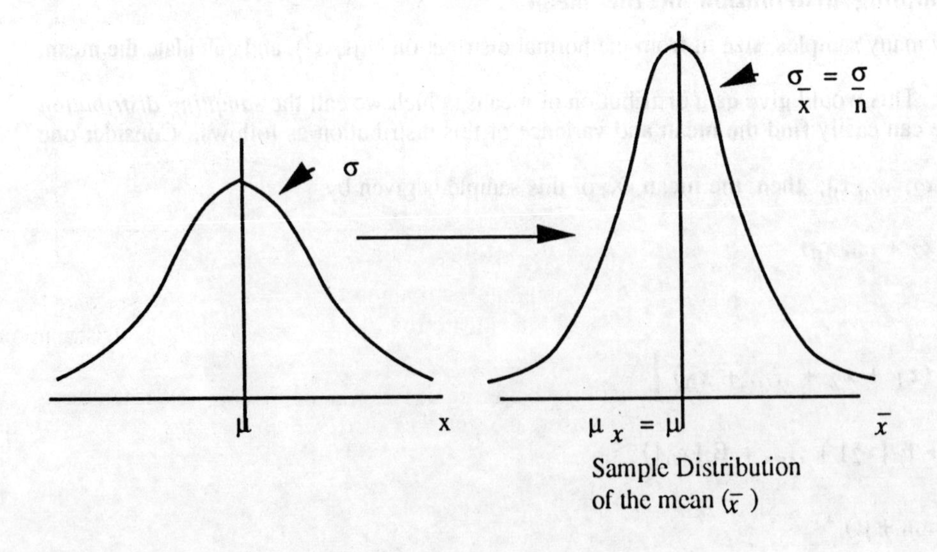

Sample Distribution
of the mean ($\bar{x}$)

Notes: i) The standard deviations are used on these diagrams, as opposed to the variances considered above.

ii) Notice that the sampling distribution of the mean has a smaller dispersion (is narrower) than the underlying distribution; this is due to the variance σ^2 being reduced by a factor of n, to become $\frac{\sigma^2}{n}$. Thus, as n increases, the means, $\bar{x}$, cluster more tightly around the population mean, μ.

Notice that this result holds whether the underlying distribution is normal or not; however, if it *is* normal, then so is the resulting sampling distribution of the mean, and so we can denote it by $N\left(\mu, \frac{\sigma^2}{n}\right)$.

Thus, we are now in a position to tackle problems involving samples drawn from normal distributions.

Example: A certain machine produces ball bearings with diameters that are normally distributed with mean 12 mm and variance 0.007 mm^2. In a routine quality control check, a sample of 25 ball bearings is taken; calculate the probability that the mean of this sample exceeds the upper tolerance limit of 12.02 mm. What size of sample should be taken to ensure that the upper tolerance limit is exceeded by at most 1 in a 100 samples?

Here, $\mu = 12$ mm, $\sigma^2 = 0.007$ mm^2 and n = 25

Thus, $\mu_{\bar{x}} = 12$ and $\sigma^2_{\bar{x}} = \frac{0.007}{25} = 0.00028$.

We require $P(\bar{x} > 12.02)$, as illustrated in the diagram below.

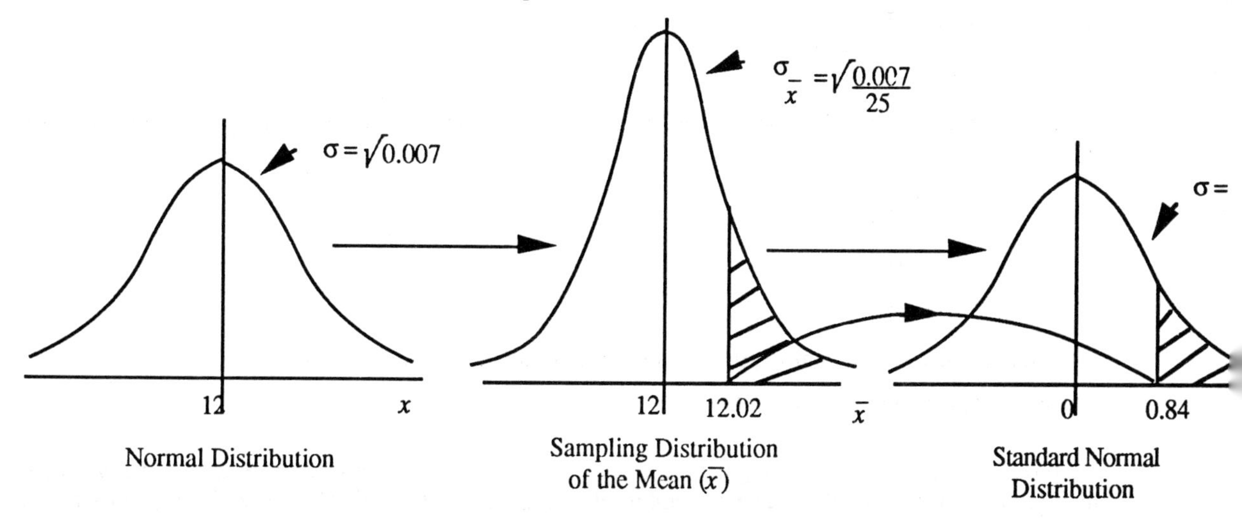

Notice that 12.02 is *not* placed on the underlying distribution since it is a *mean*, ie an $\bar{x}$ value, not an *x*-value. Problems involving *means* must be solved using the *sampling distribution of means.*

$$\text{Now, } Z_{12.02} = \frac{12.02 - 12}{\sqrt{0.00028}} = 0.84.$$

$$\begin{aligned}\text{Hence, } P(\bar{x} > 12.02) &= P(Z > 0.84) \\ &= 1 - P(Z < 0.84) \\ &= 1 - 0.7995 = 0.2005\end{aligned}$$

If we let n be the required sample size, then we require that:

$P(\bar{x} > 12.02) \leq 0.01$

as illustrated in the diagram below.

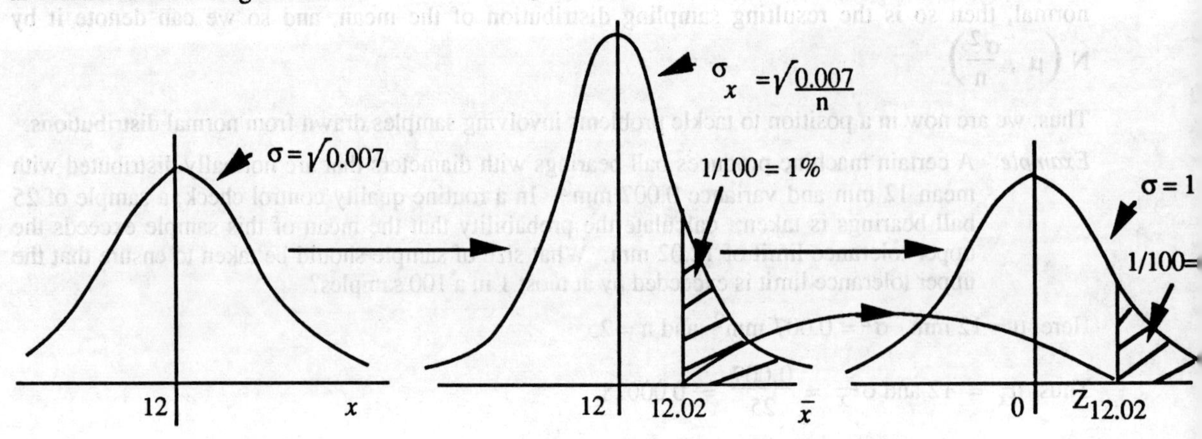

Here, $Z_{12.02} = \dfrac{12.02 - 12}{\sqrt{\dfrac{0.007}{n}}}$

From the tables, we see that the least value $Z_{12.02}$ can be is 2.33 corresponding to a probability of 0.99.

Thus, $2.33 \leq \dfrac{12.02 - 12}{\sqrt{\dfrac{0.007}{n}}}$

$\therefore \quad n \geq 95.006$

Since n must be a whole number, then n must be at least 96.

The sampling distribution of the total

Suppose that, instead of calculating the mean $\bar{X}$ of samples of size n drawn from $N(\mu, \sigma^2)$, we calculate their totals, T; we would then have a *sampling distribution of totals*. Furthermore, for a given sample, size n, whose mean is $\bar{X}$, we would have:

$$\bar{X} = \frac{T}{n}$$

and so,

$$\boxed{T = n.\bar{X}}$$

Since $\bar{X}$ is normally distributed, then, so too is T. Furthermore, the mean and variance of such totals can be easily derived, as follows:

$$E[T] = E[n\bar{X}] = n.E[\bar{X}] = n\mu$$

$$V[T] = V[n\bar{X}] = n^2 V[\bar{X}] = n^2 \frac{\sigma^2}{n} = n\sigma^2$$

Summarising, we have:

$$E[T] = n\mu$$
$$V[T] = n\sigma^2$$

This can be illustrated using the diagram below.

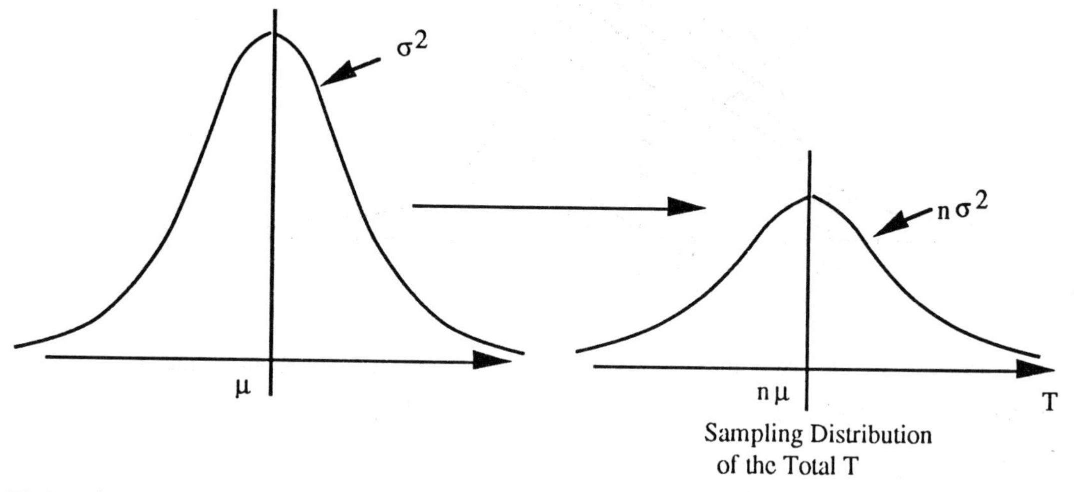

Sampling Distribution of the Total T

Notice that the distribution of the total T has a greater dispersion (is wider) than the underlying dispersion, due to the variance being increased by a factor of n.

We can now solve problems involving totals of samples drawn from normal distributions.

Example: Metal workers produce masses that are normally distributed with mean 0.5g and variance 0.003 g^2. They are packed into bags of 50.

Calculate:

a) the mean mass of a bag

b) the probability that a randomly chosen bag weighs more than 25.4g

c) the upper and lower tolerance limits between which 99% all bags lie (assuming these to be symmetrical about the means).

Now, samples of size 50 drawn from the normal distribution N(0.5, 0.003) will have masses also normally distributed; furthermore, if T is the total mass of a bag, then,

$$E[T] = n\mu = 50 \times 0.5 = 25g$$

and $V[T] = n\sigma^2 = 50 \times 0.003 = 0.15g^2$

a) Thus, the mean mass of a bag is 25g.

b) We require P(T > 25.4g).

$$\text{Now, } Z_{25.4} = \frac{25.4 - 25}{\sqrt{0.15}} = 1.03$$

$$\begin{aligned} \text{Thus, } P(T > 25.4) &= P(Z > 1.03) \\ &= 1 - P(Z < 1.03) \\ &= 1 - 0.8485 \\ &= \underline{0.1515} \end{aligned}$$

c) We require the values T_0 and T_1 between which 99% of all bags lie, as shown in the diagram below.

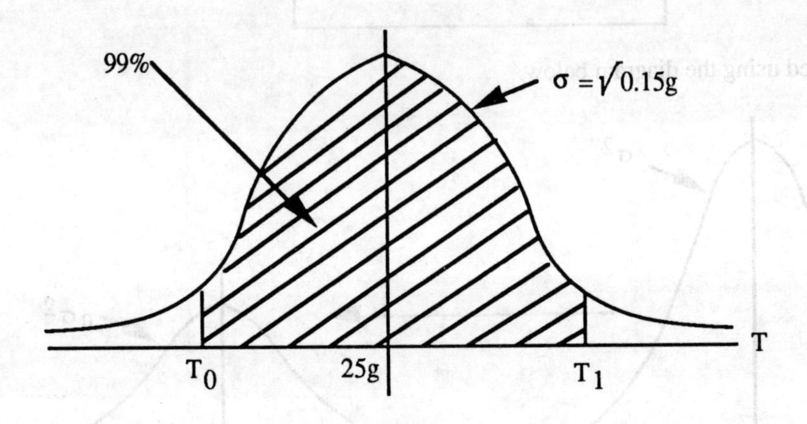

It is only necessary to find one of these, T_1, say, since they are symmetrical about the mean.

Now, $P(T < T_1) = 0.995$, and so if Z_1 is the corresponding Z-value,

$$Z_1 = 2.575$$

$$\begin{aligned} \text{Thus, } T_1 &= 25 + 2.575\sqrt{0.15} \\ &= 26\text{g} \end{aligned}$$

Hence, the upper and lower tolerance limits are 26g and 24g, respectively.

9.8 The central limit theorem

So far we have only considered samples taken from normal distributions; in these cases, both the sampling distribution of the *mean* and the sampling distribution of the *total* are automatically *normal* too. However, if the underlying distribution is *not* normal, these two sampling distributions are *not necessarily normal*; the following important result tells us when they will be normal though; you may be asked to quote this theorem.

> The Central Limit Theorem (CLT) states that the sampling distribution of the mean $(\bar{x})$ of samples (size n) drawn from *any* distribution with mean μ and variance σ^2 becomes normally distributed with mean μ and variance σ^2/n, ie becomes $N(\mu, \sigma^2/n)$, as n becomes large.

This is illustrated in the diagram below, with the underlying distribution being intentionally made to look *not* normal.

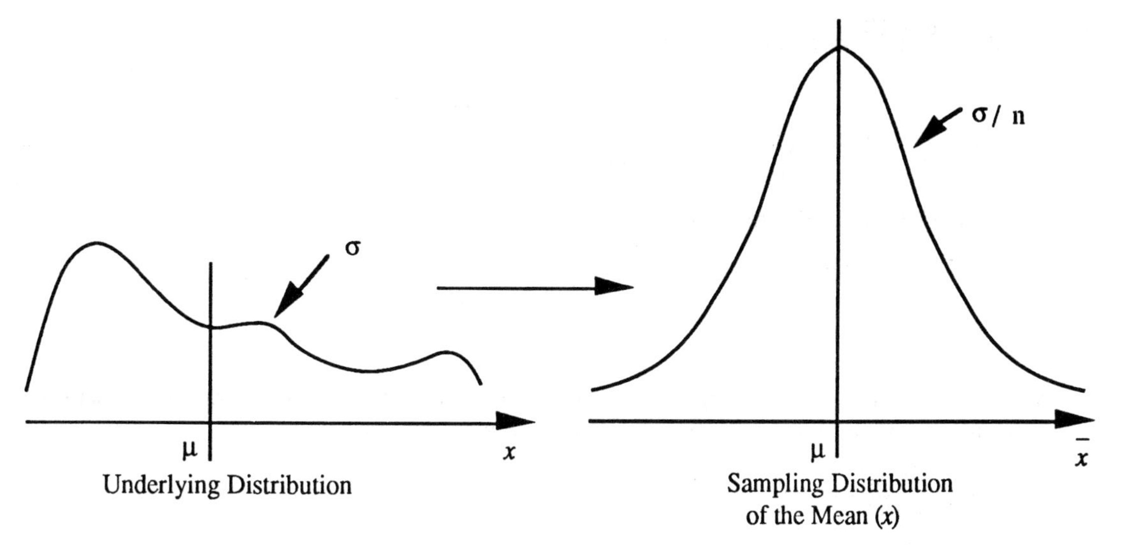

Now, since the sampling distribution of the total (T) is normal providing the sampling distribution of the mean $(\bar{x})$ is, then, we have the immediate corollary:

> Corollary:
>
> As n becomes large, the sampling distribution of the total (T) of samples, size n, drawn from *any* distribution with mean μ and variance σ^2 becomes normally distributed, with mean $n\mu$ and variance $n\sigma^2$, ie becomes $N(n\mu, n\sigma^2)$.

We can now solve the following problem.

Example: Trains depart from a station every ten minutes. A man sets out to catch one every day of the week, arriving at the station at random times in the morning, knowing that he will have to wait at most ten minutes.

a) What kind of distribution will his waiting times follow?

b) If his waiting times were averaged over a 1 month period, what is the probability that this average waiting time exceeds 6 minutes? (Assume 1 month = 30 days.)

c) Over a 3 month period, how long would you expect him to have waited on the platform altogether? What is the probability of this time being less than 7 hours?

a) Since he arrives at the station in a random fashion, then his waiting times will be uniformly distributed over the interval 0 - 10 minutes. Thus, the mean μ and variance σ^2 of this distribution will be:

$$\mu = \frac{0 + 10}{2} = 5 \text{ mins.}$$

} see page 175

$$\sigma^2 = \frac{(10 - 0)^2}{12} = \frac{100}{12} = 8.33 \text{ mins}^2$$

b) In one month he catches the train 30 times.

Thus, n = 30 and we require

$P(\bar{x} > 6)$

Now, if we assume n to be large, then the sampling distribution of the mean $(\bar{x})$ will be approximately N(5, 8.33/30), from the CLT as illustrated below, and so:

$$Z_6 = \frac{6 - 5}{\sqrt{\frac{8.33}{30}}} = 1.90$$

Thus, $P(\bar{x} > 6) = P(Z > 1.90) = \underline{0.0287}$

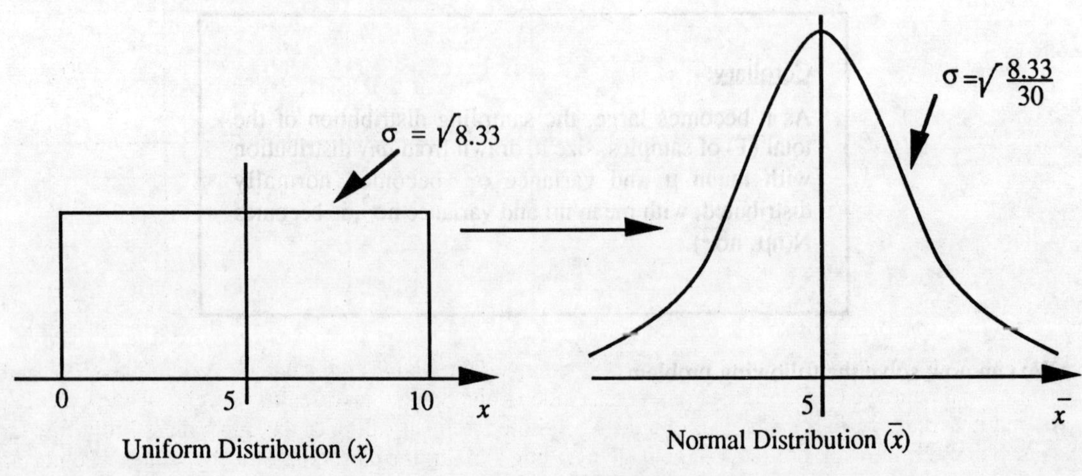

c) Assuming 3 months = 90 days, let T be the total waiting time; then n = 90 and again, assuming n to be large, the sampling distribution of totals (T) will be approximately normal with

mean = $n\mu = 90 \times 5 = 450$ mins

and variance = $n\sigma^2 = 90 \times 8.33 = 750$ mins^2

ie T will be distributed approximately N(450, 750).

We require P(T < 7 hrs) = P(T < 420 mins).

Now, $Z_{420} = \dfrac{420 - 450}{\sqrt{750}} = -1.10$, and so:

$P(T < 7 \text{ hrs}) = P(Z < -1.10) = P(Z > 1.10) = 1 - P(Z < 1.10) = 1 - .8643 = .1357$

9.9 The use of arithmetical probability paper

This is a special kind of graph paper which can be used to determine whether a given sample comes from a normally distributed underlying population; if it does, then an estimate can be obtained for the mean and standard deviation of the population. Consider the following two examples.

Example: The following table gives the heights, in inches, of 200 male students.

Height (in inches)	61-63	64-66	67-69	70-72	73-75
Observed Class Frequencies	10	40	78	56	16

Using arithmetical probability paper, determine whether this sample comes from a normal population. If so, estimate the mean and standard deviation of this population.

Method: Firstly, group the data into *continuous* classes, since the normal distribution is continuous.

Height	60.5 - 63.5	63.5 - 66.5	66.5 - 69.5	69.5 - 72.5	72.5 - 75.5
Frequency	10	40	78	56	16
Cumulative Frequency	10	50	128	184	200
Percentage Cumulative Frequency	5	25	64	92	100

Next, calculate the cumulative and, hence, percentage cumulative frequencies for each class. A scale must now be attached to the left hand vertical side of the paper; representing height. The percentage cumulative frequency for each class can now be plotted corresponding to the upper class limit; the last class, however, must be omitted, since it gives a bogus point (because 100% of the values of a normal distribution can never lie below any particular value) - (see graph opposite).

If the resulting four points approximate to a straight line, then we can assume the sample comes from a normally distributed population; in this case we see that they do.

In order to approximate the mean μ of the normal distribution we firstly draw the line of best fit, by eye, through the points, and then use this line to read off the height corresponding to a percentage cumulative frequency of 50% (since the mean of a normal distribution is also the median); in this case, we see that:

$\mu \approx 68.3"$

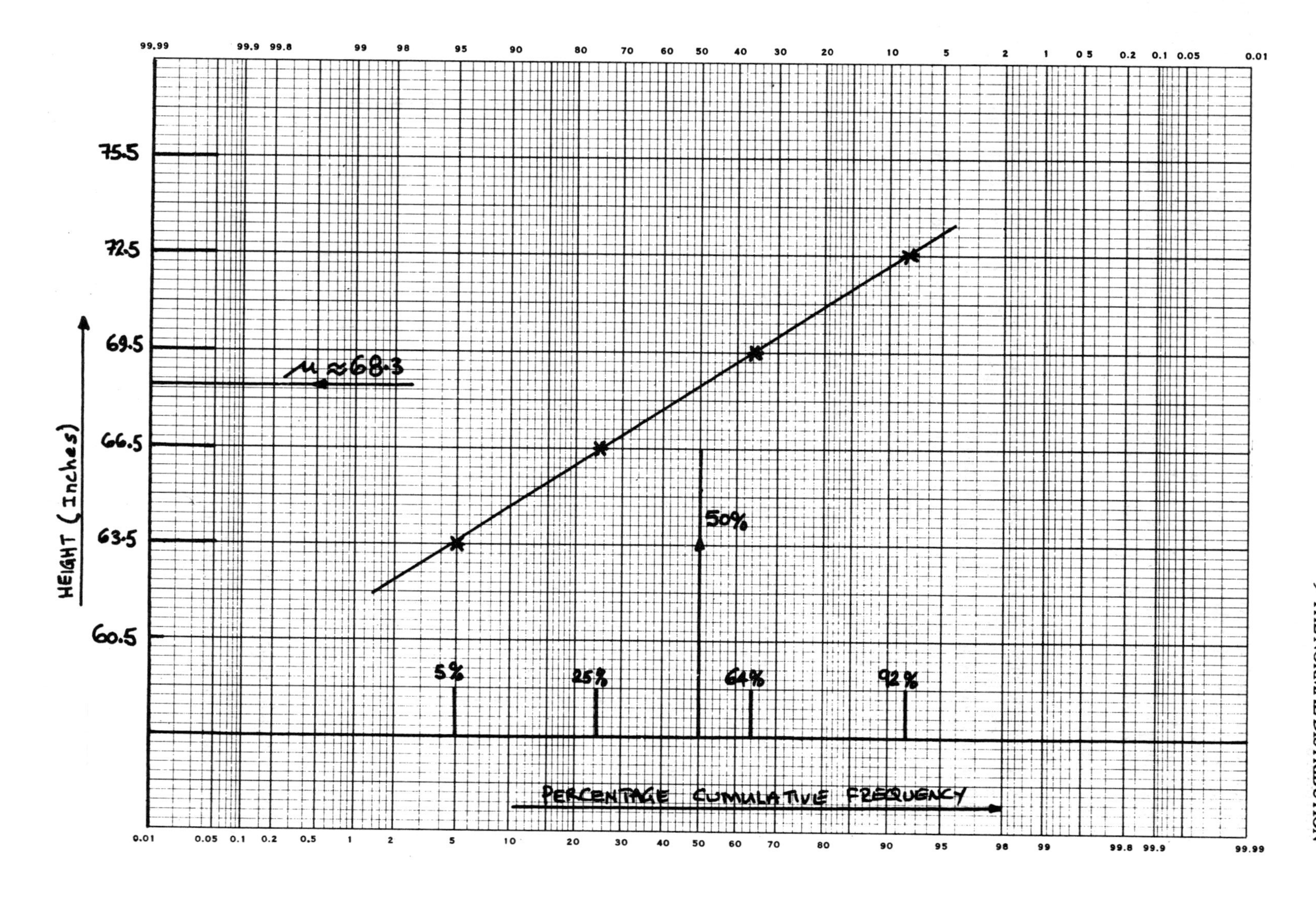

75.5
72.5
69.5
66.5
63.5
60.5
HEIGHT (Inches)
μ ≈ 68.3
50%
5%
25%
64%
92%
PERCENTAGE CUMULATIVE FREQUENCY

To estimate the standard deviation σ we need first to recall the comments on page 173/174 earlier in this chapter. Here we see that the area below one standard deviation to the *right* of the mean is 0.,8413, ie a percentage relative cumulative frequency of 84.13%. Similarly, the relative cumulative frequency for one standard deviation to the *left* of the mean is 15.87%. Thus, the difference between the two heights corresponding to 84.13% and 15.87% will be approximately two standard deviations. From the diagram, we see that these two heights are approximately 65.5" and 71.3".

Thus, $2\sigma \approx 5.8"$

ie $\sigma \approx 2.9"$

The next example shows how the *individual values* of a *small* sample can be plotted directly onto the arithmetic probability paper, avoiding the need to group the data. However, if the sample size exceeds say 10, then it may be quicker to group the data first.

Example: An investigation was conducted into the dust content in the flue gases of solid burners. Nine boilers were used over a similar period of time, and the following quantities, in grams of dust, were deposited in traps inserted in each of the nine flues:

53.0 42.4 55.8 61.5 43.8 66.6 50.0 58.4 65.4

Use arithmetic probability paper to test whether the dust deposits are normally distributed; if so, estimate the mean and standard deviation of all dust deposits.

This time, we construct a simple frequency table, and as before calculate the percentage cumulative frequencies.

Dust deposit	42.4	43.8	50.0	53.0	55.8	58.4	61.5	65.4	66.6
Frequency	1	1	1	1	1	1	1	1	1
Cumulative Frequency	1	2	3	4	5	6	7	8	9
Percentage Cumulative Frequency	11.1	22.2	33.3	44.4	55.6	66.7	77.8	88.9	100

As before, attach a scale to the left-hand vertical axis.

We can now plot these *individual* percentage cumulative frequencies on the arithmetical probability paper, again excluding the last value (see graph opposite).

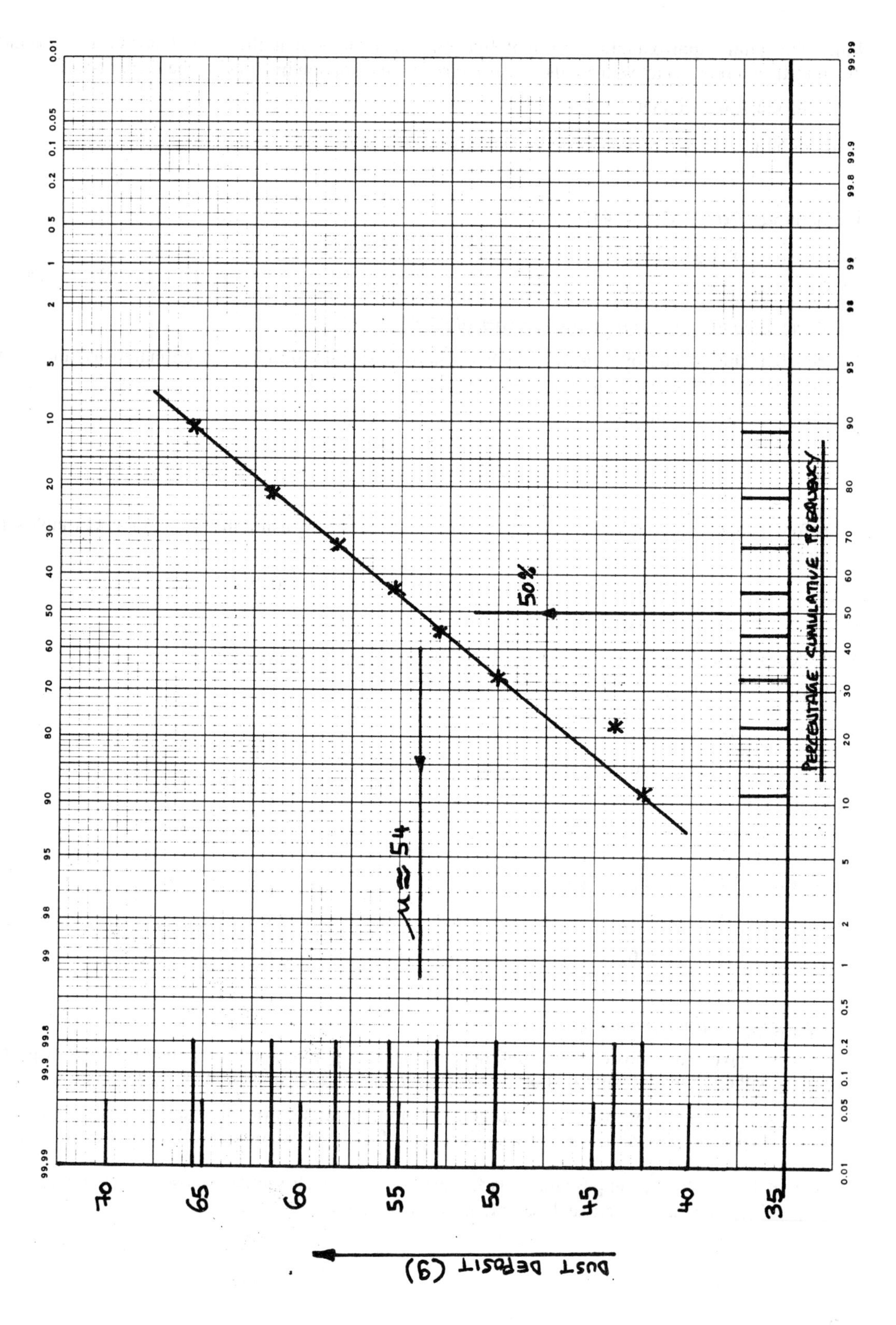

PERCENTAGE CUMULATIVE FREQUENCY
DUST DEPOSIT (g)
50%
$\mu \approx 54$
70
65
60
55
50
45
40
35

Since the points approximate a straight line we can assume that the dust deposits are normally distributed, and proceed to estimate the mean and standard deviation by drawing the line of best fit.

From the graph we see that

$\mu \approx 54$ g

Furthermore, $2\sigma \approx 63.5 - 44.7 = 18.8$

ie $\sigma \approx 9.4$ g

Choosing a random sample from a given normal distribution

We conclude this chapter by illustrating how to choose a sample from a given normal distribution; this is using the same method as demonstrated in chapter 8.8 page 183.

Example: Construct a random sample of size 5 from N(5, 2).

Here, $\mu = 5$ and $\sigma = \sqrt{2}$

As before, choose 5 probabilities, correct to 4 decimal places, say, using the random number tables, for example:

0.9008 0.1424 0.0151 0.9546 0.3032

Consider the first of these, 0.9008. We need to find the value x_0 such that:

$P(x < x_0) = 0.9008$

If Z_0 is the corresponding Z-value, then,

$P(x < x_0) = P(Z < Z_0) = 0.9008$

From the normal distribution tables, we see that

$Z_0 = 1.285$

Thus, $x_0 = 5 + 1.285\sqrt{2} = 6.82$

In a similar manner, we can obtain the other values, giving the sample:

6.82 3.49 1.93 7.39 4.27

Sampling from a finite population

So far, we have only considered sampling from *infinite* populations, or populations so large that sampling without replacement does not have any perceptible effect on the remaining population.

We now consider sampling without replacement from a population of finite size N, say. Suppose we repeatedly take samples, size n, *without* replacement, from a population of values of x of size N, with mean μ and variance σ^2; suppose that $\bar{x}$ is the mean of one such sample whose values are x_1, x_2, x_3 ....., x_n. Then the mean of all such sample means is given by:

$$E[\bar{X}] = E\left[\frac{x_1 + x_2 + \dots + x_n}{n}\right]$$

$$= \frac{1}{n} E[x_1] + \frac{1}{n} E[x_2] + \dots + \frac{1}{n} \cdot E[x_n]$$

$$= \frac{1}{n} \cdot \mu + \frac{1}{n} \cdot \mu + \dots + \frac{1}{n} \mu$$

$$= \frac{1}{n} \cdot n\mu = \mu$$

This is the same result as derived on page 187.

However, the variance of these means is *not* the same as the variance of the sampling distribution of mean drawn from an *infinite* population (which is $\frac{\sigma^2}{n}$); we state this variance here without proof (it is too complicated):

$$V[\bar{X}] = \frac{\sigma^2(N - n)}{n(N - 1)}$$

Summarising these two results, we have:

> If samples of size n are taken without replacement from a finite population of size N whose mean is μ and variance is σ^2, then the mean and variance of the means, $\bar{X}$ of these samples are:
>
> $$E[\bar{X}] = \mu \text{ and } V[\bar{X}] = \frac{\sigma^2(N - n)}{n(N - 1)}$$

Example: A school is attended by 500 pupils whose IQs have mean 105 and variance 220. Samples of size 20 are taken, without replacement, and the mean IQ of each sample is calculated. What is the mean and variance of these mean IQs? Assuming these means are normally distributed, what is the probability that the mean of one such sample exceeds 110?

Here, $N = 500$, $n = 20$, $\mu = 105$ and $\sigma^2 = 220$; thus, from the above result:

$$E[X] = 105 \text{ and } V[X] = \frac{220(500 - 20)}{20(500 - 1)} = 10.58$$

Now we require $P(\bar{X} > 110)$.

So, $Z_{110} = \dfrac{110 - 105}{\sqrt{10.58}} = 0.47$;

thus, $P(\bar{X} > 110) = P(Z > 0.47) = \underline{0.32}$

Confidence limits

The purpose of taking a sample is to learn something about the population from which it is drawn. The confidence which can be placed in estimates of mean, standard deviations, proportions etc depends on the unbiased manner of taking the sample, the size of the sample taken and the standard error of the parameter being considered. The average value of the sample will invariably differ to some extent from the true value for the population. Therefore an estimate of the true value for the population derived from samples taken should be quoted as a range of values *within which there is confidence that the true value lies*. The estimate of the true value is given as a *range* together with an *accompanying probability* which expresses the confidence shown that the true value lies within the range.

The simplest method of obtaining confidence intervals is to take a large sample, use the unbiased estimate of the parameter obtained from the sample and consider that the value for the population lies within ONE standard error (possibly), TWO standard errors (quite likely), THREE standard errors (most likely).

9.10 Confidence intervals for the mean of the population

Example: A sample of 500 items gives a mean $\bar{x}_o = 45$ with a standard deviation of $\sigma = 6$.

Find confidence intervals for the population mean.

The sample is large. The standard deviation of the sample may be taken therefore as the standard deviation of the population. Using the formula (see page 193)

$$Z = \frac{x_o - \mu}{\frac{\sigma}{\sqrt{n}}}$$

where Z = number of standard deviations

$\bar{x}_o$ = mean value of the sample

μ = mean value of the population

σ = standard deviation

n = number in sample

$\frac{\sigma}{\sqrt{n}}$ = standard error of estimate

i) If it is assumed that μ lies within ONE standard deviation either way, from the normal scale, the probability of such a deviation is 0.8413 ie 84.13%. The population mean is then given by

$$Z = 1 = \frac{45 - \mu}{\left(\frac{6}{\sqrt{500}}\right)}$$

$$\mu = 45 \pm 1\left(\frac{6}{\sqrt{500}}\right) \qquad + \text{ sign gives } \mu = 45.27 \qquad - \text{ sign gives } \mu = 44.73$$

ie there is confidence that 84.13% of the population will have a mean in the range

$44.73 \le \mu \le 45.27$.

ii) On the other hand if it is assumed that 95% of the population lies within the limits, the probability of such a deviation is P(1 - 0.025) = P(.975). From the standard table Z = 1.96. The calculation of the range of the mean of the population is

$$Z = 1.96 = \frac{45 - \mu}{\left(\frac{6}{\sqrt{500}}\right)}$$

ie $= 45 \pm 1.96 \left(\frac{6}{\sqrt{500}}\right)$

ie $44.47 \leq \mu \leq 45.53$

there is confidence that 95% of the population gives a population mean in the range $44.47 \leq \mu \leq 45.53$

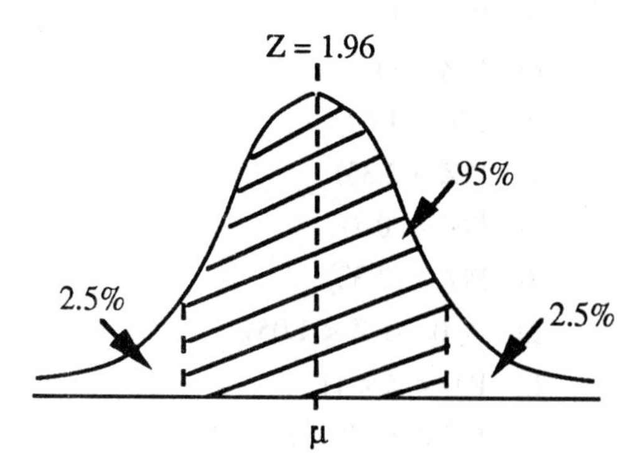

9.11 Confidence intervals for the proportion of a population

Recall from page 185/186 that the proportion of successes (p) from r successes in n repeated trials is $p = \frac{r}{n}$ (where all proportions lie between 0 and 1). The mean and variance of the proportion are given by $E[P] = \beta$ and $V[P] = \frac{\beta(1 - \beta)}{n}$

Example Before a general election an opinion poll showed that 725 people from a random sample of 1,250 people said that they would vote for the Blue party. Find 95% confidence limits for the true proportion of all voters in favour of this party.

Proportion of sample intending to vote Blue $= \frac{725}{1250} = 0.58$ (58%)

Thus: $n = 1250$ and $\beta = 0.58$

As n is large then $\mu = \beta = 0.58$ and $\sigma^2 = \frac{0.58(1 - 0.58)}{1250} = 0.00019$

From tables a 95% confidence limit gives Z = 1.96

therefore: $Z = 1.96 = \frac{P_o - 0.58}{\sqrt{0.00019}}$

$$P_o = 0.58 \pm 1.96 (\sqrt{0.00019})$$

$$0.553 \leq P_o \leq 0.607$$

Percentage of population intending to vote Blue lies between 55.3% and 60.7%.

Exercise 9

1. Find the following probabilities using the normal distribution tables:

a) $P(Z < 1)$
b) $P(Z > 2)$
c) $P(Z < 1.64)$
d) $P(Z > 2.33)$
e) $P(Z < -0.7)$
f) $P(Z \geq -2.33)$
g) $P(-0.5 \leq Z < 1.05)$
h) $P(0 < Z < 1)$
i) $P(|Z| \geq 2.33)$

2. Find the value of Z_0 which satisfies the following probabilities:

a) $P(Z < Z_0) = 0.95$
b) $P(Z \geq Z_0) = 0.90$
c) $P(Z < Z_0) = 0.4$
d) $P(Z \leq -Z_0) = 0.45$
e) $P(|Z| < Z_0) = 0.90$
f) $P(-1 \leq Z \leq Z_0) = 0.8$

3. Find the Z-values corresponding to the values given from their respective normal distributions.

a) $x = 52$, N(55, 4)
b) 95, N(100, 225)
c) 0.31, N(0.3, 0.1)
d) $x = a$, $N(2a, a^2)$
e) 2000, N (1950, 420)
f) 0.01, N(0.02, 0.003)

4. Calculate the value from the given normal distribution which corresponds to the given Z-score.

a) $Z = 1$, N(100, 225)
b) $Z = -2$, N(0.4, 0.4)
c) $Z = 2$, N(40, 52)
d) $Z = 2.33$, N(10, 2.3)
e) $Z = 0.5$, $N(a, a^2)$

5. Find the following probabilities corresponding to the given normal distribution.

a)	N(100, 225)	i) $P(x < 110)$,	ii) $P(x \geq 85)$,	iii) $P(105 < x < 120)$
b)	N(52, 4)	i) $P(50 < x < 51)$	ii) $P(x \geq 49.5)$	iii) $P(x = 52)$
c)	N(0.1, 0.1)	i) $P(x > 0)$	ii) $P(x \leq 0.1)$	iii) $P(0 \leq x < 0.05)$

6. Find the values of x_0 which satisfy the given probabilities corresponding to the given Normal distribution.

a)	N(100, 225)	i) $P(x < x_0) = 0.95$,	ii) $P(x \geq x_0) = 0.6$
b)	N(20, 16)	i) $P(x < x_0) = 0.2$	ii) $P(18 < x < x_0) = 0.5$
c)	N(1, 1)	i) $P(x < x_0) = 0.90$	ii) $P(x \geq -x_0) = 0.10$

7. Find the mean, μ, and/or standard deviation, σ, of $N(\mu, \sigma^2)$ given the following information:

i) $\mu = 10$; $P(x > 12) = 0.1$

ii) $\sigma = 3$; $P(x < 2) = 0.05$

iii) $P(x > 0.1) = 0.3051$; $P(x < 0.08) = 0.1034$

8. Billy Bouncer has achieved high jumps which are Normally distributed, with mean 2 m, variance 0.20 m^2. What is the probability that he

i) jumps higher than 2.1m;

ii) jumps between 1.9m and 2.1m;

iii) jumps less than 1.92m, in any one jump?

What is the probability that he jumps higher than 2.3m in 4 attempts? What is the probability that he jumps higher than 2.1m in each of 3 successive attempts? What height does he exceed 5% of the time? If he jumped below average height on two occasions, what is the probability that he would do so on the third occastion?

9. Acme Enterprises manufacture rubber balls. 84% of balls have diameter less than 7cm whilst 99.4% of balls have diameter greater than 3.5cm.

Find

a) the mean and variance of diameters,

b) the probability that a ball chosen at random has diameter

i) greater than 6.2cm,

ii) less than 5.5cm,

iii) between 5.1cm and 7cm.

c) 80% tolerance limits about the mean.

10. 1,000 pupils take an examination in which previous marks have been roughly normally distributed with mean 50, and variance 400. How many pupils would be expected to

    a) achieve more than 55'

    b) achieve lower than 22;

    c) achieve between 60 and 80.

    If 70% are required to pass, what is the pass mark? If 1% went to Harvard, how many would go?

11. Reaction time is normally distributed such that 12% of people respond slower than 0.7 secs. If the standard deviation is 0.05 secs, calculate the average reaction time. What percentage of people have reaction time faster than 0.72 secs? If the effect of a certain drug is to reduce reaction time by 0.15 secs, and to halve the variance, calculate the average reaction time and variance of reaction time under the influence of the drug. What is the percentage change in people reacting faster than 0.65 secs?

12. The yield (kg) of tomato plants is approximately normally distributed, with mean 10 kg, variance 3 kg. How many of 500 plants would be expected to yield less than 8 kg. If 25 out of 500 plants were discarded because of low yields, what was the approximate yield below which a plant was discarded. How many plants have yields within 1 kg of the average yield?

13. Given that X is N(2, 5), Y is N(0, 9) and W is N(-4, 12) calculate the mean and variance of the following linear combinations:

    i) $3X + 2Y$

    ii) $4X - Y + W$

    iii) $Y - W$

    iv) $2X + 3W - 4$

    v) $-X - Y - W$

    vi) $\frac{1}{2}(X + Y + W)$

    What assumption must you make in order to perform these calculations?

14. Cubic building blocks are made in three sizes, the sides of which are normally distributed with mean 4cm, 8cm, 12cm and variances $0.5cm^2$, $0.6cm^2$, $0.75cm^2$.

    a) What is the average height of a column made up of one block of each type? What is the probability that the height exceeds 25cm?

    b) What is the average height of a column made up of 2 layers of small, 1 layer of medium, and 3 layers of large blocks. What percentage of these exceed a height of 43 cm? Given that 7% lie below a certain height, what is that height?

    c) If three blocks are chosen at random from a box containing an equally large number of each type of block, and stacked on top of each other, what proportion of stacks will have a height between 13.5 cm and 22 cm?

15. Nuts and bolts are manufactured such that a nut and a bolt will fit together only if the positive difference in their diameters does not exceed 0.5mm. Given that the diameters X of holes in nuts are normally distributed with mean 10mm and standard deviation 1mm, and that the diameter Y of bolts are normally distributed with mean 9.5mm and standard deviation 0.75mm. Calculate:

    a) the mean and standard deviation of the difference in diameters $D = X - Y$;

    b) the probability that a nut and bolt, randomly chosen, will fit together;

    c) the probability that, of 2 nuts and 2 bolts randomly chosen, at least one pair will fit.

16. When painting a wall, one layer of undercoat and 3 layers of gloss paint are used. Given that undercoat and gloss paint layers have thicknesses which are normally distributed with means $\mu_x = 0.03$mm, $\mu_y = 0.01$mm and standard deviations $\sigma_x = 0.005$mm, $\sigma_y = 0.0025$mm, respectively, calculate:

    a) mean and standard deviation of the total thickness of paint on the wall;

    b) the probability that, of a randomly chosen paint on the wall, the total thickness of paint lies within 0.01mm of the mean thickness;

    c) given that one of the three layers of paint was put on with a *uniform* thickness of 0.012mm, what is the probability that the total thickness exceeds 0.0535mm at a randomly chosen point on the wall.

17. Write down

    a) the sampling distribution of the mean $(\bar{x})$, and

    b) the sampling distribution of the total (T) of samples chosen from the distributions below, and of given size. Which of these are Normal?

        i) $N(3, 9)$, $n = 5$;

        ii) $N(0, 100)$, $n = 30$;

        iii) $\sigma = 12$, $\sigma^2 = 12$, $n = 5$;

        iv) $\sigma = -2$, $\sigma^2 = 0.3$, $n = 50$;

        v) $N(100, 100)$, $n = 10$;

        vi) $\mu = a$, $\sigma^2 = a^2$, $n = 100$

18. The probability of an error occuring on a page of a book is 0.01. If there are 250 pages in a book, calculate the probability that

    i) more than 4 pages have errors, and

    ii) less than 15 pages have errors.

19. 5% of washers are defective. Find the probability that

    a) 6% or more out of a sample size 20 are defective;

    b) more than 3 out of a sample size 40 are defective;

    c) more than 4 out of each of two samples of size 40 chosen independently from each other.

    Between what limits would you expect 90% of a sample of size 100 washers to be defective?

20. Calculate the probability of obtaining a product of 12 with two dice. How often would you expect to exceed 15 such products in 100 throws?

21. A multiple choice paper comprises 100 questions, each question has 3 possible answers. If a student guesses all 100 answers what is the probability that he will pass the exam if the pass percentage is 40%? How many choices should be offered to each question to ensure that there is only a 1% or less chance of passing by guessing all the answers? Suppose a student knows the answers to 20 of the questions, what are his chances of passing if he guesses the remaining questions?

22. It is estimated that 32% of the population support a certain political party. What is the probability that of a sample of size (i) 50; (ii) 500 people, more than 35% will support the party? What is the probability that a sample of size 40 will contain 10 people, or less supporting the party? How large a sample would need to be chosen to be 95% certain of the *percentage* of supporters lying within $\frac{1}{2}$% of the true percentage of supporters?

23. A machine produces bars of chocolate with masses Normally distributed N(100g, 5g^2). Quality control checks involve choosing samples of size 20 from the production line. Calculate:

    a) the probability that the mean mass of such a sample exceeds 100.2g;

    b) 99% upper and lower quality control limits for the sample means;

    c) the percentage of sample means lying within 0.1g of the mean mass.

24. The lifetimes of certain electrical components produced by a factory is found to be Normally distributed N(400 hrs, 1000 hrs^2). The sales company guarantees its products to function for at least 330 hours. Calculate.

    a) the percentage of components that could be returned under guarantee;

    b) the proportion of components lasting longer than 500 hours;

    c) the life expectancy exceeded by 99% of components;

    d) the probability that the mean life of a sample of 5 components exceeds 415 hours;

    e) the size of sample required to be 95% confident that the mean will lie within 10 hours of the overall mean life of a component.

25. Samples of size 5 are drawn from a population of mean 20 and variance 15. State the mean and variance of the total, T, of the samples. If the underlying population is normal calculate

    i) $P(T < 98)$,

    ii) $P(T < E[T] + 4)$.

26. Treets chocolate peanuts have masses distributed with mean 4g, variance $0.5g^2$. Given that a packet contains 25 treets, calculate the probability that a randomly chosen packet has a mass, M,

    i) less than 94g,

    ii) between $\pm$ 3% of the mean mass.

    What are the limits between which 95% of masses lie? If a box of packets contains 100 packets what is:

    i) the average weight of a box,

    ii) the mass above which 90% of the boxes lie?

    If a fork lift truck can carry a maximum load of 54kg, could it safely carry 5 boxes at a time? What would be the 0.01% safety limit for the truck? What assumption do you have to make in order to carry out these calculations?

27. The pages of a book have masses distributed with mean 0.1g variance $0.0001g^2$. A book comprises 250 pages; what is the expected mass of book? What is the probability that its mass exceeds 25.8g? If a satchel contains 7 such books find the limits between which 99% of their total mass lies. What assumptions must be made here?

28. Cork tiles are stacked in piles of 50. If individual tiles have thicknesses distributed with mean 0.5cm, variance $0.0025cm^2$; find the probability that:

    i) a stack is taller than 25.7cm,

    ii) three stacks are taller than 74cm when placed on top of each other,

    iii) if a pile of height 75.2cm has 8 tiles removed, its height is still above 71.6cm.

    Do any assumptions have to be made here?

29. State the Central Limit Theorem.

    IQ is distributed with mean 100, variance 225. Calculate the probability that:

    i) a sample of 30 people have above 102 IQ on average,

    ii) the size of sample needed to be selected to be 90% certain that the mean is less than 101.

30. 20% of samples of size 5 have mean less than 9, whilst 40% of samples of size 10 have mean greater than 10.5. Calculate the mean μ and variance σ^2 of the population from which they were drawn. What assumptions do you have to make? Are these justified?

31. The uniform distribution has the pdf given below:

$$f(x) = \frac{1}{6}, \qquad 1 < x < 7.$$

Samples of size 50 are taken at random. Calculate the mean and variance of:

i) the uniform distribution,

ii) the sampling distribution of means.

Five of these samples are pooled into one larger sample; what is the probability that this has mean less than 3.9?

Find the probability that three samples of size 50 chosen successively will have each a mean of less than 3.9.

32. Battery life follows the exponential distribution with mean of 30 hours. What is the probability that

i) the mean life of 20 batteries exceeds 32 hours,

ii) each of 20 batteries has a life greater than 32 hours.

How large a sample would you advise to be taken to be 95% confident that its mean lies within 2 hours of the mean battery life?

33. Calculate confidence intervals of μ given the following information:

a) $\bar{x}_0 = 12$, $n = 5$; $N(\mu, 2)$

b) $\bar{x}_0 = 0.3$, $n = 50$; $\sigma = 0.01$

c) $\bar{x}_0 = 0$, $n = 10$; $N(\mu, 3.4)$

d) $\bar{x}_0 = -2$, $n = 25$; $\sigma = 1$.

34. Calculate confidence intervals for π given the following information:

a) $P_0 = 0.2$, $n = 50$

b) $P_0 = 25\%$, $n = 200$

c) $P_0 = \frac{1}{3}$, $n = 100$

35. a) In an examination in mathematics, 20 students at Greyfriars School had a mean mark of 52, and the sum of squares of deviations from this mark was 1312.

    At Dotheboys Hall, 17 boys sat for the same examination, obtaining a mean mark of 36, and the sum of squares of deviations from this mean was 1401.

    Calculate 95% confidence limits for the mean mark of the population, supposed normal, from which the marks of the 20 students at Greyfriars were drawn.

    b) Of 1,000 schoolboys examined, 180 were found to have flat feet. Calculate 95% confidence limits for the fraction having flat feet in the population from which the sample was drawn.

36. a) The masses, in grams, of thirteen ball bearings taken at random from a batch are:

    21.4, 23.1, 25.9, 24.7, 23.4, 24.5, 25.0, 22.5, 26.9,

    26.4, 25.8, 23.2, 21.9.

    Calculate a 95% confidence interval for the mean mass of the population, supposed normal, from which these masses were drawn, assuming the standard deviation of the population is 1.77.

    b) A random sample of fourteen cows were selected from a large dairy herd at Brookfield Farm. The milk yield in one week was recorded, in kg, for each cow. The results are given below.

    169.6, 142.0, 103.3, 111.6, 123.4, 143.5, 155.1,

    101.7, 170.7, 113.2, 130.9, 146.9, 169.3, 155.5.

    Calculate a 95% confidence interval for the mean of the population, assuming that the variance equals 289.

37. The probability of success in each of a long series of n independent trials is constant and equal to p. Explain how 95% approximate confidence limits for p may be obtained.

    In an opinion poll carried out before a local election, 501 people out of a random sample of 925 declare that they will vote for a particular one of two candidates contesting the election. Find 95% confidence limits for the true proportion of all voters in favour of this candidate.

    Do you consider there is significant evidence that this candidate will win the election?

38. a) Discuss briefly the relative merits of estimating an unknown parameter by means of either a single value of a confidence interval.

    Crates of bananas are packed in the West Indies with a nominal net mass of 55 kilograms. However, on their arrival in Liverpool, this has usually decreased due to ripening and shrinkage. A large batch of such crates has just arrived aboard SS Gauss. A random sample of twelve crates is selected and their net masses are recorded in kilograms as listed below.

    56.4, 52.1, 49.5, 56.4, 56.0, 48.1, 54.5, 47.8, 58.0,

    48.4, 53.9, 46.7.

    Assuming that this sample came from an underlying normal population with variance 16, calculate a 95% confidence interval for the mean mass of the population.

b) A random sample of fifteen workers from a vacuum flask assembly line was selected from a large number of such workers. Ivor Stopwatch, a work-study engineer, asked each of these workers to assemble a one-litre vacuum flask at their normal working speed. The times taken, in seconds, to complete these tasks are given below.

109.2, 146.2, 127.9, 92.0, 108.5, 91.1, 109.8, 114.9,
115.3, 99.0, 112.8, 130.7, 141.7, 122.6, 119.9.

Taking the variance $\sigma^2 = 11.69$ and assuming an underlying normal population again, calculate a 95% confidence interval for the mean of the population.

39. a) When a confidence interval is constructed for a parameter, one measure of its usefulness ishow narrow it is. Now if $X_1, X_2, \ldots\ldots X_n$ is a random sample from a normal distribution with mean μ and variance σ^2, both unknown, and a γ% confidence interval is constructed for μ, how does the width of the interval change (other things being equal) if

i) n becomes larger;

ii) γ becomes larger;

iii) σ^2 becomes larger?

(Justify your replies briefly.)

b) During a particular week, 13 babies were born in a maternity unit. Part of the standard procedure is to measure the length of the baby. Given below is a list of the lengths, in centimetres, of the babies born in this particular week.

49, 50, 45, 51, 47, 49, 48, 54, 53, 55, 45, 50, 48

Assuming that this sample came from an underlying normal population with $\sigma = 3.13$ calculate a 95% confidence interval for the mean of the population.

40. Derive the mean and variance of the binomial distribution.

In a survey carried out in Funville, 28 children out of a random sample of 80 said that they bought Bopper comic regularly. Find 95% approximate confidence limits for the true proportion of all children in Funville who buy this comic. A similar survey in Funville found that 45 children out of a random sample of 100 said that they bought Shooter comic regularly. Find 95% approximate confidence limits for the true proportion of all children in Funville who buy this comic.

On the basis of these surveys, is there any evidence that the sales of Shooter comic are higher than the sales of Bopper comic in Funville? Justify your reply.

41. Eight fish of a certain species are measured and their lengths are found to be:

10.6, 11.2, 10.4, 12.2, 11.3, 10.2, 10.3, and 12.5 inches respectively.

Find 95% confidence limits for the mean length of a fish of the species, assuming these lengths form a normal distribution with $\sigma = 1.12$.

Estimate the length which will be exceeded by one fish in a thousand.

42. Twelve cotton threads are taken at random from a large batch, and the breaking strengths are found to be

    7.41, 7.01, 8.34, 8.29, 8.08, 6.60, 6.59, 7.39, 4.72, 8.65, 8.51 and 8.02 ounces respectively.

    Assuming the breaking strengths of the threads form a normal distribution with $\sigma = 0.88$ find 95% confidence limits for the mean breaking strength of threads in the batch, using the unbiased estimate of the population variance.

    Estimate the standard deviation of the breaking strength from the range of the sample of twelve and comment on the agreement with the value of the standard deviation given by the unbiased estimate of the population variance.

43. The probability that a person believes in God is $\frac{1}{3}$, whilst the probability that a person has fair hair is $\frac{1}{4}$. Assuming that hair colour and faith are independent, find the probability that, in a random sample of 100 people, the number who believe in God exceeds the number who have fair hair.

    (Hint: Use binomial approximation.)

44. Given that chain links have lengths normally distributed with mean 2cm and variance 0.04cm^2, calculate the probability of a chain comprising 50 links has a length exceeding 1.014m? What is the probability that:

    a) at least one chain in five exceeds 1.014m?

    b) more than 36 chains exceed 1.014m in a random sample of 200 chains.

45. Use Arithmetical Probability paper to test whether the following samples are likely to come from underlying normal distributions; if they do, estimate their means and standard deviations.

a)	112	156	406	216	447	141	315	451
b)	12.3	13.7	10.4	11.4	14.9	12.6		
c)	15.7	11.1	6.6	14.5	13.0	13.8	11.9	
d)	22.4	23.1	25.9	24.7	23.4	24.5	25.0	
	22.5	26.9	24.4	25.8	23.2	21.9		
e)	106	100	105	103	103	102	105	
	105	102	104	104	101			
f)	22.0	23.9	20.9	23.8	25.0	24.5		
	21.7	22.8	23.4	23.0	25.5			

g)

$-\infty < x \leq 44$	$45 \leq x \leq 47$	$48 \leq x \leq 50$	$51 \leq x \leq 53$	$54 \leq x \leq 56$	$57 \leq x < \infty$
4	12	63	59	10	2

h)

x	0	1	2	3	4	5	6
f	1	4	15	21	33	18	8

i)

x	5-9	10-14	15-19	20-24	25-29	30-34
f	2	29	37	16	4	2

46. Construct samples of the given size from the following normal distributions:

a) N(3, 9), n = 5;

b) N(0, 4), n = 3;

c) N(100, 100), n = 4;

d) N(-5, 12), n = 6.

47. a) An extinct species was considered to have an average life of 25 years, with a standard deviationof only 2 years, and to be normally distributed. Of a family of 50 members of the species, born at the same time, how many would be expected to have lived

i) between 20 and 26 years,

ii) less than 22 years?

What would be the age at death of the thirty-fifth member of the family to die?

b) If x and y are independent random variables, derive expressions for the mean and variance of

i) $x + y$

ii) $x - y$

in terms of the means and variances of x and y.

48. In the assembly of a certain machine, circular bolts have to fit into circular holes. Continuous inspection has shown that the bolts are distributed normally with mean diameter 0.503 in. and standard deviation 0.004 in, while the holes are distributed normally with mean diameter 0.513 in. and standard deviation 0.006 in.

If bolts are selected at random, what proportion will not fit into the hole to which they are first applied? What is the probability that 100 successive bolts will all fit the holes to which they are first applied?

49. The independent variables x_1 and x_2 have means μ_1, μ_2 and variances $\sigma_1{}^2, \sigma_2{}^2$ respectively. State the mean and variance of the variable $m_1x_1 + m_2x_2$ where m_1 and m_2 are constants.

The scores of boys of sixteen years of age in a certain aptitude test form a normal distribution with mean 70 and standard deviation 3. The scores of boys of fifteen years of age in the same test form a normal distribution with mean 64 and standard deviation 4. What percentage of the boys of each age will score 62 or less?

Find the mean and standard deviation of the excess score of a sixteen year old boy over that of a fifteen year old boy, both taken at random.

What is the probability that the younger boy will obtain a higher score in the test?

50. Show that the variance of the rectangular distribution of a variable which lies at random between -a and +a is $\frac{a^2}{3}$.

The n numbers of a set are to be added together. Before they are added, each number is rounded off to the nearest integer, thus introducing an error whose distribution is rectangular. Find the mean and variance of the total error of the sum. Assuming that, for large enough values of n, the total error is distributed normally find the least value of n for which the total rounding-off error is more likely than not to lie outside the limits ± 10.

51. An athlete finds that in the long jump his distances form a normal distribution with mean 6.1m and standard deviation 0.03m.

Calculate the probability that he will jump more than 6.17m on a given occasion.

Find the probability that three independent jumps will all be less than 6.17m. What distance can he expect to exceed once in 500 jumps?

52. The variable X can take the values 0, 1, 2, 3, 4, 5, 6, 7, 8, 9 each with a probability of 0.1. Show that the mean of X is 4.5, and the variance of X is 8.25.

Find the mean and variance of the distribution of the sum of 20 random digits chosen from 0 to 9 inclusive.

Assuming this distribution to be approximately normal, estimate the probability that the sum of 20 random digits lies in the range 80 to 109 inclusive.

53. The independent random variables x and y have means μ_1, μ_2 and variances σ_1^2, σ_2^2 respectively. State the mean and variance of the random variable $x + y$.

The ages of persons included in a survey are given in years, as the age last birthday. The error involved in thus recording ages may be taken as having a rectangular distribution, the variable x(yr) satisfying

$0 < x < 1$. Find the mean and variance of this distribution.

Using the ages last birthday, the mean age for a set of 200 persons was found to be 31.6 yrs and the standard deviation 3.1 yrs. Assuming the birthdays to be uniformly distributed throughout the year, find corrected values for the mean and variances of the ages of the 200 persons.

54. The independent random variables X_1, X_2, X_3 have means μ_1, μ_2, μ_3 and variances $\sigma_1^2, \sigma_2^2, \sigma_3^2$ respectively. State the mean and variance of the random variable $Y = a_1X_1 + a_2X_2 + a_3X_3$ where a_1, a_2, a_3 are constants.

Norman Longlegs is a well known international athlete. His best event is the 'Hop-Step-Jump'. In this event, as the name suggests, an athlete takes a long jump up to a starting board whereupon he hops, then steps and finally jumps. His recorded distance is from the starting board to his final position. Norman has observed that the three actions of his leap all follow independent normal distributions. The hop has a mean of 4m and standard deviation 0.6m; the step has a mean of 2m and standard deviation 0.5m; the jump has a mean of 3m and standard deviation 0.5m.

The world record for this event is a distance of 10.5m. What is the probability that Norman will break this record on any given attempt?

In the European championships Norman is allowed three attempts. Assuming these attempts are independent, what is the probability that he breaks the world record at these championships? (You should assume that all leaps are fair and count.)

55. a) If X1, ..., Xn is a random sample from the rectangular distribution on (0, 1), then the Central Limit Theorem says that the random variable

$$S = \sum_{i=1}^{n} X_i \quad \text{is approximately normally distributed with mean n/2 and variance n/12.}$$

Use this result and a table of random sampling numbers in order to generate a random sample of size 7 which has approximately a normal distribution with mean 6 and variance 1. Derive the sample mean and sample variance.

b) The following is a random sample of size 15, which it is believed may be from an underlying normal population

74.1, 77.7, 74.4, 73.8, 79.3, 75.8, 82.2, 72.2,

75.2, 78.2, 77.1, 78.4, 76.3, 76.8.

Plot this sample on arithmetical probability paper. Estimate from your graph the mean and variance of this sample.

56. What is meant by the 'sampling distribution of a statistic'? If random samples of size n are drawn from an unspecified distribution with mean μ and variance $\sigma2$, what is the sampling distribution of the sample mean as n tends to infinity?

The distribution of breaking loads for certain fibres has mean 20 units and standard deviation 2 units. A rope is assumed to be made up of 64 independent fibres, and to have a breaking load which is the sum of the breaking loads of all the fibres in it. Find the probability that such a rope will support a weight of 1,300 units. The manufacturers wish to quote a breaking load for such ropes that will be satisfied by 99% of the ropes. Determine what breaking load should be quoted.

57. In Urbania selection for the Royal Flying Corps (RFC) is by means of an aptitude test based on a week's intensive military training. It is known that the scores of potential recruits on this test follow a normal distribution with mean 45 and standard deviation 10.

i) What is the probability that a randomly chosen recruit will score between 40 and 60?

ii) What percentage of the recruits is expected to score more than 30?

iii) In a particular year 100 recruits take the test. Assuming that the pass mark is 50, calculate the probability that less than 35 recruits qualify for the RFC.

58. Each time an experienced kingfisher tries to catch a fish the probability that he is successful is 1/3. Find the probability that in 5 attempts he catches:

i) exactly 2 fish,

ii) at least 2 fish.

A less experienced kingfisher is only successful, on the average, in 1 attempt out of 10.

Stating any assumptions or approximations that you make, calculate the probability that he catches more fish in 100 attempts that the experienced kingfisher catches in 20 attempts.

59. In a certain country men have masses which are normally distributed with mean 70 kg and standard deviation 8 kg. Women, however, have mean 60 kg and standard deviation 6 kg.

Find the probability that

i) three men, chosen at random, all have mass less than 66 kg;

ii) the mean mass of 3 men is greater than 72 kg;

iii) the mean mass of 5 women is less than 67 kg;

iv) the mean mass of 3 men exceeds the mean mass of 5 women by at least 7 kg.

60. A transport company finds that the amount of fuel that its lorries use during a week may be taken as a normal variable with mean 5,000 litres and standard deviation 400 litres.

    i) Calculate the probability that in a particular week less than 6,000 litres will be used.

    ii) The probability that more than L litres will be used in any week is 0.75. Find L.

    iii) Calculate the probability that in a particular 4-week period more than 21,000 litres will be used.

    iv) The company has the option to buy a garage which has an annual capacity of 260,000 litres. If the company operates a 50 week working year find the probability that the garage will be large enough to satisfy the company's demands.

61. Before joining the Egghead Society, every candidate is given an intelligence test which, applied to the general public, would give a normal distribution of IQs with mean 100 and standard deviation 20.

    The candidate is not admitted unless his IQ, as given by the test, is at least 130.

    Estimate the median IQ of the members of the Egghead Society, assuming that their IQ distribution is representative of that of the part of the population having IQs greater than, or equal to, 130.

    What IQ would be expected to be exceeded by one member in ten of the society?

62. Derive the mean and variance of the binomial distribution.

    A roulette wheel contains the numbers 0, 1, 2, ..., 36. When the wheel is spun and a ball introduced it is equally likely to come to rest on any of these numbes when the wheel stops. What is the probability that the outcome is divisible by 4 or 5 (or both)? If the wheel is spun 100 independent times what is the probability of getting more than 45 outcomes which are divisible by 4 or 5 as above?

63. Choose samples of size 2 drawn without replacement from the following population:

    5 7 8 9

    a) Calculate the mean of each sample.

    b) Calculate the variance of these means.

    c) Calculate μ and $\sigma 2$ for the population.

    d) Calculate E $[\bar{X}]$ and V $[\bar{X}]$ for the distribution of means of samples, size 2, drawn from the population without replacement using the appropriate formula. How does this compare with parts a) and b)?

64. A bag of bingo numbers, from 1 to 100, is shaken up; 20 samples of size 5 are drawn without replacement. Calculate the expected mean and variance of the means of these samples.

    If it can be assumed that the resulting distribution of means is Normal, what is the probability that a randomly chosen sample has a mean exceeding 60?

65. A finite population consists of the numbers 0, 2, 4, 6, 8, 10 and 12. Find the mean and standard deviation of this population. Make a list of all possible samples of size 2 that can be drawn (without replacement) from the population, and calculate the mean of each sample. Construct the sampling distribution of probability for the means of samples of size 2 drawn from the given population. Find the standard deviation of this probability distribution. How does this value compare with the obtained by using the appropriate formula for $\sigma_{\bar{x}}$?

66. Comment briefly upon some of the problems of obtaining information about a population by examining samples from it.

A population of size N has a mean μ and variance σ^2. Random samples of size n are taken without replacement. Write down the expectation and variance of the means of these samples.

As the sample size n gets larger and approaches N, the variance of the sample mean tends to 0. Comment on this observation.

In the Ruritanian state lottery, m tickets are drawn at a time out of n tickets numbers 1 to n ($m \leq n$). Find the expectation and variance of the random variable S denoting the sum of the numbers of the m tickets drawn.

10 THE CHI-SQUARED ($\chi2$) DISTRIBUTION

10.1 Introduction

The χ^2 distribution is a continuous distribution, (see diagram overleaf) and can be used to compare *observed data* with what would be expected by chance (*expected data*).

Consider, for example, the results of throwing a fair, six-sided die 60 times; the table below gives the frequencies *expected* by chance, denoted by E.

Die Number	1	2	3	4	5	6	Total
Expected Frequency (E)	10	10	10	10	10	10	60

However, it is unlikely that each number would occur *exactly* 10 times. In one experiment, the following frequencies were *observed*, denoted by 0:

Die Number	1	2	3	4	5	6	Total
Observed Frequency (O)	9	11	12	8	13	7	60

As can be seen, the observed frequencies (O) differ from the expected frequencies (E) above; we can evaluate this difference using the formula below:

$$\chi^2 = \Sigma \frac{(0 - E)^2}{E}$$

(This is, in fact, a value of the χ^2 distribution.)

ie $$\chi^2 = \frac{(9-10)^2}{10} + \frac{(11-10)^2}{10} + \frac{(12-10)^2}{10} + \frac{(8-10)^2}{10} + \frac{(13-10)^2}{10} + \frac{(7-10)^2}{10}$$

ie $$\chi^2 = 2.5$$

Now, providing this value of χ^2 is not too great (indicating a great difference between the observed and expected data), we can conclude that the die used in the experiment is a fair die, ie each number has the same chance of occuring (viz 1/6). We can, in fact, compare the size of our value of χ^2 with an appropriate value found in the χ^2 tables, which we will call the *critical value*; however, in order to find this, we must first consider the parameter of the χ^2 distribution, viz v, which is called the number of degrees of freedom of the χ^2 distribution. This is a natural number which can be calculated using the formula below:

Number of Degrees of Freedom, v	=	Number of Pairs of Observed and Expected Frequencies - Number of Linear Constraints on the Expected Frequencies

In the above example, the number of pairs of O and E is 6. Furthermore, the expected frequencies (E) *must* all add up to 60, ie there is a *constraint* on them - *one* constraint.

Thus, $v = 6 - 1 = 5$

Returning to the example above, we need to find the critical value of χ^2 with which to compare our value $\chi^2 = 2.5$ with. Consulting the χ^2 tables, (table 5, page 7 of the Cambridge Elementary Statistical Tables), we see that there are several values in the row corresponding to $v = 5$ (degrees of freedom, v, are on the left-hand vertical scale). Each of these corresponds with a given percentage, P (along the top row) called the 'Level of significance'. Let us consider the value of χ^2 corresponding to P = 5%, which is

$$\chi^2\text{crit.} = 11.07$$

This is the value above which only 5% of all χ^2 values lie, as shown in the diagram below. Values of χ^2 are therefore unlikely to be greater than $\chi^2_{\text{crit.}}$

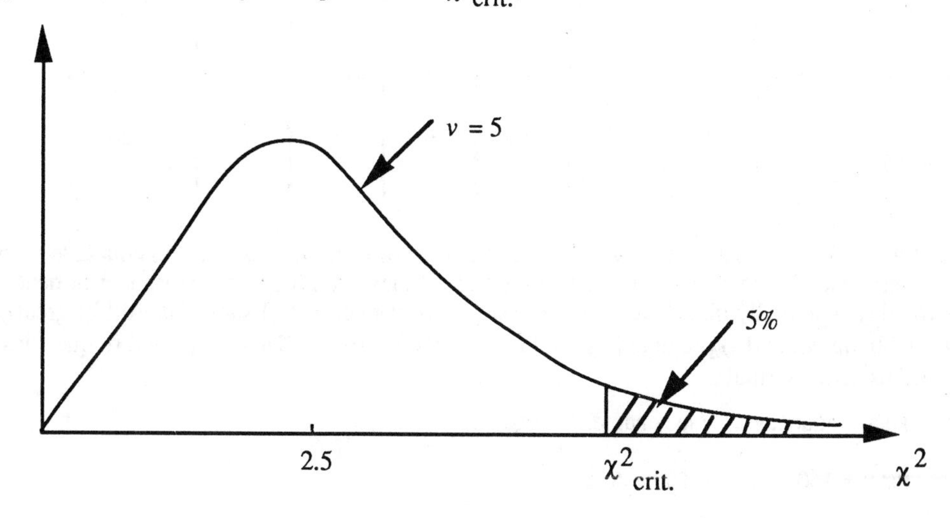

Comparing our value $\chi^2 = 2.5$ with $\chi^2_{crit.}$ we see that:

$2.5 \leq 11.07$

which is not therefore an unlikely, extremely large value, and so we can conclude that, at the 5% level of significance, the die used in the experiment is a *fair* die. We could, of course, have chosen a different level of significance - the smaller the percentage, the more stringent the test.

Note: The χ^2 test can only be used providing none of the *expected* frequencies falls below 5; see the worked example on page 222 illustrating a way around this problem.

10.2 Goodness of fit

Providing that we can find a way of calculating the expected frequencies (E) we can use the χ^2 distribution to compare with them the observed frequencies (O) from a particular situation. Consider, for example, repeated tosses of 5 coins. Clearly, the number of heads R, say, in any *one* toss is binomially distributed. Furthermore, if the 5 coins were *fair*, we can say that R is binomially distributed with parameters n = 5 and $\beta = \frac{1}{2}$, and could proceed to construct the *probability distribution* below, using the probability mass function:

$$P(r) = {}^5C_r \left(\frac{1}{2}\right)^r \left(1 - \frac{1}{2}\right)^{5-r}, \quad r = 0, 1, 2, 3, 4, 5.$$

No. of Heads (r)	0	1	2	3	4	5
Probability, P(r)	$\frac{1}{32}$	$\frac{5}{32}$	$\frac{10}{32}$	$\frac{10}{32}$	$\frac{5}{32}$	$\frac{1}{32}$

This can be thought of as a *model* of the situation (see chapter 6, page 116). Suppose we now actually take 5 coins and toss them 160 times. We may end up with the *frequency distribution* below.

No. of Heads (r)	0	1	2	3	4	5	Total
Observed Frequency (O_r)	3	20	56	38	35	8	160

Now, suppose further that we wish to test whether, on the basis of these results, we can conclude with reasonable certainty that all 5 coins are unbiased, ie are fair. If they *are* fair, then the number of heads, R, should be roughly binomially distributed, ie the observed frequencies (O_r) should not differ greatly from the *expected* frequencies (E_r) predicted by the binomial model above. These expected frequencies, E_r, are calculated using the formula:

$$Er = P(r) \times \text{Total frequency} \quad \text{(see chapter 6, page 98)}$$

eg $\quad E_0 = P(O) \times 160 = \frac{1}{32} \times 160 = 5$, etc.

Thus, we obtain the expected frequencies, E_r, below:

No. of Heads (r)	0	1	2	3	4	5	Total
Expected Frequency (E_r)	5	25	50	50	25	5	160

[Notice that the expected frequencies must total 160, ie are constrained to total 160.]

We can now compare the expected and observed frequencies shown below as before.

No. of Heads (r)	0	1	2	3	4	5
E_r	5	25	50	50	25	5
O_r	3	20	56	38	35	8

Then, $\chi^2 = \Sigma \frac{(O - E)^2}{E}$

$$= \frac{(3-5)^2}{5} + \frac{(20-25)^2}{25} + \frac{(56-50)^2}{50} + \frac{(38-50)^2}{50} + \frac{(35-25)^2}{25} + \frac{(8-5)^2}{5}$$

$= 11.2$

Now, $v = 5$ and so, at the 5% level of significance, $\chi^2_{crit} = 11.07$ again.

Comparing, we see that,

$11.2 > 11.07$

which is unlikely (the probability of exceeding χ^2_{crit} is only 0.05).

Thus, the binomial distribution is *not* a good model for our data, ie one or more of the coins is biased.

10.3 The null hypothesis, H_0

What we have shown above is that the hypothesis that the 5 coins are fair (ie the number of heads, R, is binomially distributed) is not supported by the evidence (the observed frequencies) since its assumption leads to an *unlikely* result (at the 5% level of significance). We call this kind of hypothesis the null hypothesis, written H_0, and speak of either 'accepting H_0' or 'rejecting H_0', on the basis of the evidence. Thus, in the above example we could state the null hypothesis as:

H_0 : All 5 coins are fair.

(Or, equivalently, H_0 : The number of heads, R, is binomially distributed)

Since the assumption of this hypothesis, H_0, leads to an *unlikely* result, we 'rejected H_0'. In the remainder of this chapter, this terminology will be adopted.

Now consider the following example.

Example: The production of microchips is a very difficult process, with a high proportion of defectives occuring in production. In order to monitor the current proportion of defectives, samples of size 4 are regularly made, and the number of defectives, r, recorded. The data from 200 such samples is given below. Test, at the 10% level of significance, whether the number of defectives is binomially distributed.

No. of Defectives (r)	0	1	2	3	4	Total
Observed Frequency (E)	5	36	60	68	31	200

Firstly, we must state the Null Hypothesis,

H_0 : The number of defectives is binomially distributed.

The difference between this and the last example is that we are not given the parameter β of the binomial distribution; how, then, are we to construct the binomial model to test the data against? The answer is to estimate β using the available data. Since we are testing whether a binomial model 'fits' the data well (hence the title of this section), we should choose the binomial model most likely to fit the data, which common sense tells us must be the one with the same mean as that of the data. Using a calculator, this mean is 2.42; thus, we require that:

$$n\beta = 2.42$$

But, n = 4, and so: $4\beta = 2.42$

ie $\beta = 0.605$

We can now proceed to calculate the expected frequencies predicted by this binomial model, using the probability mass function:

$$P(r) = {}^4C_r\,(0.605)^r(1 - 0.605)^{4-r},\ r = 0, 1, 2, 3, 4.$$

obtaining:

No. of defectives (r)	0	1	2	3	4	
P(r)	0.024	0.149	0.343	0.350	0.134	Total
$E_r = P(r) \times 200$	4.87	29.83	68.53	69.98	26.79	200

[Notice that the expected frequencies must not be rounded to the nearest whole number, as the total may then not be 200, which is a requirement of this test.]

We can now compare the observed and expected frequencies below:

No. of defectives (r)	0	1	2	3	4	Total
E_r	4.87	29.83	68.53	69.98	26.79	200
O_r	5	36	60	68	31	200

$$\text{Here, } \chi 2 = \Sigma \frac{(O - E)^2}{E}$$

$$= \frac{(5 - 4.87)2}{4.87} + \ldots\ldots + \frac{(31 - 26.79)^2}{26.79}$$

$$= 3.06 .$$

Now, whenever a parameter of a distribution needs to be estimated in order to calculate the expected frequencies, an extra constraint is imposed upon them; since *one* parameter, β, was estimated here, we have one extra constraint, in addition to the estimated frequencies totalling 200. Thus,

$$\nu = 5 - 1 - 1 = 3$$

From the tables, we have $\chi^2_{crit} = 6.25$, at a 10% level of significance.

Comparing, we see that: $3.06 \leq 6.25$

and so accept H_0.

The above method can also be used to test whether data is modelled by a *continuous* distribution; we simply calculate the expected frequency for each continuous class, and compare them with the observed frequencies as usual.

Example: The working lives of 200 electrical components are recorded below. Estimate the mean life of a component, assuming that the last class has width 50 hours. Test the goodness of fit of an exponential distribution to the data, using a 5% level of significance.

Life of Component (x hours)	0-50	50-100	100-150	150-200	200-250	250-300	300 or more
No. of Components	94	53	25	12	4	4	8

First we state the Null Hypothesis of the test.

H_0: the data is exponentially distributed. Now, using the mid-class values, we can estimate the mean component life of a component using the calculator, to get 80.75 hours.

Thus, the pdf of the most likely fitting exponential distribution is:

$$f(x) = \frac{1}{80.75} e^{-\frac{x}{80.75}} \quad , x \geq 0$$

This is illustrated in the diagram overleaf, together with a histogram of the above data. We can now calculate the probability of x lying in each class, as below.

$$P(O \leq x < 50) = \int_0^{50} f(x)\, dx = \int_0^{50} \frac{1}{80.75} e^{-\frac{x}{80.75}} . dx$$

$$= \left[- e^{-\frac{x}{80.75}} \right]_0^{50}$$

$$= e^{-\frac{0}{80.75}} - e^{-\frac{50}{80.75}} = 0.462$$

$$P(50 \leq x < 100) = e^{-\frac{50}{80.75}} - e^{-\frac{100}{80.75}} = 0.249$$

$$P(100 \leq x < 150) = e^{-\frac{100}{80.75}} - e^{-\frac{150}{80.75}} = 0.134$$

$$P(150 \leq x < 200) = e^{-\frac{150}{80.75}} - e^{-\frac{200}{80.75}} = 0.072$$

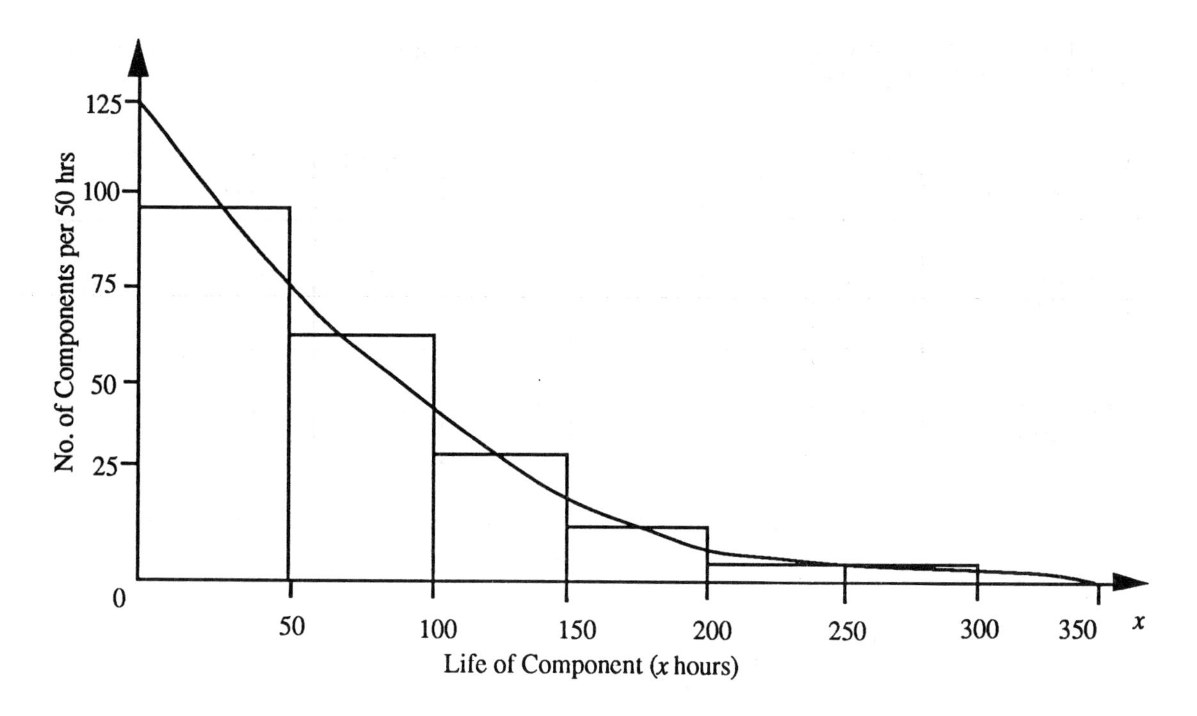

[Note: The above curve is of the function f(x) x 50 x 200, in order to ensure that the total area under the histogram (50 x 200) equals that under the curve.]

$$P(200 \le x < 250) = e^{-\frac{200}{80.75}} - e^{-\frac{250}{80.75}} = 0.039$$

$$P(250 \le x < 300) = e^{-\frac{250}{80.75}} - e^{-\frac{300}{80.75}} = 0.021$$

$$\begin{aligned} P(x \ge 300) &= 1 - P(x < 300) \\ &= 1 - (0.462 + 0.249 + 0.134 + 0.072 + 0.039 + 0.021) \\ &= 1 - 0.977 = 0.023 \end{aligned}$$

Notice how the last probability is calculated by subtracting all the others from 1; this ensures that the total probability is 1, and hence, the expected frequencies total 200. We can now find the expected frequencies for each class by multiplying each of these probabilities by the total frequency of 200; these together with the observed frequencies, are shown in the table below.

Life of Component (x hrs)	0-50	50-100	100-150	150-200	200-250	250-300	300 or more
Expected Frequency	92.41	49.8	26.8	14.4	7.8	4.2	4.6
Observed Frequency	94	53	25	12	4	4	8

Notice, however, that the last two *expected* frequencies are both less than 5; in order to remedy this situation, we must combine the last class, 300 or more, with the adjacent class, 250-300, to obtain the new class, 250 or more, as shown in the table below.

Life of Component (x hrs)	0-50	50-100	100-150	150-200	200-250	250 or more
Expected Frequency (E)	92.4	49.8	26.8	14.4	7.8	8.8
Observed Frequency (O)	94	53	25	12	4	12

This process obviously reduces the number of degrees of freedom, v, by 1, and so is an *important* step.

Now, $\chi^2 = \Sigma \frac{(0 - E)^2}{E}$

$$= \frac{(94 - 92.4)^2}{92.4} + \dots + \frac{(12 - 8.8)^2}{8.8}$$

$= 3.769$

Also, $v = 6 - 1 - 1 = 4$ (since we had to estimate a parameter)

and so, $\chi^2_{crit} = 9.49$

Comparing, we see:

$3.769 \leq 9.49$

and so accept H_o.

Goodness of fit (summary)

Tests of Goodness of fit are concerned with the comparison of some observed frequency distribution with theoretical distribution.

Most appropriate test depends on the precise form of the hypothesis being tested. This problem of fitting a distribution is most commonly tackled with the general χ^2-test. This is not always the best test available.

The degrees of freedom, v, are generally calculated as follows:

v = number of classes minus the number of parameters estimated from the sample (such as μ, λ, σ etc) minus 1

Therefore, for

i) Poisson's fit

$v = K - 2$

ii) Binomial fit

$v = K - 2$

iii) Normal fit

$\nu = K - 3$

Only large values of $\chi 2$ are considered to be significant, ie significant enough to reject H_o.

H_o is always the assumption of goodwill ie

H_o: Data to follow (fit) the proposed distribution (poisson's, binomial or normal).

10.4 Contingency tables

We now consider the use of the χ^2 distribution in testing whether two variables are related (associated with each other). For example, we may wish to test whether there is an association between (the variables) *handedness* and *colour of eyes*; are left-handed people more likely to have brown eyes, for instance? Suppose we collected data on eye colour and handedness from 500 people; this data could be tabulated as below - this configuration of data is called a *contingency* table.

		Colour of Eyes		
		Brown	Blue	Grey/Green
Handedness	L	25	5	20
	R	180	80	190

Contingency Table

Each square in the above table is called a *cell*, and at the moment, contains a single observed frequency. We wish to test whether these observed frequencies are due to random processes ie if there were *no* association between handedness and eye colour, or whether they are different to what would be expected by chance, ie if there *were* an association.

In order to do this, we must calculate those frequencies *expected* by chance, assuming there to be *no* association between handedness and eye-colour (the Null Hypothesis). If we sum the rows and columns of the above table, we obtain the so-called *marginal totals*, as shown in the table below.

		Colour of Eyes			
		Brown	Blue	Grey/Green	Marginal Totals
Handedness	L	25	5	20	50
	R	180	80	190	450
	Marginal Totals	205	85	310	Total = 500

Now, consider the 205 people with brown eyes. Since 50 out of the 500 people are left-handed, ie 0.1, we would expect that, of these 205 people with brown eyes 0.1 of them would also be left-handed, *providing* there were *no* association between handedness and eye colour. This gives us the expected frequency for the top left cell of:

$$\frac{50}{500} \times 205 = 20.5$$

Continuing in this manner, we can now construct the table of observed and expected frequencies below,

		Eye Colour		
		Brown	Blue	Grey/Green
Handedness	L	25 / 20.5	5 / 8.5	20 / 31
	R	180 / 184.5	80 / 76.5	190 / 209

which can be explained using the key:

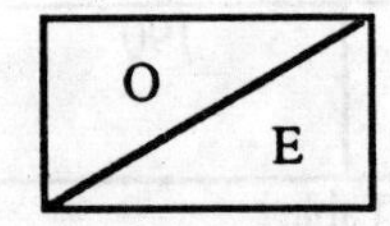

Before proceeding further, we must state the Null Hypothesis, which is always as follows.

H_0: there is no association between handedness and eye colour.

We can now calculate the value of χ^2

$$\chi^2 = \Sigma \frac{(0 - E)^2}{E}$$

$$= \frac{(25 - 20.5)^2}{20.5} + \ldots + \frac{(190 - 209)^2}{209}$$

$$= 8.33.$$

Now, the number of degrees of freedom for a contingency table can be calculated using the formula:

$$\nu = (\text{no. of rows - 1})(\text{no. of columns - 1})$$

[Note: In the case of there being only one row or column, we simply replace the appropriate bracket by 1.]

In this case, there are 2 rows and 3 columns, giving

$\nu = (2 - 1)(3 - 1) = 2.$

From the tables, $\chi^2_{crit} = 5.99$, at a 5% level of significance.

Comparing, we see that: 8.33 > 5.99

which is unlikely, and so reject H_o, ie there is an association between handedness and eye colour, on the basis of this data.

10.5 Yates' correction

In the special case of a 2 x 2 table (2 rows and 2 columns), we must use a slightly different formula to calculate $\chi 2$ by, as below:

$$\chi^2 = \frac{\left(| O - E | - \frac{1}{2} \right)^2}{E}$$

which involves subtracting $\frac{1}{2}$ from the positive difference between O and E, before squaring. The use of this formula is illustrated in the following example.

Example: In a survey, 100 people were asked the following questions:

a) Do you play any kind of sport?

b) Are you a vegetarian?

The results are given below. Using the $\chi 2$ test of association, with a 5% level of significance, test whether sporting inclination is related to dietary habit.

		Plays Sport	
		Yes	No
Vegetarian	Yes	10	10
	No	30	50

H_o: there is no association between sporting inclination and dietary habit.

Construct the table of expected frequencies below as usual.

		Plays Sport: Yes	Plays Sport: No	Marginal Totals
Vegetarian	Yes	10 / 8	10 / 12	20
	No	30 / 32	50 / 48	80
	Marginal Totals	40	60	Total = 100

(Each cell: O / E)

Then, using the new formula for χ^2, we have:

$$\chi^2 = \Sigma \frac{\left(| O - E | - \frac{1}{2}\right)^2}{E}$$

$$= \frac{\left(| 10 - 8 | - \frac{1}{2}\right)^2}{8} + \frac{\left(| 10 - 12 | - \frac{1}{2}\right)^2}{12} + \frac{\left(| 30 - 32 | - \frac{1}{2}\right)^2}{32} + \frac{\left(| 50 - 48 | - \frac{1}{2}\right)^2}{48}$$

$$= \frac{\left(1\frac{1}{2}\right)^2}{8} + \frac{\left(1\frac{1}{2}\right)^2}{12} + \frac{\left(1\frac{1}{2}\right)^2}{32} + \frac{\left(1\frac{1}{2}\right)^2}{48}$$

$$= 0.59$$

Using the formula for the number of degrees of freedom, we have:

$\nu = (2 - 1)(2 - 1) = 1.$

From the tables, $\chi^2_{crit} = 3.84$, at the 5% level of significance.

Comparing, we see that:

$0.59 \leq 3.84$

and so we accept H_o.

Exercise 10

1. Using χ^2 tests at 5% level of significance, test whether the following sets of digits were constructed using random number tables.

a)

digit	0	1	2	3	4	5	6	7	8	9	Total
frequency	9	7	12	8	15	10	11	8	11	9	100

b)

digit	0	1	2	3	Total
frequency	15	23	20	22	80

c)

digit	3	5	9	Total
frequency	25	70	25	120

2. A coin is flipped 100 times, and turns up 72 heads. Is it biased?

3. Two dice are thrown 360 times, their totals being recorded in the table below. Ascertain whether the dice are fair.

Total of Dice	2	3	4	5	6	7	8	9	10	11	12
Frequency	7	24	35	38	50	55	54	40	28	16	13

4. A sociologist claims that factory strikes occur more frequently in the summer months than the winter months. Data concerning strikes during the last year is given below. Is his claim justified?

Month	Jan	Feb	Mar	Apr	May	Jun	Jul	Aug	Sept	Oct	Nov	Dec	Total
No. of Strikes	12	10	10	17	20	25	12	10	18	15	16	15	100

5. The English examination papers for a class of 50 students are marked by two independent teachers, A and B. The number of passes and fails given by each teacher are given in the table below. Do these teachers mark consistently?

		Teacher A	
		Pass	Fail
Teacher B	Pass	10	18
	Fail	12	10

6. A botanist believes that a certain species of plant will flower 70% of the time. In an experiment it was found that 125 out of 200 plants flowered. Is this idea consistent with the botanist's beliefs?

7. A survey was carried out by a company making TV sets. 600 viewers were asked what kind of set they had and how long they spent viewing. The data is given below.

		Type of TV Set			
		Large Colour	Large Black and White	Portable Colour	Portable Black and White
Amount of Viewing	High	120	50	30	20
	Medium	110	60	25	25
	Low	45	30	45	40

Perform a 5% χ^2 test of association between Type of Set and Amount of Viewing.

By combining the appropriate categories of data, make another 50% χ^2 test of association to determine whether the amount of viewing is related to whether a colour or black and white set is used.

Finally, by eliminating those viewers with low viewing time, carry out the same test as above.

8. A die is tossed repeatedly until a six shows; this trial is repeated 100 times, with the number of tosses needed to produce a six being recorded each time, as in the table below. Use a χ^2 test with a 10% level of significance to determine if the die is fair.

No. of throws required	1	2	3	4	5	6	7 or more	Total
No. of trials	14	18	17	8	8	12	23	100

9. In a seed viability test, 600 seeds were planted in 100 rows of 6. The number of seeds that germinated in each row was counted and the results are shown in the table below:

No. of seeds germinating per row	0	1	2	3	4	5	6
Observed no. of rows	1	4	7	29	33	18	8

Calculate

i) the mean number of seeds germinating per row, and

ii) the expected frequencies corresponding to these observed values for a binomial distribution with the same mean as that found in i).

10. If P(x) denotes the probability of x successes in a binomial distribution for n trials for which the probability of a success in each trial is p, show that

$$P(x + 1)/P(x) = p(n - x)/(1 - p)(x + 1)$$

A bag contains a very large number of black marbles and white marbles. 8,192 random samples of 6 marbles are drawn from the bag. The frequencies of the numbers of black marbles in these samples are tabulated below:

No. of black marbles per sample	0	1	2	3	4	5	6	Total
Frequencies	3	42	255	1115	2505	2863	1409	8192

Test the hypothesis that the ratio of the numbers of black to white marbles in the bag is 3 : 1.

11. Show that the mean and the variance of a binomial distribution of x trials are np and $np(1 - p)$ where p is the probability of success in any trial. In an experimental trial, a biased coin is tossed 5 times. 500 such trials are carried out and the number of tails in each trial is recorded. The results are summarised in the following table:

No. of tails per trial	0	1	2	3	4	5	
Observed frequency	19	76	165	148	77	15	Total 500

Fit a binomial distribution to these data.

12. The following frequency table gives the heights in inches of 100 male students, classified in intervals of 2", with the corresponding class frequencies.

Height (inches)	61-63	64-66	67-69	70-72	73-75	Total
Observed Class Frequencies	5	20	39	28	8	100

Show that the mean and standard deviation are 68.42 and 2.97 inches, respectively. Fit a normal curve to the above data. Compare the theoretical frequencies with the observed values. Use a χ^2 test to check that the fit is very good.

13. Four coins are thrown 160 times, and the distribution of the number of heads is observed to be

x (number of heads)	0	1	2	3	4
f (frequency)	5	35	67	41	12

Find the expected frequencies if the coins are unbiased. Compare the observed and expected frequencies and apply the χ^2 test. Is there any evidence that the coins are biased?

14. The masses of 500 vegetable marrows were recorded during a season, as in the table below. Use a $\chi 2$ 'goodness of fit' test at the 5% level of significance to test if the masses of marrows are normally distributed. (Assume the last class has width 100 when estimating the parameters.)

Mass (g)	100-200	200-250	250-300	300-400	400-500	500 or more	Total
No. of Marrows	20	50	180	210	25	15	500

15. Four players meet weekly and play eight hands of cards. In a year one of the players finds that he has won x of the eight hands with a frequency given in the following table.

x	0	1	2	3	4	5	6	7	8
f	4	13	12	12	6	3	2	0	0

Find the frequencies of the number of hands he would expect to win if the probability of winning any hand were 1/4.

Use the χ^2 distribution to test this hypothesis.

16. At St Trinian's College for Young Ladies there are 1,000 pupils. Of these 75 have represented the College at both hockey and netball, 10 have represented the College at hockey but do not play netball, 35 have represented the College at netball but do not play hockey and 100 do not play games at all. In all 100 girls have represented the College at hockey, and 150 at netball. The number who do not play hockey is 200 and the number who do not play netball is 125.

Arrange the above data in the form of a 3 x 3 contingency table, and state how many pupils play both hockey and netball but have not represented the College in either.

Apply the χ^2 test to your 3 x 3 table, and state the hypothesis which it tests.

17. Two fair dice are thrown 432 times. Find the expected frequencies of scores of 2, 3, 4, ..., 12. Two players A and B are each given two dice and told to throw them 432 times, recording the results.

The frequencies reported are:

Scores	2	3	4	5	6	7	8	9	10	11	12
A's frequency	18	33	28	54	62	65	66	42	30	27	7
B's frequency	14	22	34	51	58	73	63	45	38	25	9

Is there any evidence that either pair of dice is biased?

What can be said about B's alleged results?

18. A large haulage company employs drivers to work a 'round-the-clock' system. The drivers' union is concerned that some periods are more dangerous than others. The management contests this, claiming that no one period is more dangerous than any other. The following data on the incidence of traffic accidents by time of day, for this group of drivers over the previous two years is presented.

Time of day (24-hr clock)	00.01 -04.00	04.01 -08.00	08.01 -12.00	12.01 -16.00	16.01 -20.00	20.01 -24.00	Total
Number of accidents	14	16	24	22	24	20	120

Are these results consistent with the management's claim?

19. Derive the mean and variance of the binomial distribution. Mass production of miniature hearing aids is a particularly difficult process and so the quality of these products is monitored carefully. Samples of size six are selected regularly and tested for correct operation. The number of defectives in each sample is recorded. During one particular week 140 samples are taken and the distribution of the number of defectives per sample is given in the following table.

No. of defectives per sample (x)	0	1	2	3	4	5	6
No. of samples with x defectives (f)	27	36	39	22	10	4	2

Find the frequencies of the number of defectives per sample given by a binomial distribution having the same mean and total as the observed distribution.

20. The table below summarises the incidence of cerebral tumours in 141 neurosurgical patients.

		Type of Tumour		
		Benign	Malignant	Others
Site of Tumour	Frontal Lobes	23	9	6
	Temporal Lobes	21	4	3
	Elsewhere	34	24	17

Find the expected frequencies on the hypothesis that there is no association between the type and site of a tumour. Use the χ^2 distribution to test this hypothesis.

21. The government of Cashmania is in the process of making drastic public spending cuts in order to reduce the national debt. One controversial proposal is to change primary school hours from the existing 9 am to 3.30 pm to a new 8 am to 1.30 pm, thus saving large amounts of both heating costs and staff costs. In order to gauge public reaction to this proposal, a survey of 200 people was undertaken, and their replies are summarised in the table below accounting for the political party that they support.

	Wurkuz Party (B_1)	Bossis Party (B_2)	Freedom Party (B_3)	Totals
For proposals (A_1)	24	36	10	70
Against proposals (A_2)	52	50	28	130
Totals	76	86	38	200

i) On the hypothesis that there is no association between reaction to the proposal and politcal party supported,

the value of $\chi^2 = \Sigma\frac{(O - E)^2}{E}$

has been calculated to be 3.43. Use a χ^2 distribution and a 5% level of significance to complete the test of the above hypothesis.

ii) It is also felt that the sex of people surveyed may be an important factor. Thus the following additional information is obtained.

In cell (A_1, B_1) 16 were male;

in cell (A_1, B_2) 17 were male;

in cell (A_1, B_3) 4 were male;

in cell (A_2, B_1) 17 were male;

in cell (A_2, B_2) 35 were male;

in cell (A_2, B_3) 11 were male.

Construct two new 2 x 3 contingency tables, one for males and one for females. Use the χ^2 distribution and a 5% level of significance to test the hypothesis of no association between political party and reaction to the proposal, both for males and for females.

iii) Explain briefly the results obtained.

22. The shape of the human head is the subject of an international project financed by the World Council for Health and Welfare. Observations were taken in many countries and the nose lengths, to the nearest millimetre, of 150 Italians are summarised below.

Nose lengths (mm) x	$-\infty<x<44$	$45<x<47$	$48<x<50$	$51<x<53$	$54<x<56$	$57<x<\infty$
frequency f	4	12	63	59	10	2

Estimate the mean and standard deviation of the population from which these observations were taken. (For these calculations you should assume that the lower and upper classes have the same range as the other classes.)

Use the χ^2 distribution and a 1% level of significance to test the adequacy of the normal distribution as a model for these data.

11 THE POISSON DISTRIBUTION

11.1 Introduction

This is a discrete distribution arising from the binomial distribution. We will introduce it by considering the following situation.

Suppose we record the number of cars passing a fixed point on the roadside during many 10 minute periods, and find that, on average, 4 cars pass in 10 minutes. We could model this situation by dividing up the 10 minute period into 10 one minute intervals, say, and specifying that at most one car can pass in any one of these intervals. The diagram below shows 4 cars in 4 separate intervals.

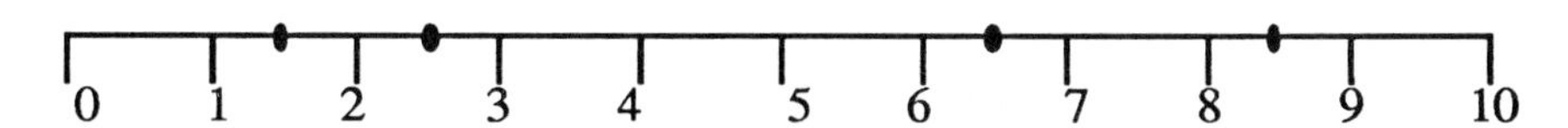

Now, the probability that any one minute interval containing a car is clearly $\frac{4}{10} = 0.4$. Furthermore, the number of intervals in any 10 minute period that contains a car will be binomially distributed with n = 10 and $\beta = 0.4$. Thus, we have a binomial model of the number, R, say of cars passing in any 10 minute period, with probability mass function:

$$P(r) = {}^{10}C_r\ (0.4)^r\ (1 - 0.4)^{10-r},\ r = 0, 1, \ldots\ldots, 10.$$

However, this is obviously a crude model since it does not allow for the possibility of more than one car passing in a single minute.

We can refine the above model by dividing the 10 minute period into smaller intervals, of say $\frac{1}{2}$ minute, and specifying that at most one car can pass in any single interval; this model is illustrated in the diagram below:

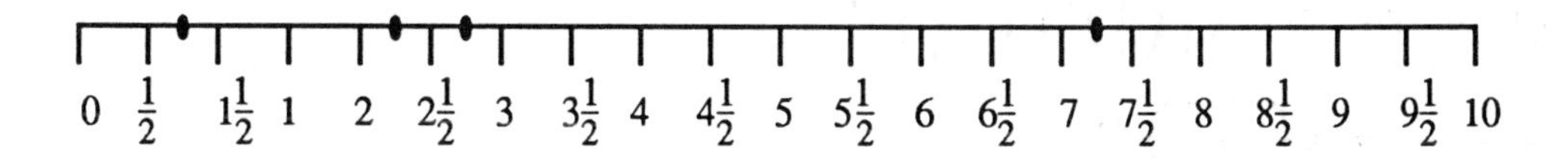

Again, since 4 cars pass every 10 minutes, on average, then the probability of a car passing in a single $\frac{1}{2}$ minute interval is

$$\frac{4}{20} = 0.2$$

Furthermore, the number of $\frac{1}{2}$ minute intervals occupied in a 10 minute period will be binomially distributed with $n = 20$ and $\beta = 0.2$. We thus have the more refined model:

$$P(r) = {}^{20}C_r (0.2)^r (1 - 0.2)^{20-r}, r = 0, 1, ..., 20.$$

Notice that in both models, $n\beta = 4$, since this is the mean of the binomial distribution which is 4 cars every 10 minutes in both cases.

Continuing in this manner, further subdividing the 10 period into smaller intervals we obtain more refined (realistic) models of the number of cars passing; as we do this, n becomes greater (the number of intervals) and β becomes correspondingly smaller (the probability of a car passing in one interval), but always $n\beta$ remaining constantly 4.

We can now proceed with the derivation of the probability mass function for the poisson distribution.

Now, the probability mass function for the binomial distribution with parameters n and β is:

$$P(r) = {}^nC_r \beta^r (1 - \beta)^{n-r}, r = 0, 1, ..., n.$$

Suppose we let $n\beta = \lambda$ (in the above case, $\lambda = 4$); then λ is the mean of the distribution and, furthermore,

$$\beta = \frac{\lambda}{n}$$

Substituting for β in the above expression, we have:

$$P(r) = {}^nC_r \left(\frac{\lambda}{n}\right)^r \left(1 - \frac{\lambda}{n}\right)^{n-r}, r = 0, 1, ..., n.$$

Let us consider each of these probabilities in turn.

<u>When r = 0,</u> $\quad P(r = 0) = {}^nC_o \left(\frac{\lambda}{n}\right)^0 \left(1 - \frac{\lambda}{n}\right)^{n-0}$

$$= \left(1 - \frac{\lambda}{n}\right)^n$$

Now, as $n \to \infty$, the expression $\left(1 - \frac{\lambda}{n}\right)^n \to e^{-\lambda}$

(this is a result which must be assumed).

Thus, $\quad P(r = 0) = e^{-\lambda}$

<u>When r = 1,</u> $\quad P(r = 1) = {}^nC_1 \left(\frac{\lambda}{n}\right)^1 \left(1 - \frac{\lambda}{n}\right)^{n-1}$

$$= \lambda \left(1 - \frac{\lambda}{n}\right)^{n-1}$$

Again, as $n \to \infty$, $\left(1 - \frac{\lambda}{n}\right)^{n-1} \to e^{-\lambda}$, and so:

$P(r = 1) = \lambda . e^{-\lambda}$

When r = 2, $\quad P(r = 2) = {}^{n}C_2 \left(\frac{\lambda}{2}\right)^2 \left(1 - \frac{\lambda}{n}\right)^{n-2}$

$$= \frac{\lambda^2}{2!} \frac{(n-1)}{n} \left(1 - \frac{\lambda}{n}\right)^{n-2}$$

As $n \to \infty$, $\frac{n-1}{n} \to 1$ and $\left(1 - \frac{\lambda}{2}\right)^{n-2} \to e^{-\lambda}$, and so:

$$P(r = 2) = \frac{\lambda^2}{2!} e^{-\lambda}$$

Continuing in this manner, we obtain the probability mass function for the poisson distribution:

$$P(r) = \frac{\lambda^r}{r!} e^{-\lambda}, \quad r = 0, 1, ...$$

Notice that, unlike the binomial distribution, r has no upper limit, since we have let $n \to \infty$; however, in practice there obviously must be - in the above example, for instance, n is the maximum possible number of cars passing in a 10 minute period, which must obviously be finite.

Returning to the above example, we can now specify the best model for the number of cars passing in a 10 minute period by simply knowing the mean number of cars passing per 10 minutes. Since the mean = 4 cars per 10 minutes, then, the probability mass function for this model is:

$$P(r) = \frac{4}{r!} e^{-4}, \quad r = 0, 1, ...$$

where r is the number of cars passing in a 10 minute period.

As with the binomial distribution, certain conditions must be met in order for a poisson model to be applicable; these are as follows.

1. The events being observed (eg passing cars) must occur independently of each other; this is the same as for a binomial distribution.
2. The variable being modelled is the number of events occuring in a given *interval* (this could be an interval of time, height, volume etc.)
3. Events cannot occur *simultaneously*; this precludes the possibility of modelling a two-lane road, for example.
4. The events must occur as the result of random processes; this condition is not particularly satisfied by traffic flow because of rush hours, traffic lights, etc.
5. The mean number of events occuring per unit interval must remain constant; in the above example, this was 4 cars per 10 minutes (or, equivalently 2 cars per 5 minutes, etc.)

Providing the above conditions are sufficiently well satisfied, a poisson model can be used; typical situations which can be modelled this way are:

a) the number of blades of a rare grass in $1m^2$ of grounds;

b) the number of rare blood cells of type X in a cm^3 of blood;

c) the number of particles emitted from a weak radioactive source per second;

d) the number of deaths per hour in a city.

Notice that each of the events considered is relatively rare; this is because if n is large then β will be small, since $n\beta = \lambda$ remains constant.

11.2 Poisson approximation of the binomial distribution

Providing $n\beta$ remains constant, we have seen that the binomial distribution becomes closer to the poisson distribution as n becomes large; thus, providing n is large, we can use the poisson distribution to approximate the probabilities of the binomial distribution. This is sometimes necessary since the evaluation of ${}^nC_r$ in the proability mass function of the binomial distributin can be extremely difficult for large values of n (indeed, the calculator cannot handle n! for large n).

Consider the following example.

Example: It is known that 2.5% of all new cars will need repairs within the first week after sale. What is the probability that, of 200 cars sold, more than 3 will be returned for repair within a week.

Here, the number of cars needing repair is binomially distributed with $n = 200$ and $\beta = 0.025$. Since n is large, we can use a poisson approximation with $\lambda = n\beta = 200 \times 0.5 = 5$, to estimate $P(r > 3)$ as follows:

$$P(r > 3) = 1 - P(r \leq 3)$$

$$= 1 - P(r = 0) - P(r = 1) - P(r = 2) - P(r = 3)$$

Now $P(r = 0) = e^{-5} = 0.0067$

and $P(r = 1) = 5\,e^{-5} = 0.0337$

$$P(r = 2) = \frac{5^2}{2!}e^{-5} = 0.0842$$

$$P(r = 3) = \frac{5^3}{3!}e^{-5} = 0.1404$$

Thus, $P(r > 3) = 1 - 0.0067 - 0.0337 - 0.0842 - 0.1404$
$= 0.735$

Before we can solve other problems we need to know the mean and variance of the poisson distribution.

11.3 Mean of the poisson distribution

As shown in the derivation of the probability mass function of the poisson distribution, the parameter λ was the mean of all the binomial distribution approximating to it; it is of no surprise then (and as already intimated on page 264) that the mean of the poisson distribution with parameter λ is indeed λ itself. We now prove this as follows.

Now, $P(r) = \frac{\lambda^r}{r!}e^{-\lambda}$, $r = 0, 1, \ldots$

Thus, $E[R] = \sum_{r=0}^{\infty} r.P(r)$

Now, when $r = 0$, $r.P(r) = 0$ and so we may as well begin the sum at $r = 1$; thus:

$$E[R] = \sum_{r=1}^{\infty} r.P(r)$$

$$= \sum_{r=1}^{\infty} r.\frac{\lambda^r}{r!}e^{-\lambda}$$

$$= e^{-\lambda}\sum_{r=1}^{\infty}\frac{\lambda^r}{(r-1)!} = e^{-\lambda}\sum_{r=1}^{\infty}\frac{\lambda.\lambda^{r-1}}{(r-1)!} = \lambda.e^{-\lambda}\sum_{r=1}^{\infty}\frac{\lambda^{r-1}}{(r-1)!}$$

Replace r - 1 by s; then, when r = 1, s = 0 and when r = ∞, s = ∞.

Thus, $$E[R] = \lambda.e^{-\lambda}\sum_{s=0}^{\infty}\frac{\lambda^s}{s!} = \lambda e^{-\lambda}.e^{\lambda} = \lambda$$

$$\left(\text{since } e^{\lambda} = \sum_{s=0}^{\infty}\frac{\lambda^s}{s!} \text{ by definition}\right)$$

11.4 Variance of the poisson distribution

Now, $V[R] = E[R^2] - (E[R])^2$

$= E[R^2] - \lambda^2$

Also, $E[R^2] = E[R(R-1) + R]$

$= E[R(R-1)] + E[R]$

$= E[R(R-1)] + \lambda$

Now, $$E[R(R-1) = \sum_{r=1}^{\infty} r(r-1)P(r)$$

$$= \sum_{r=2}^{\infty} r(r-1)P(r)$$

since when r = 0 or r = 1, r(r - 1) P(r) = 0.

Then, $$E[R(R-1)] = \sum_{r=2}^{\infty} r(r-1)\frac{\lambda^r}{r!}e^{-\lambda}$$

$$= e^{-\lambda}\sum_{r=2}^{\infty}\frac{\lambda^r}{(r-2)!}$$

$$= e^{-\lambda}\sum_{r=2}^{\infty}\frac{\lambda^2.\lambda^{r-2}}{(r-2)!}$$

$$= \lambda^2 e^{-\lambda} \sum_{r=2}^{\infty} \frac{\lambda^{r-2}}{(r-2)!}$$

Replace r - 2 by s; then, when r = 2, s = 0 and

when r = ∞, s = ∞.

$$\text{Thus,} \quad E[R(R-1)] = \lambda^2 e.^{-\lambda} \sum_{s=0}^{\infty} \frac{\lambda^s}{s!} = \lambda^2 e^{-\lambda}.e^{\lambda} = \lambda^2$$

Hence, $E[R^2] = \lambda^2 + \lambda$, and so:

$V[R] = (\lambda^2 + \lambda) - \lambda^2 = \underline{\lambda}$

Thus, we see that the mean and variance of the poisson distribution are both λ.

> If R follows the poisson distribution, then:
>
> $E[R] = \lambda$
>
> and $V[R] = \lambda$

In order to completely specify a poisson distribution we simply need to find its mean λ, since this is its only parameter; this fact is illustrated in the example below.

Example: The mean number of blood cells of type X is known to be 3.2 per mm^3 of blood.

Calculate the probability of finding:

a) no blood cells of type X in one mm^3 of blood;

b) more than one cell of type X in one mm^3 of blood;

c) 4 blood cells of type X in *two* mm^3 of blood.

Firstly, this situation does satisfy the conditions of the Poisson distribution.

1. Blood cells are distributed independently of each other throughout a volume.
2. The variable being modelled is the number of blood cells per mm^3.
3. No two blood cells can occupy the same position in the volume.
4. The blood cells are distributed randomly throughout a volume.
5. The mean number of blood cells of type X is constantly 4 per mm^3.

a) Thus, we can model r, the number of blood cells of type X, using a Poisson distribution with λ = 3.2.

$$\text{Hence,} \qquad P(r) = \frac{(3.2)^r}{r!} e^{-3.2}, \quad r = 0, 1, ...$$

Thus, P(no blood cells in one mm^3) = P(r = 0)

$$= \frac{(3.2)^0}{0!} e^{-3.2} = 0.041$$

b) Also, P(more than one blood cell in one mm^3)

= 1 - P(no blood cells or one blood cell in one mm^3)

= 1 - P(r = 0) - P(r = 1) = 1 - 0.041 - 0.130 = <u>0.829</u>

c) Here we are asked for the probability of 4 blood cells of type X in *two* mm^3 of blood; we must therefore construct a *new* Poisson distribution to model the number of blood cells of type X in *two* mm^3 of blood (which is a different variable to the above one).

Here, $\lambda = 2 \times 3.2 = 6.4$ blood cells per two mm^3.

And so, $P(r) = \frac{(6.4)^r}{r!} e^{-6.4}, \quad r = 0, 1, \ldots.$

Thus, $P(r = 4) = \frac{(6.4)^4}{4!} e^{-6.4} = \underline{0.116}$

Example: A certain disease is known to affect 2 trees per km^2 of forest on average. Calculate the probability of

a) finding 2 diseased trees in a km^2 of forest;

b) finding more than 3 diseased trees in a km^2 of forest;

c) a disease free area of 5 km^2.

A brief run through the conditions to be satisfied for a Poisson model will show that the number of diseased trees per unit area can be modelled in this way.

a) Here, $\lambda = 2$ diseased trees per km^2 and so:

$$P(r) = \frac{(2)^r}{r!} e^{-2}, \quad r = 0, 1, \ldots.$$

Thus, $P(r = 2) = \frac{(2)^2}{2!} e^{-2} = 0.271$

b) Also, $P(r > 3) = 1 - P(r \le 3)$

$= 1 - P(r=0) - P(r=1) - P(r=2) - P(r=3)$

$= 1 - e^{-2} - 2e^{-2} - 0.271 - \frac{2^3}{3!} e^{-2}$

$= 1 - 0.135 - 0.271 - 0.271 - 0.180$

$= \underline{0.143}$

c) Here, $\lambda = 5 \times 2 = 10$ diseased trees per 5 km^2, and so:

$$P(r) = \frac{(10)^r}{r!} e^{-10}, \quad r = 0, 1, \ldots.$$

Thus, P(disease free) = $P(r = 0) = e^{-10} = 0.00005$

Example: If 3 people in 1,000 are ambidextrous on average, what size sample would have to be chosen in order to be 95% confident of finding at least one such person?

Let n be the required sample size; then, the number of ambidextrous people in such a size of sample will be binomially distributed with

$\beta = \frac{3}{1,000} = 0.003.$ Using a Poisson approximation, we have:

$\lambda = n\beta = 0.003n$, and:

$$P(r) = \frac{(0.003n)^r}{r!} e^{-0.003n}, \quad r = 0, 1, ...$$

We require that:

Probability of at least one ambidextrous person = 0.95

ie $P(r \geq 1) = 0.95$

ie $1 - P(r < 1) = 0.95$

ie $1 - P(r = 0) = 0.95$

ie $1 - e^{-0.003n} = 0.95$

ie $e^{-0.003n} = 0.05$

ie $-0.003n = \log_e(0.05) = 12.996$

ie $n = \frac{2.996}{0.003} = 998.6$

Thus, we would have to take a sample of size 999 to be 95% confident of finding at least one ambidextrous person.

11.5 Normal approximation to the poisson distribution

Recall from chapter 9, page 207, that the binomial distribution becomes normal as n becomes large. But, as n becomes large, such that $n\beta$ remains constant, the binomial distribution becomes Poisson distributed.

Thus, the Poisson distribution is also approximately normal; the approximation is good only for large values of λ, however ($\lambda \geq 10$).

Just as the normal approximation to the binomial distribution saved tedious calculations, so can its approximation to the Poisson distribution, as demonstrated in the following example.

Example: The letter k occurs 11 times per page of a certain book, on average. What is the probability of there being more than 7 of these letters on a page.

Here, $\lambda = 11$ letters per page, and so:

$$P(r) = \frac{(11)^r}{r!} e^{-11}, \quad r = 0, 1, ...$$

we require: $P(r > 7)$.

Using the Poisson distribution, we would have to calculate:

$P(r > 7) = 1 - P(r{=}0) - P(r{=}1) - P(r{=}2) - P(r{=}3) - P(r{=}4) - P(r{-}5) - P(r{=}6) - P(r{=}7)$

which is obviously tedious.

However, we can approximate this Poisson distribution by the normal distribution with same mean and variance as that of the Poisson distribution (since λ is large), ie

$\mu = \lambda = 11$ and $\sigma^2 = \lambda = 11$

We can calculate P(r > 7) as for a Normal distribution as follows.

$$Z_7 = \frac{7 - 11}{\sqrt{11}} = -1.21$$

So, $P(r > 7) = P(Z > -1.21) = \underline{0.8869}$

Naturally, this normal approximation must have the same mean and variance as the Poisson distribution, ie

$\mu = \lambda$ and $\sigma^2 = \lambda$

Summarising this result, we have:

Providing $\lambda \geq 10$, the Poisson distribution with parameter can be approximated by $n(\lambda, \lambda)$

Thus, providing the right conditions are met, the Poisson, normal and binomial distributions are approximately equal, and so it is important to cover all three distributions if you decide to attempt a question on any *one* of them in the examination. The diagram below shows when to use what approximation.

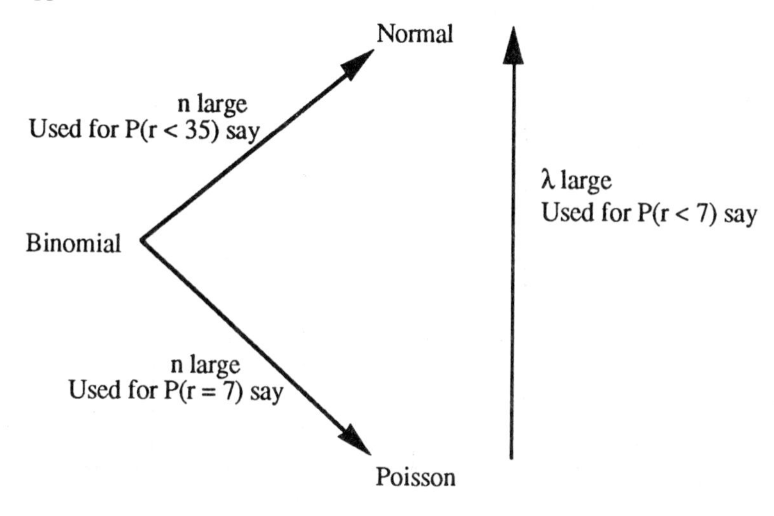

11.6 χ^2 goodness of fit of the poisson distribution as a model

We can apply the theory of the last chapter to the new distribution in testing for goodness of fit, as illustrated in the following example.

Example: The number of emissions, r, from a radioactive source are recorded in 200 successive intervals of 1 millisecond, as in the table below. Use a χ^2 test of goodness of fit at the 5% level of significance to test whether the variable R follows a Poisson distribution.

No. of emissions per millisecond	0	1	2	3	4	5 or more	Total
No. of milliseconds	107	63	20	7	1	2	200

H_0: the data follows a Poisson distribution.

Firstly, we must estimate the parameter λ of the Poisson distribution; this must be the mean of the above distribution.

From the calculator, $\lambda = 0.69$ emissions per millisecond.

Then, the Poisson distribution has probability mass function.

$$P(r) = \frac{(0.69)^r}{r!} e^{-0.69}, \ r = 0, 1, ...$$

from which we can calculate the expected frequencies as follows.

$$P(r=0) = \frac{(0.69)^0}{0!} e^{-0.69} = 0.502$$

$\therefore$ Expected frequency = 0.502 x 200 = 100.3

Continuing in this manner, we obtain the expected frequencies below.

No. of Emissions per millisecond	0	1	2	3	4	5 or more	Total
Observed Frequency (O)	107	63	20	7	1	2	200
Expected Frequency (E)	100.3	69.2	23.8	5.4	0.93	0.37	200

As can be seen, the last two expected frequencies are well below 5, and so we must combine the last three classes together, giving:

No. of Emissions per millisecond	0	1	2	3 or more	Total
Observed Frequency (O)	107	63	20	10	200
Expected Frequency (E)	100.3	69.2	23.8	6.7	200

We can now calculate the value of χ^2

$$\chi^2 = \Sigma \frac{(0 - E)^2}{E}$$

$$= \frac{(107 - 100.3)^2}{100.3} + \ldots\ldots + \frac{(10 - 6.7)^2}{6.7}$$

$$= 3.24$$

Comparing with $\chi^2_{crit} = 5.99$ for $v = 4 - 1 - 1 = 2$

and at a 5% level of significance, we have:

$$3.24 \leq 5.99$$

and so accept H_o, ie the Poisson model is a good fit.

11.7 The use of poisson probability paper

This graph paper can be used to see whether a given sample is likely to have come from a Poisson distribution, and, if so, to estimate its mean, λ. Consider the following example.

Example: The number of bees, r, visiting a square metre of meadow is thought to follow a Poissondistribution. The data below was collected on a summer day from an area of $150m^2$ of meadow. Use Poisson probability paper to verify this hypothesis, and estimate the mean number of bees visiting a square metre of meadow. Furthermore, use your graph to estimate the probability of more than 8 bees visiting a square metre of meadow land.

No. of bees, r, visiting a m^2	0	1	2	3	4	5	6 or more	Total
No. of m^2	18	45	40	28	15	3	1	150

Method: Firstly, calculate the reverse relative cumulative frequency, P, for each value of r, ie P = 1 - relative cumulative frequency, as in the table below.

No. of bees, r, visiting a m^2	0	1	2	3	4	5	6 or more	Total
No. of m^2	18	45	40	28	15	3	1	150
Cumulative Frequency	18	63	103	131	146	149	150	
Relative Cumulative	0.12	0.42	0.69	0.87	0.97	0.99	1.00	
P	0.88	0.58	0.31	0.13	0.03	0.01	0.00	
c = r + 1	1	2	3	4	5	6	7 or more	

The bottom row is simply found by adding 1 to each value of r; this is necessary, since the values on the Poisson probability graph paper begin at c = 1, as opposed to r = 0.
We can now plot values of P (left hand vertical axis) against values of c (right hand vertical axis) by following a curved line from each value of c until the correct level of P is reached (see diagram opposite). Notice that the last value is missed out.

Draw the vertical line of best fit through these points, extending it downward to cross the bottom horizontal axis; as can be seen, the crosses do approximate to a straight line, and so we can assume that the data does follow a Poisson distribution.

Furthermore, reading the value where the line crosses the bottom axis, we obtain the mean value of:

$\lambda \approx$ 1.95 bees per m^2.

Finally, since P is the probability of exceeding any particular value of c, we can find the probability of r exceeding 8 by reading off the value of P corresponding to c = 9; from the graph, this is P = 0.00024 (as shown by the arrow).

Thus, $P(r > 8) \approx 0.00024$

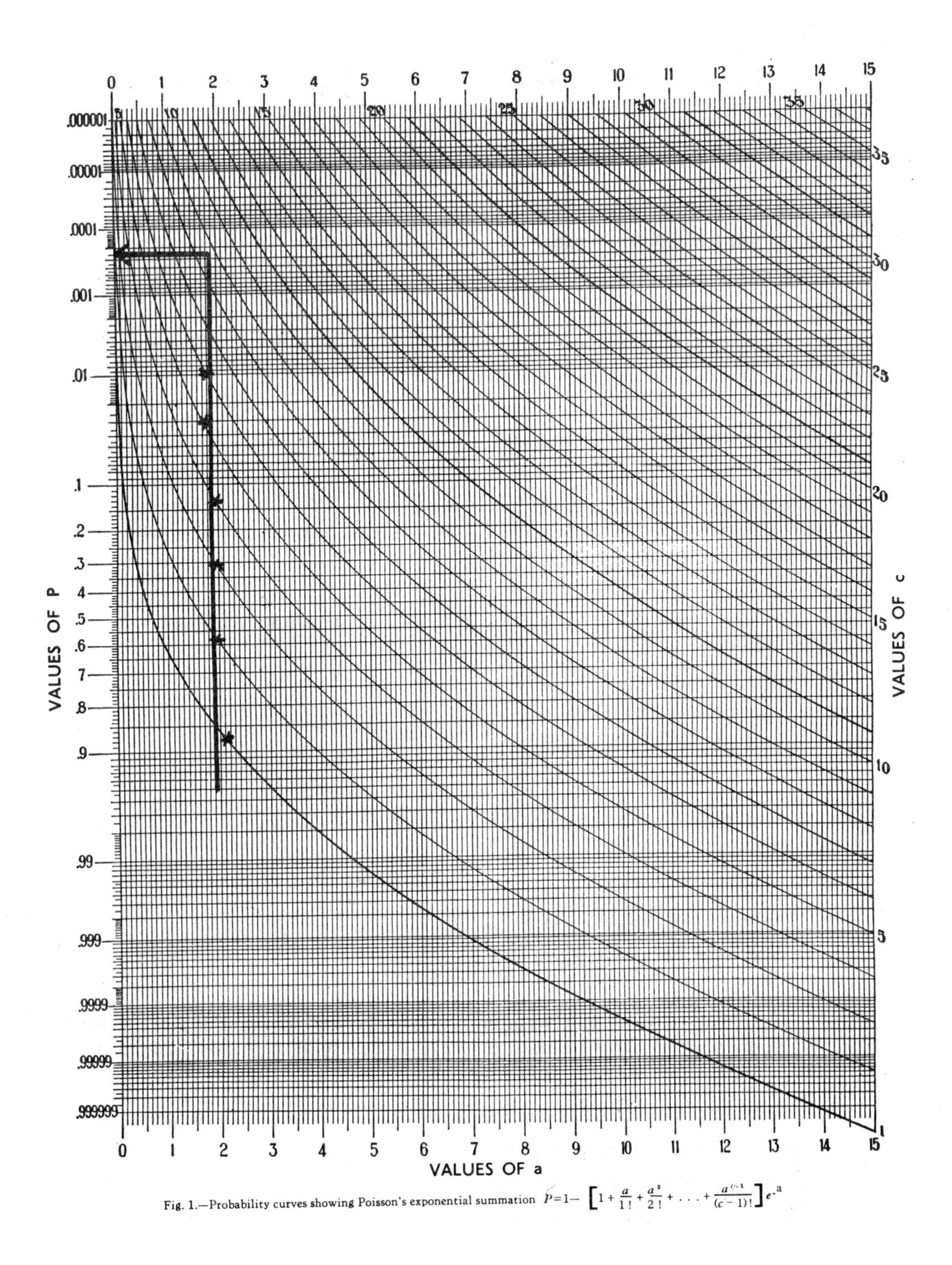

Fig. 1.—Probability curves showing Poisson's exponential summation $P = 1 - \left[1 + \frac{a}{1!} + \frac{a^2}{2!} + \ldots + \frac{a^{c-1}}{(c-1)!}\right] e^{-a}$

Exercise 11

1. Sketch the shape of the Poisson distribution for $\lambda = \frac{1}{2}$, 1, $1\frac{1}{2}$ and 2. What happens as λ becomes large?

   The number of spelling mistakes in a child's written work averages 2 per page. What is the probability of:

   a) a page with no spelling mistakes;

   b) more than 5 mistakes on a page;

   c) less than 3 mistakes on 2 pages.

2. A loaded die shows the digit 1 when tossed only two times in 25 throws. What is the probability that, in 100 throws

   a) no 1s occur

   b) three 1s occur.

3. The probability of a transistor being defective in production is 0.1. What is the probability that a randomly chosen sample of 75 transistors, in a control check, contains between 3 and 5 faulty ones (inclusive)?

4. On a certain road, the number of cars passing a town hall is approximately 3.2 per minute. Calculate the probability of

   a) more than 4 passing in a given minute;

   b) at least 3 passing between 12-01 and 12-03;

   c) 10 cars passing in a 5 minute interval.

5. In a certain hospital ward, the number of dust particles per cm^3 averages 7.2. What is the probability of there occuring:

   a) no dust particles in a cm^3;

   b) more than 4 dust particles in a cm^3;

   c) at least 5 dust particles in $0.5cm^3$.

6. How large a sample of people must be taken in order to be 99% confident of it containing at least one with a certain rare bone marrow found in only 3 out of every 200 people on average?

7. On average, 4 out of every 100 eggs is unsaleable on a certain farm. What is the probability of:

   a) more than 3 unsaleable eggs in 100;

   b) at least one unsaleable egg in 3 dozen;

   c) no unsaleable eggs in a box of 10 dozen.

8. A rare type of grass occurs with a frequency of only 1 plant per m^2. What is the probability of

   a) finding no plants in one m^2;

   b) finding more than 5 plants in one m^2;

   c) finding less than 6 plants in $10m^2$.

   What area would need to be examined in order to be 95% certain of finding at least one specimen of this type of grass?

9. In a certain grade of gravel there occurs the odd large stone which has escaped the grading mesh; this occurs 25 times per m^3 of gravel passing through the mesh. Calculate the probability of:

   a) a m^3 of gravel containing more than 18 large stones;

   b) less than 40 large stones in $1.5m^3$ of gravel.

   If a further sieving reduced the number of large stones by 20%, what is the probability that $10m^3$ of gravel contains more than 180 stones?

10. A certain species of plant contains a medicinal compound such that from 4 kg of plants, 52g of the compound can be extracted on average. Calculate the probability that:

    a) more than 75g can be extracted from 4 kg of plants;

    b) less than 30g can be extracted from 5 kg.

11. A certain aphid is attracted to French dward bean plants. The data below gives the number of aphids found on 200 plants in a test area. Use the χ^2 goodness of fit test at a 5% level of significance to see if the data can be modelled by a Poisson distribution.

No. of aphids	0	1	2	3	4	5	6	7 or more
No. of plants	28	94	41	18	11	4	4	0

12. In a different test area to that in question 11, a further 200 plants were inspected.

    Use Poisson probability paper to test whether this distribution is Poisson. If so, estimate the mean no. of aphids per plant, and the probability of more than 8 aphids occuring on a plant.

No. of aphids	0	1	2	3	4	5	6	7 or more
No. of plants	42	63	51	29	11	3	1	0

13. Specify the conditions under which a binomial distribution reduces to a Poison distribution and derive the expression for the Poisson distribution from the expression for the binomial distribution. Hence derive expressions for the mean and variance of the Poisson distribution.

A clerk in the ticket issuing office of a suburban railway station noted that the number of tickets issued for journeys away from London was equal to the number of first class tickets issued for all journeys, and the total of the number of tickets issued in both categories was in the proportion of one in five hundred of all tickets issued.

Determine the probability that, of the 2,000 travellers departing from that station between 8 am and 9 am

a) exactly 3 required first class seats,

b) more than 2 travelled away from London.

c) Determine the probability that 2 travellers, journeying away from London required first class seats,

d) the probability that one person, with a first class ticket, joined the train travelling away from London.

14. The results of twenty football matches, played on the same occasion, revealed the following frequency distribution for the numbers of goals scored by the forty teams:

Goals scored	0	1	2	3	4	5	6	7	8
Teams scoring	3	5	14	9	3	4	1	1	0

Use the χ^2 test to compare these observed frequencies with the frequencies which would be expected from the Poisson distribution having the same mean and total frequency.

Comment on the agreement.

15. At a busy intersection of roads, accidents requiring the summoning of an ambulance occur with a frequency, on average, of 1.8 per week. These accidents occur randomly, so that it may be assumed that they follow a Poisson distribution.

i) Calculate the probability that there will not be an accident in a given week.

ii) Calculate the smallest integer n such that the probability of more than n accidents in a week is less than 0.02.

iii) Calculate the probability that there will not be an accident in a given fortnight.

iv) Calculate the largest integer k such that the probability that there will not be an accident in k successive weeks is greater than 0.0001.

16. State the conditions under which the binomial distribution approximates to the Poisson distribution. Hence drive the Poisson distribution of mean m and show that its variance is also m.

    Tests for defects are carried out in a textile factory on a lot comprising 400 pieces of cloth. The results of the tests are tabulated below:

No. of faults per piece	0	1	2	3	4	5	6	Total
No. of pieces	92	142	96	46	18	6	0	400

    Show that this is approximately a Poisson distribution and calculate the frequencies on this assumption.

    How many pieces from a sample of 1,000 pieces may be expected to have 4 or more faults?

17. If $p(x)$ is the probability of x successes in a Poisson distribution with mean μ, show that

    $$p(x+1)/p(x) = \mu/(x+1)$$

    A car hire firm has 3 cars to be hired out by the day. Assuming the number of cars ordered has a Poisson distribution with mean 2, calculate, for a period of 100 days, the expected number of days when

    a) no car will be ordered;

    b) some orders cannot be met;

    c) one particular car is not used, assuming each car is used equally frequently.

18. State the conditions in which a binomial distribution approximates to the Poisson distribution given by the coefficients of powers of t in the expansion of $e^{\mu t-\mu}$.

    Find the mean and variance of this Poisson distribution.

    A batsman kept a record of the numbers of 'sixes' he hit in sixty innings, with the following result, x being the number of 'sixes' in an innings and f the frequency of each value of x.

x	0	1	2	3	4	5
f	13	22	14	5	5	1

    Find the frequencies of 'sixes' given by the Poisson distribution having the same mean and total as the observed distribution.

    Comment on the agreement between the observed and calculated frequencies.

19. Explain the conditions in which the frequencies of an observed variable may be expected to approximate to the Poisson distribution given by the coefficients of powers of t in the expansion of $e^{\mu(t-1)}$.

At the approach to a large town there is a notice showing the number of persons injured in road accidents since the beginning of the year. For the first 100 days of a given year, the figures were, in order:

5, 8, 9, 18, 18, 24, 28, 30, 32, 37, 45, 48, 52, 59, 66, 71, 74, 75, 82, 90, 96, 98,

102, 107, 114, 117, 119, 125, 133, 144, 151, 159, 163, 167, 168, 168, 174, 181,

185, 189, 194, 200, 201, 204, 208, 216, 217, 222, 227, 233, 236, 238, 239, 247,

254, 258, 261, 263, 277, 283, 290, 292, 295, 302, 310, 313, 317, 322, 327, 329,

330, 333, 338, 344, 352, 360, 362, 366, 372, 383, 389, 394, 397, 399, 402, 408,

415, 420, 428, 430, 436, 444, 455, 465, 476, 483, 488, 494, 500.

Find the mean number of persons injured per day and compare the frequencies of days with 0, 1, 2, ... 8 and 9 or more persons injured with the frequencies of a Poisson distribution having the same mean and total frequency (there is no need for a χ^2 test).

20. The independent variables x_1 and x_2 have means μ_1, μ_2 and variances $\sigma_1{}^2$, $\sigma_2{}^2$ respectively. State the mean and variance of the variable $m_1x_1 + m_2x_2$ where m_1 and m_2 are constants.

The scores of boys of sixteen years of age in a certain aptitude test form a normal distribution with mean 70 and standard deviation 3. The scores of boys of fifteen years of age in the same test form a normal distribution with mean 64 and standard deviation 4. What percentage of the boys of each age will score 62 or less?

Find the mean and standard deviation of the excess score of a sixteen year old boy over that of a fifteen year old boy, both taken at random.

What is the probability that the younger boy will obtain a higher score in the test?

21. The table below gives the distribution for the number of heavy rainstorms reported by 330 weather stations in the United States of America over a one year period.

No. of rainstorms (x)	0	1	2	3	4	5	more than 5
No. of stations (f) reporting x rainstorms	102	114	74	28	10	2	0

i) Find the expected frequencies of rainstorms given by the Poisson distribution having the same mean and total as the observed distribution.

ii) Use the χ^2 distribution to test the adequacy of the Poisson distribution as a model for these data.

22. In the Growmore Market Garden plants are inspected for the presence of the deadly red angus leaf bug. The number of bugs per leaf is known to follow a Poisson distribution with mean one. What is the probability that any one leaf on a given plant will have been attacked (at least one bug is found on it)?

    A random sample of twelve plants is taken. For each plant ten leaves are selected at random and inspected for these bugs. If more than eight leaves on any particular plant have been attacked then the plant is destroyed. What is the probability that exactly two of these twelve plants are destroyed?

23. Paper manufacture is a continuous production process, with the only planned machine stoppage occuring when a new mix of colour is introduced, or for routine maintenance. However, unplanned stoppages do occur, for example a power cut, or a paper break. The table below gives the distribution of the number of unplanned stoppages per week, for a certain machine over the last two working years.

No. of unplanned stoppages per week (x)	0	1	2	3	4	5	6	more than 6
No. of weeks with x stoppages (f)	5	15	21	24	17	11	5	2

Plot the information given by this table on Poisson probability paper. Draw a line through your plotted points and hence estimate the mean. Assuming that the number of unplanned stoppages per week does follow a Poisson distribution with mean equal to your estimate, what is

i) the probability of more than 2 unplanned stoppages in any given week,

ii) the probability of an even number of unplanned stoppages in any given week (take 0 as even),

iii) the probability of an odd number of unplanned stoppages in any given week?

24. The table below gives the distribution of the number of hits by flying bombs in 450 equally sized areas in South London during World War II.

No. of hits x	0	1	2	3	4	5	6 or more
Frequency f	180	173	69	20	6	2	0

i) Find the expected frequencies of hits given by a Poisson distribution having the same mean and total as the observed distribution.

ii) Use the χ^2 distribution and a 10% level of significance to test the adequacy of the Poisson distribution as a model for these data.

25. In an agricultural research station the number of caterpillars on a cabbage may be assumed to be an observation from a Poisson distribution with mean θ. An inspector visits the cabbages every day and records the number of caterpillars on each cabbage, but only if caterpillars are present. (Zeros are not recorded.) What is the probability that for any given cabbage he makes a recording? In this case write down the probability that x caterpillars are recorded as present on any given cabbage.

26. The number of defective microchips in 100 control samples was recorded below. Use Poisson probability paper to test whether this variable follows a Poisson distribution.

No. of defectives	0	1	2	3	4	5	Total
No. of samples	6	54	25	12	2	1	100

27. Use Poisson probability paper to test whether the number of roof top birds' nests per house can be modelled by a Poisson distribution. Estimate the mean number of nests per house. What is the probability of there being more than 5 nests in a house?

No. of nests	0	1	2	3	4	Total
No. of houses	40	30	10	5	1	86

12 CORRELATION AND REGRESSION

12.1 Introduction

This chapter is concerned with how strongly two variables are related, eg are 'height' and 'weight' of men related? Consider the table of heights and weights of 15 men, aged 30-40 years, below.

Height (x cm)	175	168	170	171	190	192	165	168	162	170	175	172	175	180	181
Weight (y kg)	65	62	66	63	70	68	60	64	60	69	68	64	64	72	74

We can obtain a good visual representation of this data by plotting weight (y) against height (x) as below; such a diagram is called a *scattergram*.

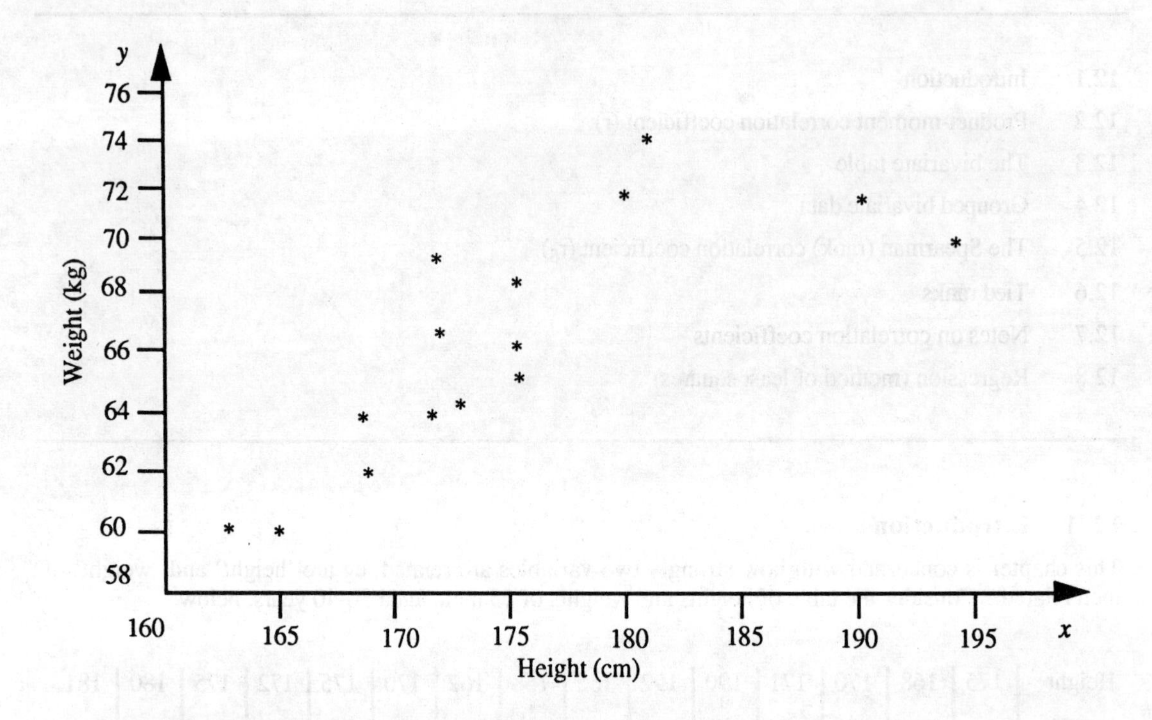

Now, if x and y were related, we would expect that, on average, as x increased, y would also increase (or, as x increases, y decreases); as can be seen below, the points of the scattergram form an upward trend, indicating just this (as would be expected, taller men weigh more on average). Such an upward trend could be indicated by drawing a straight line (the *line of best fit*) through the middle of the array of points in an upward direction, as shown on the diagram below. *Note*: the line of best fit must pass through $(\bar{x}, \bar{y})$, which is the point (174.27, 65.93) in this case, and represents the 'middle' of the scattergram and is drawn such that the perpendicular distances (*deviations*) of the points from the line are a minimum overall; this ensures that it truly passes through the middle of the scattergram and is the *best* fit (see diagram below).

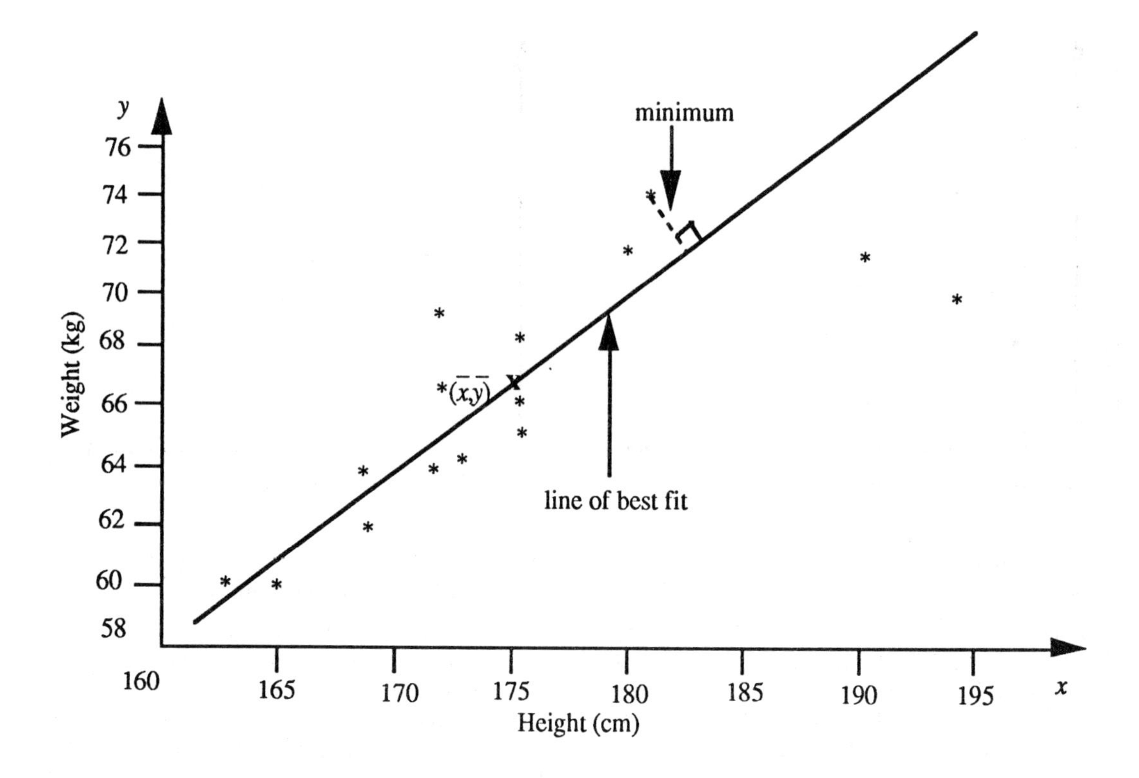

The closer the points approximate to a straight line, the stronger the relationship between the two variables; the diagram below shows examples of differing degrees of relatedness.

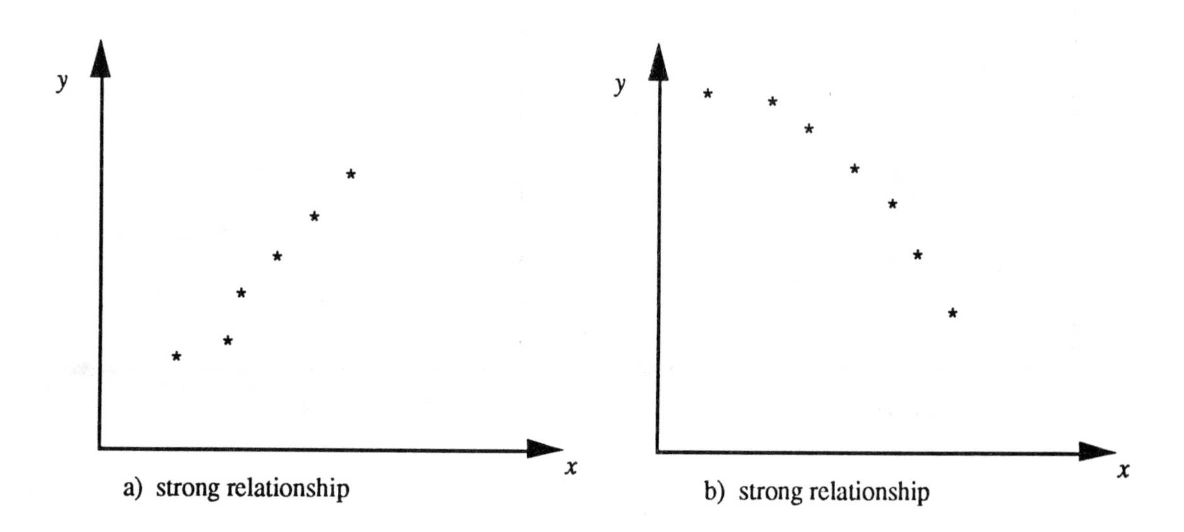

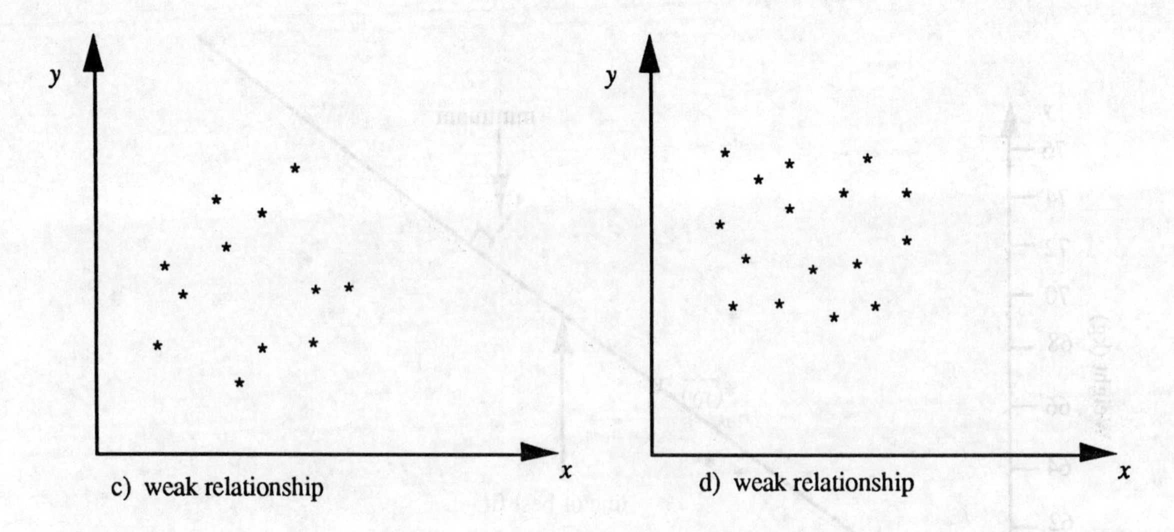

c) weak relationship

d) weak relationship

What we require is a way of *measuring* the strength of the relationship between two variables; the *correlation coefficient* (r) does just this, ie it is a measure of how co-related two variables are.

Correlation coefficients are numbers which always lie between -1 and +1. The larger the *absolute* value of the correlation coefficient, the stronger the relationship. We need both *negative* and *positive* values values since there are two types of relationships between variables as mentioned above, viz an x-increases, y-increases (diagram a) *and* as x-decreases, y-decreases (diagram b) above) a *positive* correlation coefficient indicates the *former* (a positive correlation), whilst a *negative* correlation coefficient indicates the *latter* (a negative *correlation*), and we say that variables are *positively* or *negatively correlated.* We can now attach typical correlation coefficients to the diagrams above.

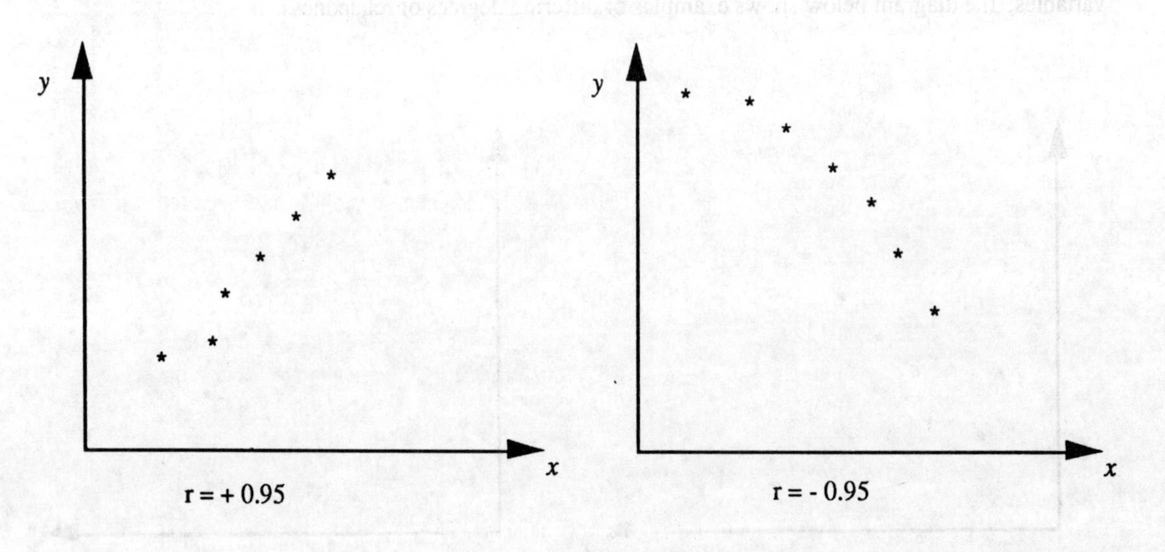

r = + 0.95

r = - 0.95

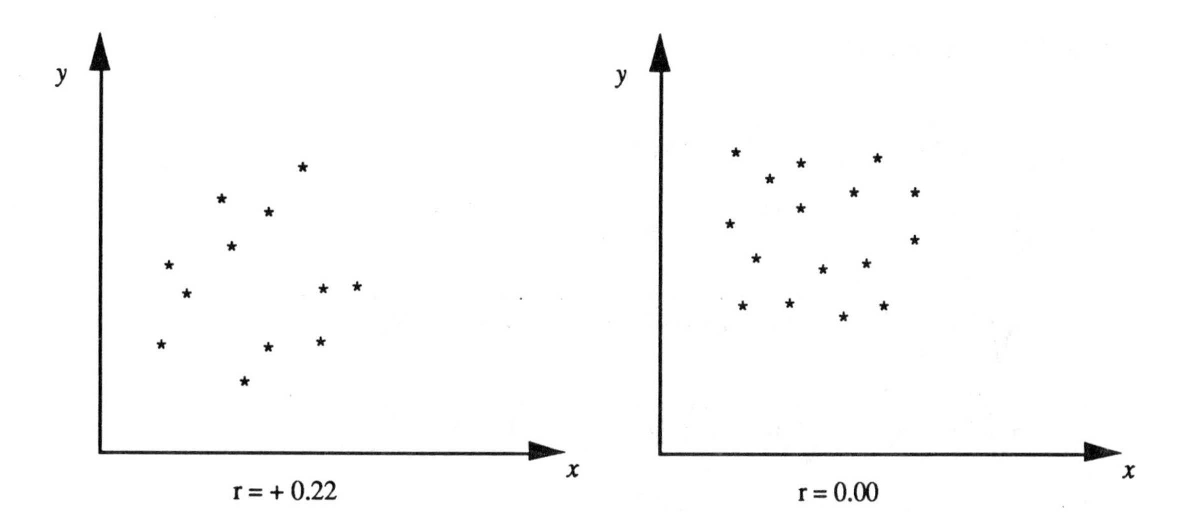

There are two main ways of calculating correlation coefficients, resulting in two different types of correlation coefficient, viz the *Product-Moment Correlation Coefficient* (r) and the *Spearman Rank Correlation Coefficient* (r_S); the need for two kinds of correlation coefficient arises from the different types of data involved, and will be discussed later on page 291.

12.2 Product-moment correlation coefficient (r)

Consider again the data concerning height and weight of men above.

Height (x cm)	175	168	170	171	190	192	165	168	162	170	175	172	175	180	181
Weight (y kg)	65	62	66	63	70	68	60	64	60	69	68	64	64	72	74

The formula below produces a value, r, which always lies between -1 and +1, and is called the *product-moment* correlation coefficient.

$$r = \frac{\Sigma xy - n\bar{x}\bar{y}}{n.S_x.\ S_y}$$

Here, n is the number of pairs of values.

S_x is the standard deviation of the x-values

S_y is the standard deviation of the y-values

$\bar{x}$ is the mean of the y-values

$\bar{y}$ is the mean of the y-values

Σxy is the mean of the products of each pair of values.

For the data above, we have:

$$n = 15$$

$$\bar{x} = \frac{\Sigma x}{n} = \frac{175 + \ldots + 181}{15} = 174.27$$

$$\bar{y} = \frac{\Sigma y}{n} = \frac{65 + \ldots + 74}{15} = 65.93$$

$$S_x = \sqrt{\frac{\Sigma(x - \bar{x})^2}{n}} = \sqrt{\frac{(175 - 174.27)^2 + \ldots + (181 - 174.27)^2}{15}} = 8.20$$

$$S_y = \sqrt{\frac{\Sigma(y - \bar{y}^2)}{n}} = \sqrt{\frac{(65 - 65.93)^2 + \ldots + (74 - 65.93)^2}{15}} = 4.02$$

$$\Sigma xy = 175 \times 65 + \ldots + 181 \times 74 = 172704$$

Thus, we have

$$r = \frac{172704 - 15(174.27)(65.93)}{15(8.20)(4.02)}$$

$$= \underline{0.73}$$

Thus, the heights and weights of this sample of men are fairly strongly positively correlated, consistent with the diagram on page 252.

12.3 The bivariate table

Where larger amounts of data are being analysed, it is sometimes more convenient to record them in the form of a *bivariate table*; this is when there are many repetitions of the same pair of values occuring. The table below gives the age (x yrs) and fastest reading speed (lines/min) of 100 girls. As can be seen, the pair of values x = 12 years and y = 7 lines/min occurs 8 times (indicated by the number of the corresponding cell). If we simply wrote down all the pairs of values in a row (as in the previous example) we would obtain many unnecessary repetitions.

		Age (x yrs)					
		10	11	12	13	14	15
Speed (y lines/min)	6	7	5	1	0	1	0
	7	4	5	8	4	3	2
	8	2	3	3	7	7	5
	9	0	1	2	3	5	8
	10	0	0	1	2	2	9

In order to calculate r for such data, we simply include the frequency, f, of each pair of values (x, y) into the above formula, which thus becomes:

$$r = \frac{\Sigma xy - n\bar{x}\bar{y}}{n.Sx.\ Sy}$$

where: $n = \Sigma f$, and $\bar{x} = \frac{\Sigma fx}{n}$, $\bar{y} = \frac{\Sigma fy}{n}$,

$$S_x = \sqrt{\frac{\Sigma f(x - \bar{x})^2}{n}}, \quad Sy = \sqrt{\frac{\Sigma f(y - \bar{y})^2}{n}}$$

The calculations of $\bar{x}$, $\bar{y}$, S_x and S_y can be simplified, however, by firstly summing the frequencies of each value of x and y, as shown in the table below; these 'marginal' frequencies can then be used in place of f in the above formulae.

		Age (x yrs)						
		10	11	12	13	14	15	Marginal Frequencies
Speed (y lines/min)	6	7	5	1	0	1	0	14
	7	4	5	8	4	3	2	26
	8	2	3	3	7	7	5	27
	9	0	1	2	3	5	8	19
	10	0	0	1	2	2	9	14
	Marginal Frequencies	13	14	15	16	18	24	Total = 100

In the above example, we have:

n = total number of pairs = Σf = 100.

$$\bar{x} = \frac{\Sigma fx}{n} = \frac{13 \times 10 + \ldots + 24 \times 15}{100} = 12.84$$

$$\bar{y} = \frac{\Sigma fy}{n} = \frac{14 \times 6 + \ldots + 14 \times 10}{100} = 7.93$$

$$S_x = \sqrt{\frac{\Sigma f(x - \bar{x})^2}{n}} = \sqrt{\frac{13(7 - 12.84)^2 + \ldots + 24(15 - 12.84)^2}{100}} = 1.73$$

$$S_y = \sqrt{\frac{\Sigma f(y - \bar{y})^2}{n}} = \sqrt{\frac{14(6 - 7.93)^2 + \ldots + 14(10 - 7.93)^2}{100}} = 1.25$$

$$\Sigma fxy = 7 \times 10 \times 6 + \ldots + 9 \times 15 \times 10 = 10{,}321$$

Thus, we have:

$$r = \frac{103121 - 100 \times 12.84 \times 7.93}{100 \times 1.73 \times 1.25}$$

$$= \underline{0.64}$$

12.4 Grouped bivariate data

As with frequency distributions, it is often necessary to group larger amounts of data, obtaining *grouped bivariate tables*; the method of calculating r is the same as above, except that the mid-class values are used in place of the individual values of x and y. This is illustrated in the example below.

Example: In a psychological experiment, 200 students were asked to take a reaction time test wherein a button must be pressed as soon as a light appears on a screen in front of the student. They were then given standard IQ tests. The data is given below in the form of a grouped bivariate table. Calculate the product-moment correlation coefficient for the data, and comment on your result.

Reaction Time (x milliseconds)

IQ (y)	100-150	150-200	200-250	250-300	300-350	Marginal Frequencies
95-105	12	11	12	7	9	51
105-115	13	14	9	11	2	49
115-125	12	8	13	10	3	46
125-135	14	12	11	10	7	54
Marginal Frequencies	51	45	45	38	21	200

Firstly, calculate the MCV for each class, eg replace the class 100-150 by the MCV 125. Then, using the formulae above, we have:

$$n = \Sigma f = 12 + 11 + \ldots + 7 = 200$$

$$\bar{x} = \frac{\Sigma fx}{n} = \frac{51 \times 125 + \ldots + 21 \times 325}{200} = \frac{41650}{200} = 208.25$$

$$\bar{y} = \frac{\Sigma fy}{n} = \frac{51 \times 100 + \ldots + 54 \times 130}{200} = \frac{23030}{200} = 115.15$$

$$S_x = \sqrt{\frac{\Sigma f(x - \bar{x})^2}{n}} = \sqrt{\frac{51(125 - 208.25)^2 + \ldots + 21(325 - 208.25)^2}{200}} = 66.01$$

$$S_y = \sqrt{\frac{\Sigma f(y - \bar{y})^2}{n}} = \sqrt{\frac{51(100 - 115.15)^2 + \ldots + 54(130 - 115.15)^2}{2}} = 11.40$$

$$\Sigma fxy = 12 \times 125 \times 100 + \ldots + 7 \times 325 \times 130 = 4794250$$

Thus, we have:

$$r = \frac{4794250 - 200 \times 208.25 \times 115.15}{200 \times 66.01 \times 11.40} = \frac{-1747.5}{150502.8} = -.012$$

The almost negligible correlation suggests that there is no relationship between IQ and Reaction Time.

12.5 The Spearman (Rank) correlation coefficient (r_s)

The need for a second kind of correlation coefficient arises from the different types of data to be analysed; some kinds of data cannot be analysed unless they are first converted into ranked data. The most obvious examples of this type of data are actual rankings or ratings made by humans, eg rank ordering 10 pieces of prose in terms of style. Alternatively, a teacher may mark the 10 pieces out of 20, say; however, the fact that a person obtained 18 does not mean to say that their piece of prose had twice as much style as one obtaining 9. All we can say realistically is that the one piece had more style than the other, ie place them in a relative order (ie rank them). Consider the data below concerning two teachers A and B who marked the same 10 pieces of prose out of a possible maximum of 20.

	1	2	3	4	5	6	7	8	9	10
Teacher A	18	10	11	20	19	7	8	12	9	17
Teacher B	16	13	11	17	14	4	12	15	9	18

We may wish to calculate a measure of how consistent (related) the two teachers' marking is. As mentioned above, we must first *rank order* the data; this can be from largest to smallest or vice-versa. However, if the marks of teacher A are ranked from smallest to largest, say, then so must those of teacher B. Thus, the lowest mark has the rank of 1, the next lowest the rank 2, and so on. (See table below.)

	1	2	3	4	5	6	7	8	9	10
Rank A	8	4	5	10	9	1	2	6	3	7
Rank B	8	5	3	9	6	1	4	7	2	10

Now,if the two teachers' marks were similar, then we would expect that their rankings, r_A and r_B, would also be similar; if this were the case then the differences, d, in these ranks would be small, as would the squares of their differences, d^2. The next step, therefore, is to calculate all the differences $d = r_A - r_B$ and their squares, d^2 as in the table below.

	1	2	3	4	5	6	7	8	9	10
r_A	8	4	5	10	9	1	2	6	3	7
r_B	8	5	3	9	6	1	4	7	2	10
d	0	-1	2	1	3	0	-2	-1	1	-3
d^2	0	1	4	1	9	0	4	1	1	9

We can now calculate the Spearman (Rank) correlation coefficient, r_s, using the formula:

$$r_s = 1 - \frac{6\Sigma d^2}{n(n^2-1)}$$

As can be seen, the smaller the differences, d, and hence, the squares of differences, d^2, the smaller will be Σd^2 and so, the *larger* r_s will be, thereby reflecting the strength of relationship between the two sets of rankings.

In the above example, we see that:

$$\Sigma d^2 = 0 + \ldots + 9 = 30$$

and so:

$$r_s = 1 - \frac{6 \times 30}{10(10^2 - 1)} = \underline{0.818}$$

12.6 Tied ranks

A slight modification must be made to the above method in the case of two values, and hence two ranks being the same. Consider the example below.

Example: A student guessed the weights (g) of a random order series of 15 weights, firstly with the left and then with the right hand; the results are shown below. Calculate the Spearman Rank Correlation Coefficient for the data and comment on the results.

Weight Number

	1	2	3	4	5	6	7	8	9	10	11	12	13	14	15
Left Hand	12	11	9	4	25	15	11	17	12	13	11	12	13	18	15
Right Hand	14	13	8	7	20	18	19	14	12	9	14	12	18	13	19

First, notice that we could use the product-moment correlation coefficient as a measure of how related the Left and Right Hand guesses are; however, in view of the (*subjective*) nature of guessing, the rank correlation coefficient is also appropriate. Proceeding as before, we can rank the first few Left Hand values with no difficulty, ie 4, 7. However, the next largest value is 11, which occurs three times. What rank should we ascribe to each? Clearly, the fairest way is to give them *equal* ranks; in fact, we simply average what their ranks would have been, and give each of them this 'averaged rank'. In this case, the three 11s would have had the ranks of 3, 4 and 5; thus, averaging, we get:

$$\frac{3 + 4 + 5}{3} = 4$$

and so give each of them the rank 4 (the reason for choosing the average rank is so that the ranks still have the same *sum* as before). Continuing this process, we arrive at the ranks below.

	1	2	3	4	5	6	7	8	9	10	11	12	13	14	15
r_L	7	4	2	1	15	11.5	4	13	7	9.5	4	7	9.5	14	11.5
r_R	9	6.5	2	1	15	11.5	13.5	9	4.5	3	9	4.5	11.4	6.5	13.5
$d=r_L-r_R$	-2	-2.5	0	0	0	0	-9.5	4	2.5	6.5	-5	2.5	-2	7.5	-2
d^2	4	6.25	0	0	0	0	90.25	16	6.25	42.25	25	6.25	4	56.25	4

Again, we have: $\Sigma d^2 = 260.5$

and so: $r_s = 1 - \dfrac{6\Sigma d^2}{n(n^2 - 1)}$

$$= 1 - \frac{6 \times 260.5}{15(15^2 - 1)} = \underline{0.535}$$

The positive correlation suggests that, to some extent, guesses with the Left Hand are related to those of the Right Hand.

12.7 Notes on correlation coefficients

The following notes concern possible *false inferences* which may be made on the basis of a given correlation coefficient, *which simply measures how strongly two variables are related to each other.*

1. *Correlation does not imply cause*

Suppose that we recorded the output of 50 similar factories together with the number of man-hours lost through sickness and found these two variables to be negatively correlated; we could *not* infer that absence through sickness causes decreased output. Other factors may be giving rise to the decreased output of some factories, eg market forces, which may far outweigh the effects of absence through sickness. It *may*, of course, be true that staff sickness significantly reduces output; however, we cannot *infer* this from a *correlational* analysis.

2. *Correlation due to a third cause*

Two variables may be correlated through the causal effect of a third variable. For example, in a reaction time experiment, a student's blood pressure was found to correlate positively to his speed of response to a flashing light; however, it cannot be concluded that a rise in blood pressure increased his speed of response, or vice-versa. In actual fact, a third variable, namely *concentration*, was causing this correlation - the more he concentrated, the higher his blood pressure became and also the faster his speed of response.

3. *Correlation coefficients only measure linear relations*

The correlation coefficients described above only have the power to detect *linear* relationships between variables; other non-linear relationships such as those depicted below may yield small or zero correlations.

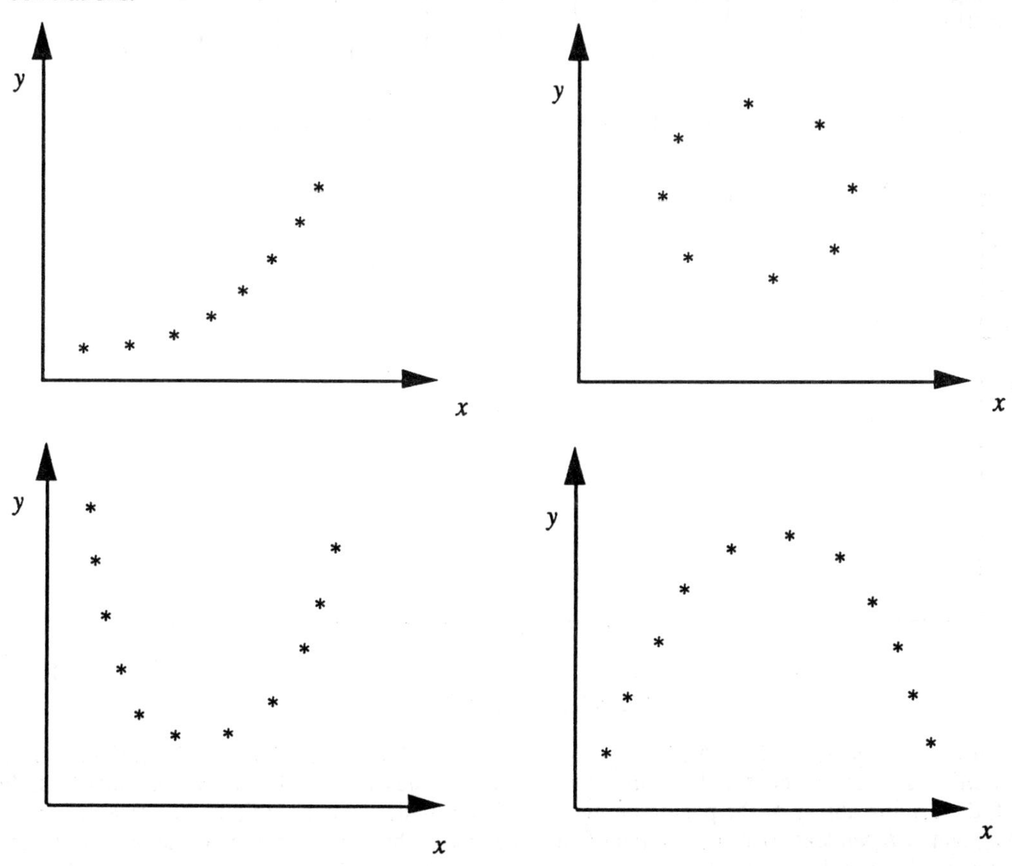

12.8 Regression (method of least squares)

As seen at the beginning of the chapter, the scattergram of two variables may approximate to a straight line fairly well, indicating a strong linear relationship. What we are now concerned with is to find what this relationship is, ie find an equation linking the two variables. Naturally, since the relationship is linear, this equation will be that of a straight line. This equation can then be used to estimate or predict pairs of values not given in the data.

Now, we have already shown one particular straight line (the lines of best fit) indicating the trend of the points of a scattergram (see page 284); however, this is only drawn by eye and so is inaccurate. What we require is a formula which can be used to give the best possible straight line *for our purposes*, viz to predict or estimate further values; such a line is called a *line of regression*. Consider the data below, which comes from an experiment designed to investigate the relationship between 'sleep loss' and 'driving ability'. A person was required to take a simulated driving test after varying amounts of sleep deprivation; marks for driving ability were out of a best possible 50.

Sleep Deprivation (x hours)	1	2	3	4	5	6	7	8	9	10
Driving Ability (y marks)	45	42	43	38	38	30	31	25	28	25

The scattergram below shows a possible linear relationship between the two variables.

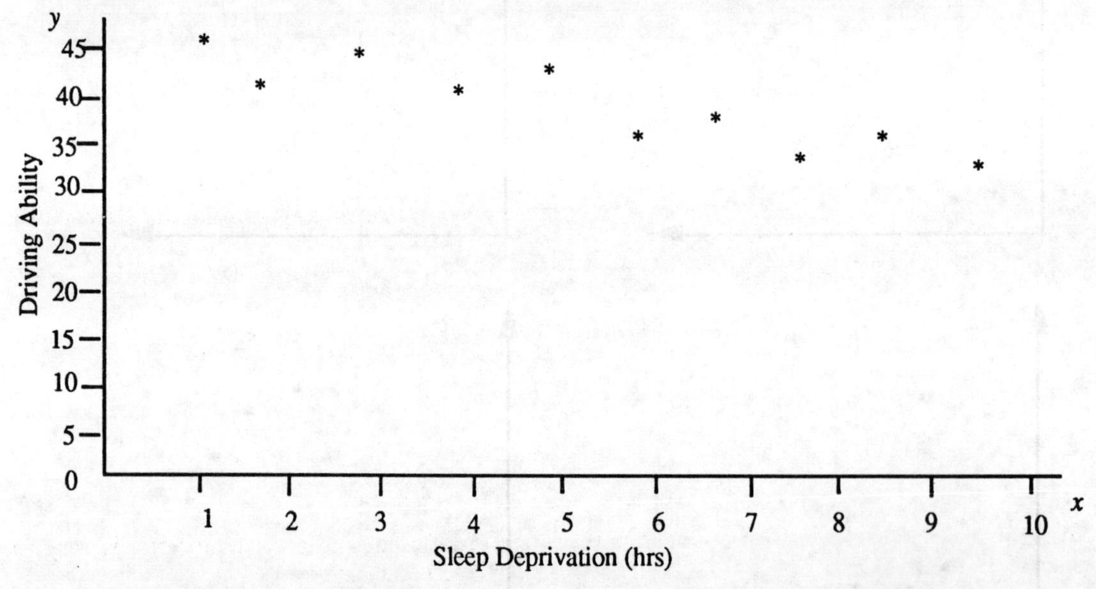

Now, in this experiment, the driving test was taken at precise intervals of 1 hour, ie the variable 'sleep deprivation' was precisely measured or *controlled*; such variables are called *independent variables*. In contrast, the driving ability of the person being tested is not under control; this is the variable we are *observing* and is *dependent* upon the amount of sleep the person has been deprived of, ie this *dependent variable* depends upon the *independent variable* (sleep deprivation). What we are interested in doing is estimating the *driving ability* of the person after say, 15 hours sleep deprivation, ie we are interested in estimating values of the *dependent variable* for given values of the independent variable (eg 15 hours). Thus, when considering how to construct the straight line (the line of regression) most suited for this purpose, we must *minimise the errors* in such estimations ie we must minimise the errors or deviations in the *dependent variable* from the straight line.

As can be seen below, the value of the *dependent* variable of each point of the scattergram deviates vertically to some extent, from that of the best line of regression. We require these (vertical) deviations, or precisely the *squares of the deviations*, to be a *minimum* (hence, the title 'method of least squares').

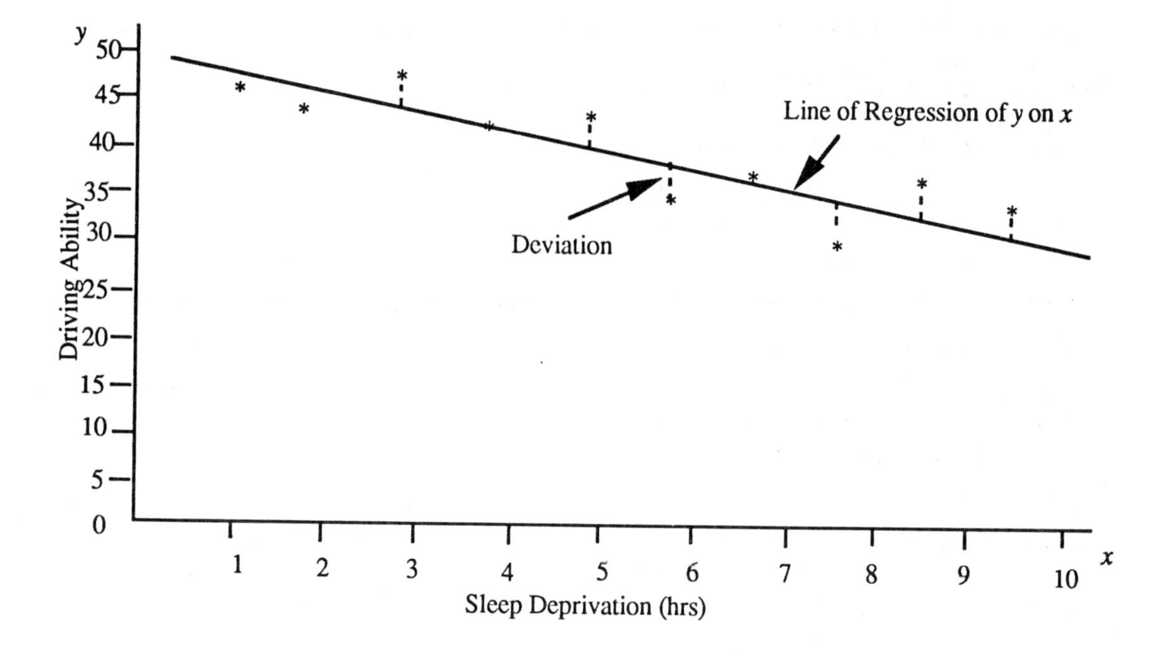

The formula below does just this, and is called the *equation of line of regression* of y on x; the latter part of this, 'y on x', refers to the fact that we are estimating y (the dependent variable) for given values of x (the independent variable).

> Equation of the line of regression of y on x.
>
> $$(y - \bar{y}) = (x - \bar{x})\, r \cdot \frac{S_y}{S_x}$$

[*Note*: Again, this formula is the most suitable one to be used with a calculator and is easy to remember; it differs from that in the formula book, and so must be memorised.]

From the above data, we have:

$$\bar{x} = \frac{\Sigma x}{n} = \frac{1 + \ldots + 10}{10} = 5.5$$

$$\bar{y} = \frac{\Sigma y}{n} = \frac{45 + \ldots + 25}{10} = 34.5$$

$$S_x = \sqrt{\frac{\Sigma(x - \bar{x})^2}{n}} = \sqrt{\frac{(1\text{-}5.5)^2 + \ldots + (10\text{-}5.5)^2}{10}} = 2.87$$

$$S_y = \sqrt{\frac{\Sigma(y - \bar{y})^2}{n}} = \sqrt{\frac{(45\text{-}34.5)^2 + \ldots + (25\text{-}34.5)^2}{10}} = 7.20$$

$\Sigma xy = 1 \times 45 + ... + 10 \times 25 = 1699$

Hence, $r = \dfrac{1699 - 10 \times 5.5 \times 34.5}{10 \times 2.87 \times 7.20} = \underline{-0.960}$

So, the equation of regression of y on x is:

$$(y - 34.5) = (x - 5.5)\ (-0.960)\left(\frac{7.20}{2.87}\right)$$

ie $\quad y = -2.41x + 47.8$

We call the gradient of the (straight) line of regression the *coefficient of regression*; here, we have:

Coefficient of regression = $\underline{-2.41}$

We can now plot this line of regression on the scattergram (as above) and use it to estimate the driving ability after 15 hours sleep deprivation by putting $x = 15$, obtaining:

$$y = -2.41 \times 15 + 47.8 = \underline{11.65}$$

There is another line of regression whose equation can be calculated, viz the line of regression of x on y; this equation can be used to estimate values of x given the value of y. The equation of the line of regresion of x on y is found by simply interchanging x and y in the above formula, giving:

Equation of the line of regression of x on y

$$(x - \bar{x}) = (y - \bar{y})\ r \bullet \frac{S_x}{S_y}$$

The following example illustrates the use of this other line of regression, viz of x on y. [Clearly, the above example is not amenable to swapping x and y around, since it would not make sense to say that the now 'dependent' variable x (sleep deprivation) depends upon the now 'independent' variable y (driving ability).]

Example: The prices of butter (y) and eggs (x) were recorded for 12 consecutive months, as shown below. Plot a scattergram of the data and draw a line of best fit by eye. Calculate and plot the lines of regression of (i) y on x, (ii) x on y.

Estimate the price of butter when the price of eggs is 97p per dozen. If the price of butter increased by 20%, what percentage increase in the price of eggs would you expect?

Butter (y pence per lb)	100	105	104	108	110	107	107	112	120	115	118	120
Eggs (x pence per doz)	90	93	95	95	94	95	98	100	100	99	102	101

The scattergram of the data is as below. In order to draw the line of best fit, we must first calculate the 'middle' point of the scattergram, ie $(\bar{x}, \bar{y})$.

Here, $\bar{x} = \frac{\Sigma x}{n} = \frac{90 + ... + 101}{12} = 96.83$

and $\bar{y} = \frac{\Sigma x}{n} = \frac{100 + ... + 120}{12} = 110.5$

Thus, the middle point is (96.83, 110.5), and is shown on the scattergram.

The line of best fit can now be drawn by eye, trying to make sure that the sum of the squares of the *perpendicular* deviations of the points from the line is a minimum (see page 284).

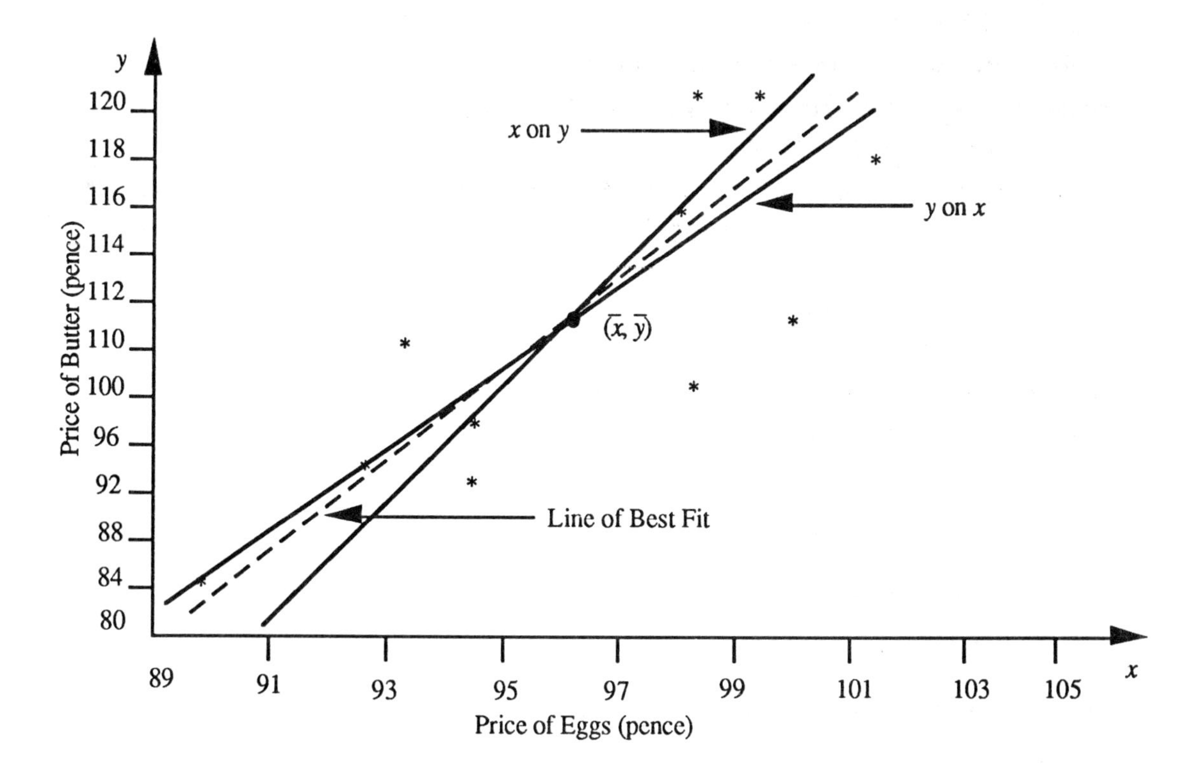

Continuing with the calculations, we have:

$$S_x = \sqrt{\frac{\Sigma(x - \bar{x})^2}{n}} = \sqrt{\frac{(90 - 96.83)^2 + ... + (101 - 96.83)^2}{12}} = 3.53$$

$$S_y = \sqrt{\frac{\Sigma(y - \bar{y})^2}{n}} = \sqrt{\frac{(100 - 110.5)^2 + ... + (120 - 110.5)^2}{12}} = 6.28$$

$\Sigma xy = 90 \times 100 + ... + 101 \times 120 = 128637$

Thus, $r = \frac{128637 - 12 \times 96.83 \times 110.5}{12 \times 3.53 \times 6.28} = 0.89$

i) Hence, line of regression of y on x is:

$$(y - 110.5) = (x - 96.83)\ 0.89\left(\frac{6.28}{3.53}\right)$$

ie $\quad y = 1.58x - 42.81$ -------------------- (1)

ii) Also, line of regression of x on y is:

$$(x - 96.83) = (y - 110.5)\ 0.89\left(\frac{3.53}{6.28}\right)$$

ie $\quad y = 2.00x - 83.05$ -------------------- (2)

See the diagram above for the plots of these straight lines.

We can now use equation 1) to estimate y when $x = 97$; thus, putting $x = 97$:

$$y = 1.58 \times 97 - 42.81 = 110.45$$

Furthermore, we can use equation 2) to estimate the percentage increase in x when y increases by 20%.

Before the increase, $y = 2.00x - 83.05$

$$\text{ie } x = \frac{y}{2.00} + \frac{83.05}{2.00}$$

After the increase, y becomes y + 20% of y, ie $1.2y$

ie $\quad 1.2y = 2.00x - 83.05$

$$\text{ie} \quad x = \frac{1.2\ y}{2.00} + \frac{83.05}{2.00}$$

Hence, the increase in x is:

$$\frac{1.2y}{2.00} - \frac{y}{2.00} = 0.1\ y$$

Thus, the percentage increase in x is:

$$\frac{0.1y}{y} \times 100 = \underline{10\%}$$

As can be seen in the above diagram, the line of best fit lies between the two lines of regression. Notice that all three lines pass through the middle point $(\overline{x}, \overline{y})$.

Here, it *is* possible to make predictions about the values of y given x (using y on x) *and* values of x given y (using x on y) since the two variables are both dependent upon a third variable, viz inflation; neither x or y can be considered the independent variable - they are *both dependent* (as opposed to the first example considered above).

Example 12

1. Draw scattergrams and a line of best fit (by eye) for the following sets of data:

a)

x	12	15	11	11	8	7	19	21	23
y	1.2	1.1	1.7	1.3	1.9	1.9	1.3	1.1	1.0

b)

x	101	105	103	108	110	115	112	113	112	120	123
y	52	53	55	58	57	59	62	63	61	61	68

c)

x	0.30	0.35	0.36	0.31	0.31	0.33	0.38	0.35	0.37	0.40	0.41
y	28	28	24	23	22	22	23	22	21	18	23

d)

x	3	2	2	3	5	5	1	8	13	11	4	9	12	15	1	16	2	8	5.
y	0.2	0.2	0.5	0.3	0.7	0.8	0.5	0.9	1.3	1.4	0.6	0.8	0.8	0.9	0.3	0.9	0.1	0.8	0.5

Comment on the type of correlation, if any, between the two variables.

2. Calculate the product-moment correlation coefficient in each of the above cases.

3. Calculate the product moment correlation coefficient for each of the following sets of data:

a)

y \ x	1	2	3	4	5	6	7
5	0	0	1	2	3	5	9
6	0	3	5	6	4	6	2
7	1	8	9	6	4	3	1
8	2	1	4	3	2	1	0
9	5	0	0	1	0	0	0

b)

y \ x	0.1	0.2	0.3	0.4	0.5	0.6	0.7	0.8	0.9
21	7	9	4	3	0	2	1	0	0
22	3	2	8	9	1	5	1	1	1
23	1	5	9	9	8	9	3	3	0
24	2	1	4	9	25	14	8	3	1
25	5	3	4	2	13	14	9	7	5

c)

y \ x	-1	-2	-3
4	0	1	5
5	1	4	1
6	3	2	0

d)

y \ x	0	1	2	3
105	5	4	6	4
106	2	3	1	4
107	4	9	3	8

e)

y \ x	3-5	6-8	9-11	12-14
1-2	0	1	7	20
2-3	5	8	19	3
3-4	7	4	5	2
4-5	18	2	1	3

f)

y \ x	100-110	110-120	120-130	130-140	150-160
1	24	20	10	5	1
2	20	24	31	14	9
3	15	13	28	38	24
4	4	5	15	21	42
5	1	8	7	15	31

g)

y \ x	1-3	4-6	7-9
0-2	1	9	8
2-4	3	1	0
4-6	4	3	7

h)

y \ x	100-200	200-300	300-350	350-400
0.1-0.2	1	10	20	18
0.2-0.4	8	9	10	5
0.4-0.8	9	3	2	0

4. Calculate the Spearman Rank correlation coefficient for the following sets of data.

a)

x	2	3	4	5	6	7
y	20	22	21	28	29	29

b)

x	-1	0	1	2	3
y	-5	-8	-8	-9	-10

c)

x	0.02	0.03	0.04	0.04	0.03	0.05	0.06
y	15.0	12.1	12.0	12.3	11.8	11.1	11.3

5. Calculate the Spearman Rank correlation coefficient for each set of data in question 1.

6. Calculate the equations of the lines of regression of

   i) y on x ,

   ii) x on y from the data in questions 1, 3 and 4.

   Plot both of these on the scattergrams on questions 1 and 4.

7. The gestation periods (days) and number of offspring of 10 rodents were recorded as below.

Gestation period (x)	100	100	105	106	107	108	108	109	118	120
No. of offspring (y)	1	2	1	2	2	2	1	3	3	2

Draw a scattergram of the data, and fit a straight line by eye.

Calculate:

a) Product-moment correlation coefficient;

b) Spearman Rank correlation coefficient;

c) Lines of regression of

i) y on x,

ii) x on y.

Plot the lines of regression on the scattergram. State the coefficient of regression in each case. Use your lines to estimate:

iii) the expected gestation period of a rodent carrying 4 offspring;

iv) the expected number of offspring after a gestation period of 125 days.

What is the value of these estimates?

8. Three judges of 10 gymnasts gave the following ratings out of 10.

	1	2	3	4	5	6	7	8	9	10
Judge A	5	5	6	3	8	9	10	10	4	7
Judge B	4	6	5	5	8	6	5	9	3	8
Judge C	2	5	9	5	8	4	3	9	8	7

Calculate the Spearman Rank correlation coefficients between the three pairs of judges, and comment on your results.

9. Fifteen competitors took part in two races: the 5000m and the 1500m. Each had a letter pinned to their shirts, from A to O, and the order in which they passed the winning post is recorded below.

Place	1st	2nd	3rd	4th	5th	6th	7th	8th	9th	10th	11th	12th	13th	14th	15th
5000m	D	A	G	K	M	E	C	F	I	O	H	L	N	B	J
1500m	K	I	N	D	G	A	H	C	E	L	F	B	J	O	M

Calculate a suitable measure of correlation between the two races.

10. The price of different makes of car and the numbers sold per year are recorded below.

	Mini	Jaguar	Ford	Citroen	Toyota	Rolls	BMW	Fiat
Price (£1000s)	4.1	11.8	4.5	5.2	4.2	15.3	6.1	3.7
No. sold (1000s)	490	7	941	53	17	8	92	42

Calculate

a) product-moment correlation coefficient

b) Spearman Rank correlation coefficient

c) Regression lines of

i) y on x

ii) x on y.

Use your regression lines to estimate

iii) the price of a car which sells 520,000 per year;

iv) the expected sales of a car costing £7,200.

11. In a study of the effectiveness of a weight-reducing drug, the following data was obtained:

Loss in weight (lbs)	74	45	30	28	20	15	16	10	39	19	26	34
Length of treatment (months)	11	5	14	3	2	4	1	1	3	6	24	18

Calculate the product-moment coefficient of correlation for the data and

a) test whether, at the 5% level, it provides evidence of correlation between the two variables, and

b) test whether, again at the 5% level, this value is consistent with the hypothesis that the true value of the correlation coefficient in the bivariate population is 0.6.

12. The heights of fathers and their eldest sons are tabulated below:

Height of father x(in)	63	68	70	64	66	72	67	71	68	62
Height of son y(in)	65	66	72	66	69	74	69	73	65	66

a) Find the regression lines of y on x and of x on y.

b) Plot a scatter diagram for the above data and draw and label the two regression lines on the same diagram.

c) Calculate the correlation coefficient.

13. The systolic blood pressure of 10 men of various ages are tabulated as follows:

Age (years (x))	37	35	41	43	42	50	49	54	60	65
Systolic blood pressure (y) (mm of mercury)	110	117	125	130	138	146	148	150	154	160

a) Find the linear regression line of y on x.

b) Find the correlation coefficient for these data.

14. The scores of twelve candidates in an examination in mathematics and physics are

Candidate	A	B	C	D	E	F	G	H	I	J	K	L
Maths (x)	68	77	52	78	90	42	33	61	25	46	63	37
Physics (y)	70	62	60	71	82	55	48	51	35	62	32	20

Fit straight lines by the method least squares (the regression lines) making

i) x,

ii) y the independent variable.

A drawing is *not* required.

Find the product-moment coefficient of correlation between the two sets of marks.

15. In Utopia the wholesale food price index (x) and the retail food price index (y) over twelve years had the following value in order.

x	100	98	96	97	95	92	89	87	88	90	86	88
y	100	97	95	98	94	95	90	89	91	93	89	87

Find the equations of the regression lines

i) of y on x,

ii) of x on y.

Calculate the product-moment coefficient of correlation.

When the wholesale food price index is 84, what is the estimated value of the retail price index?

16. Ten boys compete in throwing a cricket ball, and the following table shows the height of each boy (x cm) to the nearest cm and the distance (y m) to which he can throw the ball.

Boy	A	B	C	D	E	F	G	H	I	J
x	122	124	133	138	144	156	158	161	164	168
y	41	38	52	56	29	54	59	61	63	67

Find the equations of the regression lines of y on x, and x on y. No diagram is needed. Calculate also the coefficient of correlation.

Estimate the distance to which a cricket ball can be thrown by a boy 150 cm in height.

17. The body and heart masses of fourteen 10-month-old male mice are tabulated below:

Body mass (x) (grams)	27	30	37	38	32	36	32
Heart mass (y) (milligrams)	118	136	156	150	140	155	157

Body mass (x) (grams)	32	38	42	36	44	33	38
Heart mass (y) (milligrams)	114	144	159	149	170	131	160

i) Draw a scatter diagram of these data.

ii) Calculate the equation of the regression line of y on x and draw this line on the scatter diagram.

iii) Calculate the product-moment coefficient of correlation.

18. Explain clearly what is meant by the statistical term 'correlation'. Vegboost Industries, a small chemical firm specialising in garden fertilisers, set up an experiment to study the relationship between a new fertiliser compound and the yield from tomato plants. Eight similar plants were selected and treated regularly throughout their life with x grams of fertiliser diluted in a standard volume of water. The yield y, in kilograms, of good tomatoes was measured for each plant. The following table summarises the results.

Plant	A	B	C	D	E	F	G	H
Amount of fertiliser x (g)	1.2	1.8	3.1	4.9	5.7	7.1	8.6	9.8
Yield y (kg)	4.5	5.9	7.0	7.8	7.2	6.8	4.5	2.7

i) Calculate the product-moment correlation coefficient for these data.

ii) Calculate Spearman's rank correlation coefficient for these data.

iii) Is there any evidence of a relationship between these variables? Justify your answer. (No formal test is required.)

19. A scientist, working in an agricultural research station, believes there is a relationship between the hardness of the shells of eggs laid by chickens and the amount of a certain food supplement put into the diet of the chicken. He selects ten chickens of the same breed and collects the following data:

Chicken	A	B	C	D	E	F	G	H	I	J
Amount of food supplement x (g)	7.0	9.8	11.6	17.5	7.6	8.2	12.4	17.5	9.5	19.5
Hardness of shells y	1.2	2.1	3.4	6.1	1.3	1.7	3.4	6.2	2.1	7.1

(Hardness is measured on a 0-10 scale, 10 being the hardest. There are no units attached.)

i) Calculate the equation of the regression line of y on x.

ii) Calculate the product-moment correlation coefficient.

iii) Do you believe that this linear model will continue to be appropriate no matter how large or small x becomes? Justify your reply.

20. Ten athletes have best performances at the High Jump and Long Jump as follows:

Athlete	A	B	C	D	E	F	G	H	I	J
High Jump x m	1.8	2.1	1.9	2.0	1.8	1.8	1.6	1.8	1.9	2.3
Long Jump ym	6.7	7.6	6.3	6.8	5.9	7.9	5.5	5.6	6.5	7.2

Calculate the coefficient of correlation between x and y.

21. Miss Claire Voyant claims that she possesses 'extra-sensory perception'. A small room is divided by a metal screen, with Claire seated on one side and an experimental assistant on the other. Each is given a set of cards numbered 1 to 5. At each trial of the experiments the assistant selects one of these cards and thinks of that number; Claire then records her 'guess' by showing the appropriate card. 100 trials are completed and the results are given in the following bivariate table:

		Claire's 'guess'				
		1	2	3	4	5
Assistant's choice	1	11	4	0	0	0
	2	1	16	8	1	0
	3	0	2	18	6	1
	4	0	0	2	13	5
	5	0	0	0	4	8

Calculate the product-moment correlation coefficient for these data.

22. At the University of Batsula, students for the degree in Chemical Engineering take two examinations. They have a Part I examination at the end of the second year of their studies, and a Part II examination at the end of the third year. Degree classification depends upon an average of these marks. The following is a table of the Part I and Part II marks for a particular set of 100 students.

		Part I marks					
		40-49	50-59	60-69	70-79	80-89	90-99
Part II marks	40-49	3	5	4	0	0	0
	50-59	3	6	6	2	0	0
	60-69	0	5	9	5	2	0
	70-79	0	0	5	10	8	1
	80-89	0	0	0	5	6	5
	90-99	0	0	0	2	4	4

Calculate the product-moment correlation coefficient for these data. (Assumed means of 64.5 for Part I marks and 74.5 for Part II marks should be used.)

23. A toy company specialises in high precision scale models of vintage cars. Before being allowed on to the assembly line employees must complete a training course and receive an assessment, in the range 1-20, of their aptitude for this kind of work. The better suited to this work would expect a high course assessment. The table below gives the number of complete cars assembled in one day by 80 employees, together with their training course assessment figure.

Number of complete cars assembled in one day

		1-3	4-6	7-9	10-12	13-15
Training course assessment	1-5	5	7	4	0	0
	6-10	1	7	9	6	0
	11-15	0	2	8	11	5
	16-20	0	1	1	4	9

Calculate the product-moment correlation coefficient for these data.

24. As part of his research into the behaviour of the human memory, Hugo Nutts, a leading psychologist, asked 15 schoolgirls to talk for five minutes on 'my day at school'. Hugo then asked each girl to record how many times she thought that she had used the word *nice* during this period. The table below gives their replies together with the true values.

Girl	A	B	C	D	E	F	G	H	I	J	K	L	M	M	O
True value x	12	20	1	8	0	12	12	17	6	5	24	23	10	18	16
Recorded no. y	9	19	3	14	4	12	16	14	5	9	20	16	11	17	19

i) Draw a scatter diagram on these data.

ii) Fit a line *by eye* to these data, and label this line clearly.

iii) Calculate the equation of the regression line of y on x and draw this line on your scatter diagram also.

Discuss briefly the usefulness or otherwise of lines drawn by eye in situations such as the above.

25. The following table is taken from the International Journal of Marriage Statistics, editor Ms Honey Moon. It gives the heights, measured to the nearest centimetre, of both the husband and wife for 100 couples married during September 1980.

Heights of husbands (x)

		160-164	165-169	170-174	175-179	180-184
Heights of wives (y)	155-159	2	6	2	0	0
	160-164	5	6	8	5	0
	165-169	0	11	12	8	4
	170-174	0	0	14	5	3
	175-179	0	0	0	7	2

Calculate the product-moment correlation coefficient for these data. (If the coding method is applied, you should use assumed means of 172 for husbands' heights and 167 for wives' heights.)

26. At a certain location the percentage of sand in soil at different depths is given by the table

x (depth in mm)	25	35	45	55	65
y (% of sand)	92	86	84	79	70

i) Find the equation of the y on x regression line. Use your equation to estimate the value of y at the surface. Comment on this result.

ii) Find the product moment correlation coefficient.

13 TIME SERIES AND WEIGHTED AVERAGES

This last chapter covers a few isolated 'O' level topics that crop up occasionally. Because of their simplicity, it is advisable to learn them, as well as, of course, the other 'O' level topics already considered, eg histograms, cumulative frequency diagrams.

13.1 Time series

This topic is concerned with variables which change over time, eg sales figures; if we plotted a graph of such a variable over time we would obtain a *Time Series.* Consider the data below concerning the sales figures for a growing company producing coats over the years 1975 to 1979.

Year	1975				1976				1977			
Quarter	1	2	3	4	1	2	3	4	1	2	3	4
Sales (£1000s)x	9.1	5.2	3.1	10.5	10.0	5.7	3.9	13.0	11.3	6.1	3.9	18.3

Year	1978				1979			
Quarter	1	2	3	4	1	2	3	4
Sales (£1000s)x	15.2	6.8	4.8	13.7	11.8	7.1	5.2	14.2

Notice that each year has been split into 4 quarters, since it would be expected that coat sales would vary from season to season; if sales figures per *year* were used, then such important variations would be lost.

Plotting sales against time, we obtain the time series below.

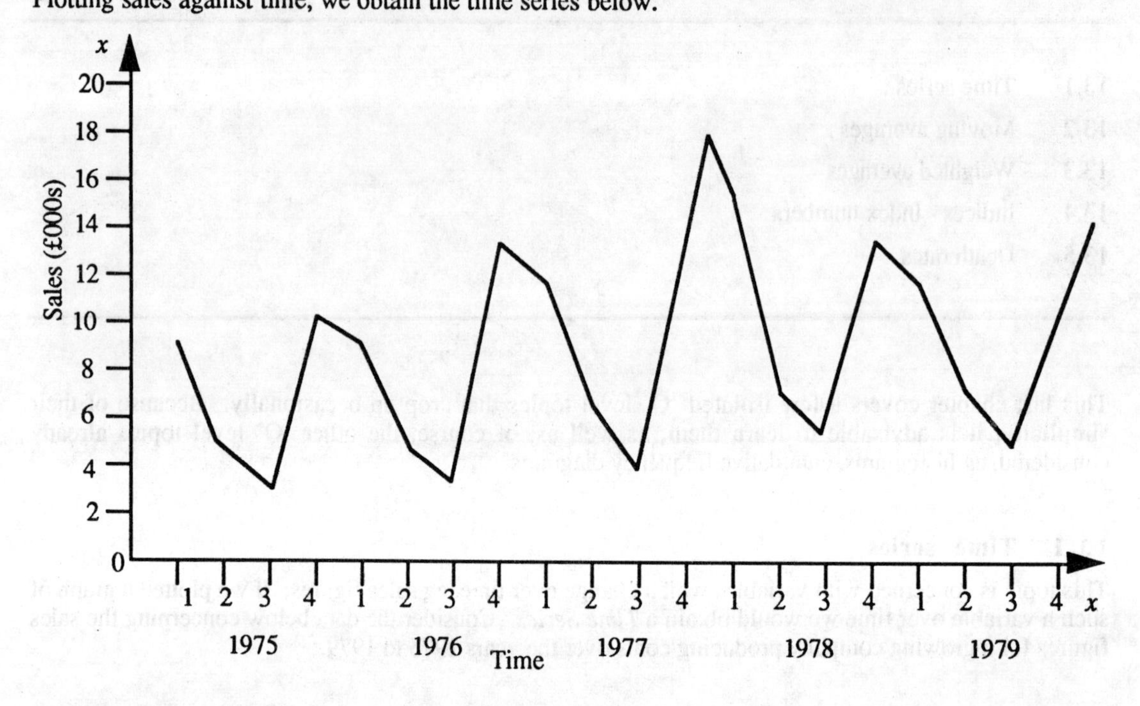

We now consider the various components that go to make up a time series, using the above example as an illustration.

Secular trend

As can be seen in the diagram above, the sales figures are increasing overall, even though the graph is not continuously rising; this is indicated by the upward inclination of the whole graph, and is termed the *secular trend* of the time series. The above time series has an *upward* trend.

Seasonal variation

In any one year, it can be seen that the sales figures vary from quarter (season) to quarter (season); this is obviously due to the climate during our seasons, and is referred to as *seasonal variation*. As expected, coat sales decline in the summer and increase in the winter.

Irregular fluctuations

These are irregular fluctuations in the variable as a result of exceptional circumstances. In the above example, it can be seen that the coat sales for the end of 1977 and beginning of 1978 were exceptionally high compared to similar quarters of other years; this may have been due to a particularly harsh winter, or a revolutionary coat design that set the fashion.

Cyclical fluctuations

These are regular fluctuations that occur over longer periods of time than seasonal fluctuations, eg every 5 years. The diagram below is a possible time series for coat sales over 25 years, and shows cyclical fluctuations that repeat every 5 years. In this case, such cyclical fluctuations may be due to broad economic factors, changing governments, etc.

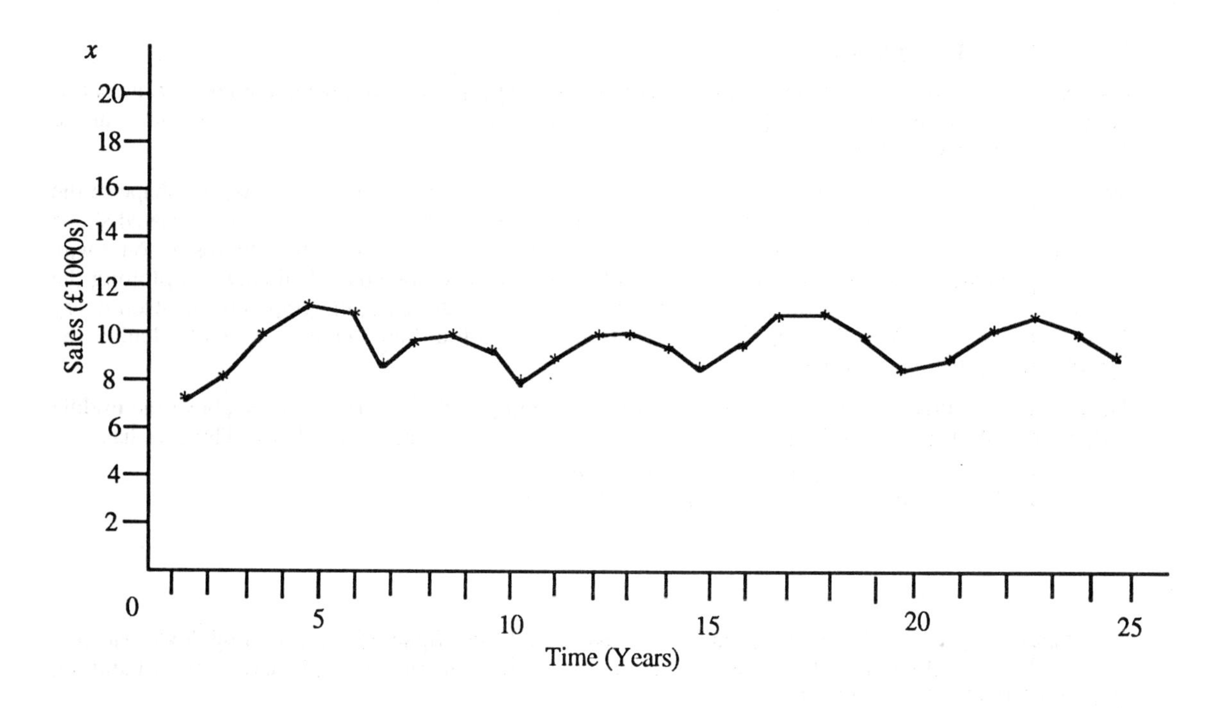

The first three of these components are labelled on the diagram below.

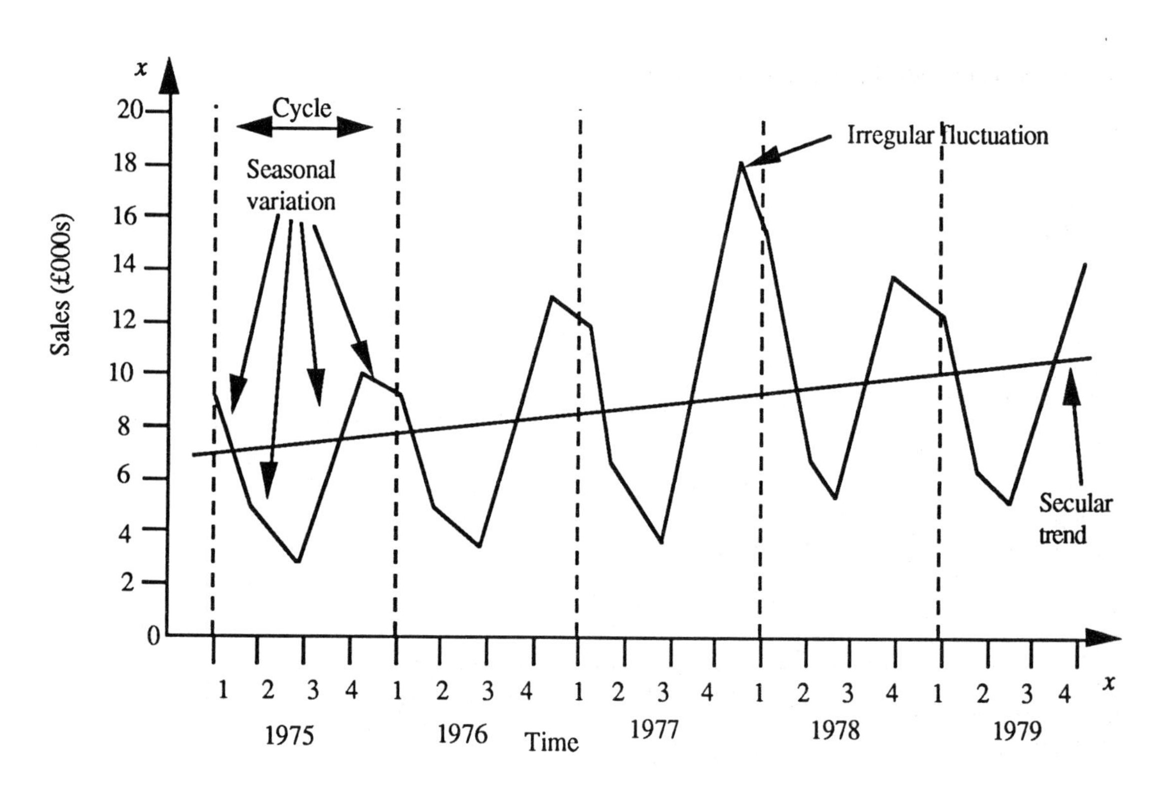

13.2 Moving averages

It would be particularly difficult to predict coat sales in the year 1980, say, using the above time series because of the jerkiness of the graph; the plotting of *moving averages* is a way of smoothing out the graph, making such predictions easier.

Looking at the time series for coat sales again, we see that every year (every 4 quarters) the shape of the graph repeats itself (due to seasonal variation) as indicated in the diagram below; such a repeated shape is called a *cycle*. (See diagram above.) The aim here is to 'average out' the sales figures in some way so as to eliminate or at least smooth out these cycles. This can be done by calculating and plotting the so-called *moving averages* for the data; more precisely, since in this case, the graph repeats itself every 4 quarters or, equivalently, the shape of the cycle is made up of four line segments, we will calculate the *4-point moving averages* for the data.

The method is simple. Average the first 4 sales figures, and plot this *moving average* above the middle (mean) of the 4 quarters used, ie above the middle of the 2nd and 3rd quarters of 1975. Thus, we have

$$\text{1st Moving Average} = \frac{9.1 + 5.2 + 3.1 + 1.05}{4} = 6.975$$

Now repeat the process, this time beginning with the second sales figure (5.2), ie shifting down one, and plotting this second moving average above the middle of the 4 quarters used, ie above the 3rd and 4th quarters of 1975. Thus, we have:

$$\text{2nd Moving Average} = \underline{5.2 + 3.1 + 10.5 + 10.0} = 7.2$$

These calculations are illustrated below.

Year	Quarter	Sales Figures (x)	Moving averages
	1	9.1	
	2	5.2	
1975	3	3.1	6.975
	4	10.5	7.2
	1	10.0	7.325
	2	5.7	7.525
1976	3	3.9	8.15
	4	13.0	8.475
	1	11.3	8.575
	2	6.1	8.575
1977	3	3.9	9.9
	4	18.3	10.875
	1	15.2	11.05
	2	6.8	11.275
1978	3	4.8	10.125
	4	13.7	9.275
	1	11.8	9.35
	2	7.1	9.45
1979	3	5.2	9.575
	4	14.2	

Plotting these moving averages above the mid-points of the corresponding 4-quarters used, we obtain the graph below, which is considerably smoother than the original.

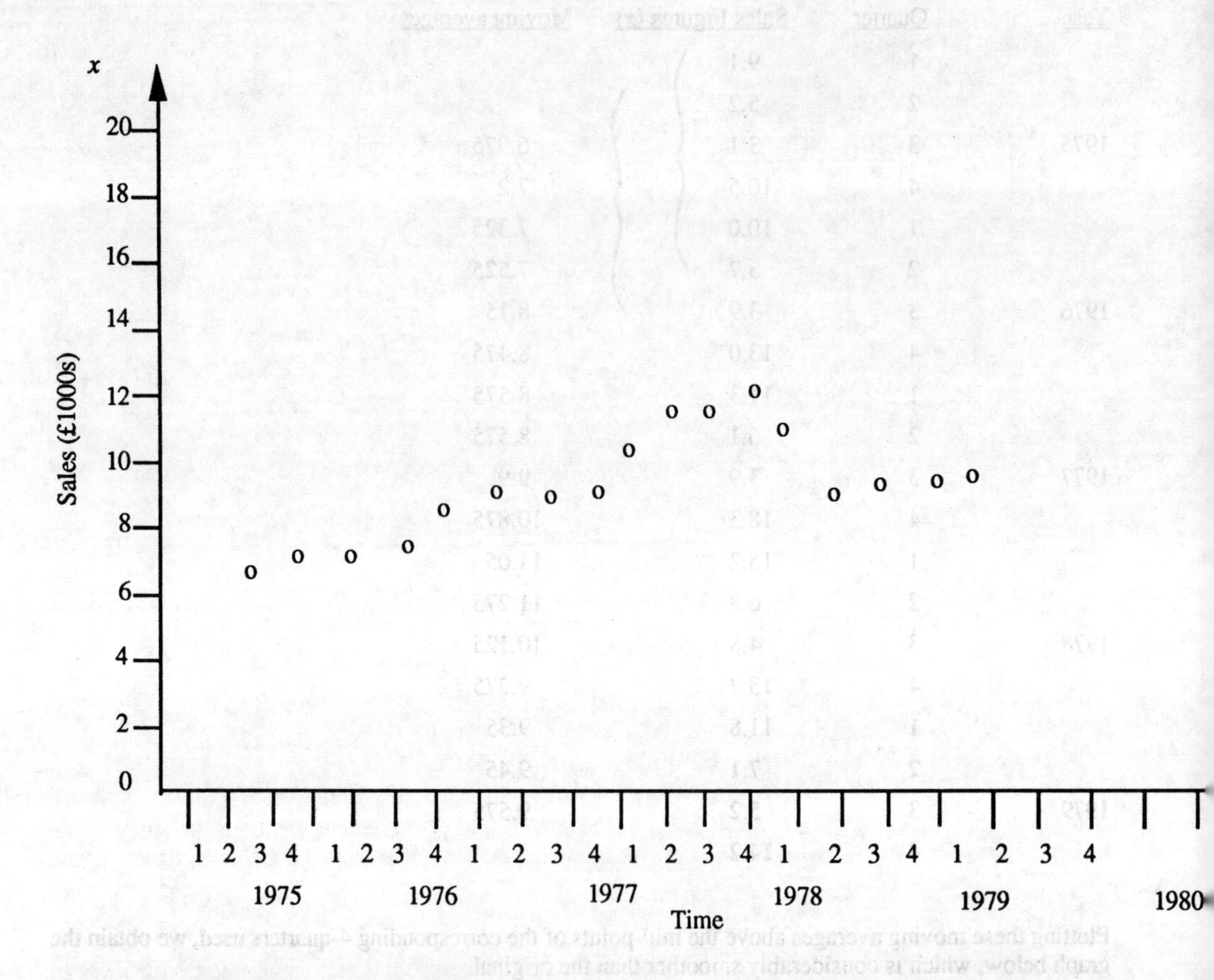

We are now able to draw a line of best fit (by eye) through these moving averages (which would have been difficult on the original graph), and can use this to predict future sales figures.

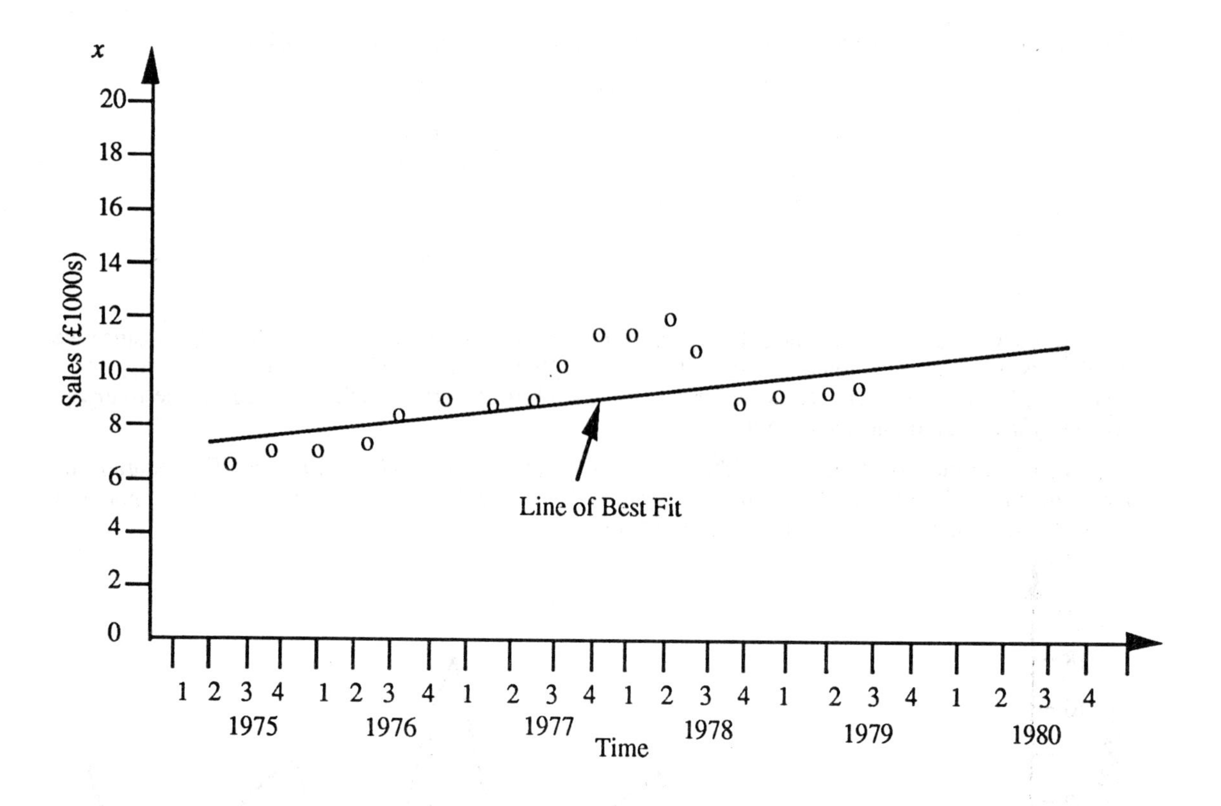

Suppose we wish to predict the sales figures for the year 1980. First, we must continue the straight line of best fit into 1980. We can then read off the expected moving average for the first quarter; in this case,

moving average = 10

Now, if we let x be the *actual* sales for this first quarter of 1980, then this moving average is the average of x and the last 3 sales figures, ie

$$10 = \frac{7.1 + 5.2 + 14.2 + x}{4}$$

Thus, $x = 4 \times 10 - 7.1 - 5.2 - 14.2$

$= 13.5$

We can now repeat this process, beginning by reading off the expected moving average for the second quarter of 1980, which is 10.15. Then,

$$10.15 = \frac{5.2 + 14.2 + 13.5 + x}{4}$$

Thus, $x = 4 \times 10.15 - 5.2 - 14.2 - 13.5$

$= \underline{7.7}$

Continuing in this way, we obtain the sales figures for the four quarters of 1980 below:

Year	Quarter	Sales
	1	13.5
1980	2	7.7
	3	5.8
	4	14.8

Now, if we wished to estimate the sales figures for 1985, the above method would be tedious, since we would firstly have to estimate the sales figures for 1980, 1981, 1982, 1983 and 1984 (easy for a computer, but not for us!). Thus, we must use a slightly different method, based on the *average seasonal fluctuation* of the available data.

Let us consider the first quarter of 1985. Notice that, in the years 1975 to 1979, the first quarters lie above the line of best fit, as shown in the diagram below. The distance above the line is the amount of seasonal variation, and is indicated by the dotted lines.

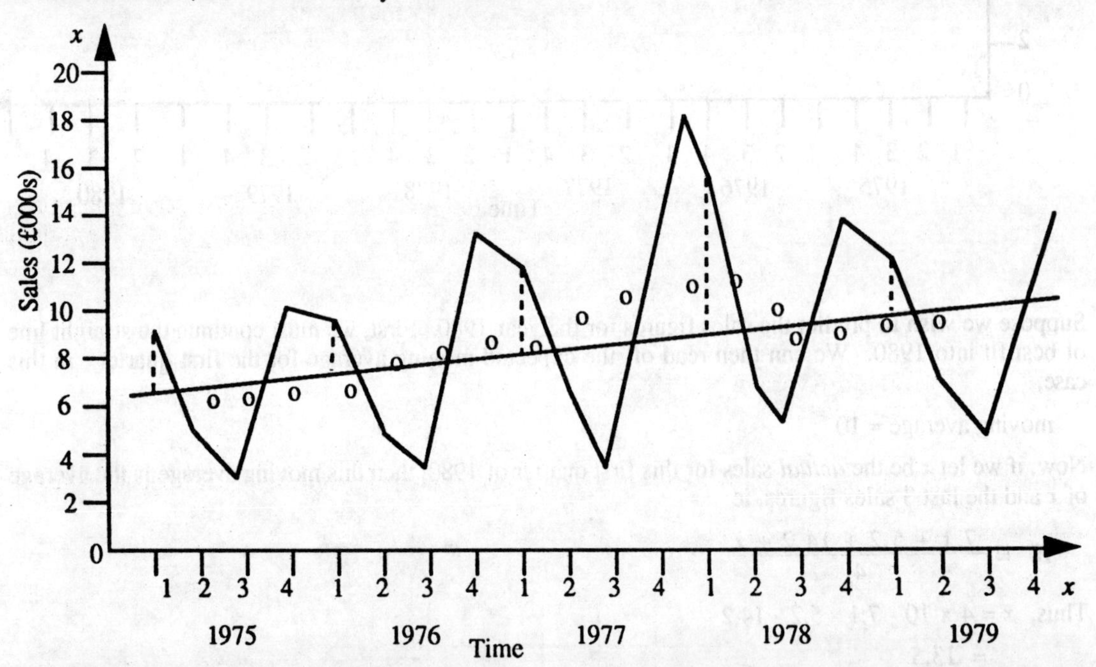

If we average these seasonal variations, we can use this as an estimate of the seasonal variation of the *same* first quarter of succeeding years, 1985 in particular. Reading from the graph above, the 5 seasonal variations are 2, 2.2, 3, 6, 2.1.

Thus, the average seasonal variation for the 1st quarter

$$= \frac{2 + 2.2 + 3 + 6 + 2.1}{5}$$

$$= 3.06$$

Now, the sales figures of the line of best fit above the first quarter of 1985 can be found by first calculating the gradient of this line; from the graph, this is 0.6 per year. Since 1985 is 10 years later than 1975, we add

10 x 0.6 to the sales figures of the line of best fit above the first quarter of 1975, to get:

$7 + 10 \times 0.6 = 13$

Finally, the required estimate of the sales figure for the first quarter of 1985 is found by adding the average seasonal variation for this quarter to the above, ie

Sales for 1st quarter of 1985 = 13 + 3.06 = 16.06

Repeating this process, we obtain the sales figures for 1985 below:

Year	Quarter	Sales figures
	1	16.06
1985	2	10.35
	3	8.96
	4	17.89

13.3 Weighted averages

Recall that the average (mean) of a discrete frequency distribution is calculated using the formula:

$$\bar{x} = \frac{\Sigma fx}{\Sigma f}$$

where f is the frequency of the value x.

In fact, this formula is a special case of a more general formula used to 'average' a set of values of x, say. Instead of each value having a given frequency f, each value x has a given *weight*, w. The formula thus becomes:

$$\text{Weighted Average} = \frac{\Sigma wx}{\Sigma w}$$

where w is the ascribed *weight* of the value x. Thus, one *possible* kind of weight is frequency, and the weighted average simply becomes the mean, $\bar{x}$-values.

Consider the data below.

x	1	2	3	4	5	6	7
w	2	1	8	7	1	3	2

The weighted average of this data is:

$$\frac{\Sigma wx}{\Sigma w} = \frac{2 \times 1 + 1 \times 2 + 8 \times 3 + 7 \times 4 + 1 \times 5 + 3 \times 6 + 2 \times 7}{2 + 1 + 8 + 7 + 1 + 3 + 2}$$

$$= \underline{3.875}$$

We now consider two applications of weighted averages with weights that are *not* frequency.

13.4 Indices - Index numbers

An *index* (plural: indices) is a number which indicates the change of a quantity over time, eg a 'cost of living' index indicates the cost of living compared with that of an earlier year. Consider the data below concerning the prices of several kinds of food for the years 1983 and 1984.

	Cost (pence)	
Kind of food	1983	1984
Meat (per lb)	150	175
Fish (per lb)	120	128
Eggs (per doz)	85	95
Butter (per lb)	98	105
Vegetables (per lb)	28	31

As can be seen (and as expected) the prices of all the food commodities increased between 1983 and 1984. For example, meat increased from 150p per lb to 175p per lb, an increase of 16.7%. What we wish to do is embody in a single figure - let us call it the 'cost-of-food index' - the cost of food to the consumer in 1984 *as compared with 1983*. In order to do this, we express the price of each commodity in 1984 as a percentage of the price of it in 1983; naturally, since all of the prices *increased* from 1983 to 1984, these percentages will all be more than 100% - what they were in 1983, *before* the increase. Thus, for example, the cost of meat in 1984 is 116.7% of the cost of meat in 1983; we call these percentages 'percentage relatives' (see table below).

	Percentage Relatives	
Kind of food	1983	1984
Meat	100	116.7
Fish	100	106.7
Eggs	100	111.8
Butter	100	107.1
Vegetables	100	110.7

Notice that each percentage relative for 1983 is 100; this is obvious since each price expressed as a percentage of itself is 100%. We call 1983 the *base year*, since all of the prices are expressed as percentages of the 1983 prices; we also call 100 the *base*, since this is the percentage relative of each 1983 value.

Clearly, it would be unfair to take *one* of these percentages, say that for meat, and use this as an index of the Cost of Eating, since these ignore different rises in prices of other commodities. Also, it would not be wise to simply average the five percentages; the reason for this is that different amounts of money are spent on the different kinds of food. For example, maybe as much as half the money spent by a family on food may go to purchase meat, as compared with maybe only a tenth on eggs. Clearly, it would be unrepresentative to give the percentage relative for eggs (110.7) as much influence on the final index as that for meat (116.7). In other words, the different kinds of food have different relative importance, or *weight*, viz. the amount of the food budget spent on them. We must take these weights into consideration when calculating the final index, ie we must calculate the *weighted average* of the percentage relatives, using the proportions spent on the different kinds of food as weights.

Suppose the average family spends each £1 of the food budget as in the table below; the amount (price) spent, per £1, on each kind of food will be the weight (w) for its corresponding price relative of 1984.

	Percentage Relative		Weight (w)
			Amount (pence) per £1
Kind of Food	1983	1984	
Meat	100	116.7	45
Fish	100	106.7	18
Eggs	100	111.8	10
Butter	100	107.1	12
Vegetables	100	110.7	15

We can now calculate the weighted average of the 1984 percentage relatives using the formula:

$$\text{Weighted Average} = \frac{\Sigma wx}{\Sigma w}$$

where w is the amount (pence) spent per £1, and x is the percentage relative for 1984.

$$\text{Thus, Cost of Eating Index} = \frac{\Sigma wx}{\Sigma w}$$

$$= \frac{45 \times 116.7 + 18 \times 106.7 + 10 \times 111.8 + 12 \times 107.1 + 15 \times 110.7}{45 + 18 + 10 + 12 + 15}$$

$$= 112.4$$

Notice that the final index is also a number above 100, and represents the overall percentage increase in the cost of eating, ie in this case, a 12.4% increase.

The following example uses different weights; finding which weights to use is the main problem in such questions.

Example: The average middle-class family spends their income per month in the ways shown in the table below. Calculate a cost of living index for 1984 based on 1983.

Type of Expense	Amount Spent (£)	
	1983	1984
Housing	150	180
Heating	25	30
Food	60	75
Clothes	25	29
Entertainment	30	36
Transport	40	55

Again, we first calculate the percentage relatives for 1984 based on 1983, as shown in the table below.

Type of Expense	Percentage Relatives		Weights (w)
	1983	1984	
Housing	100	120	180
Heating	100	120	30
Food	100	125	75
Clothes	100	116	29
Entertainment	100	120	36
Transport	100	137.5	55

The problem now is to find suitable weights which indicate the relative importance of each type of expense; in the last example, these were the relative amounts spent per £1 on the different kinds of food. But, in this example, such relative amounts are the very amounts given in the table of data. Thus, we simply use the 1983 or 1984 column of monthly expenditures; since 1984 is the most recent data, we will use this, although it is perfectly legitimate to use the 1983 data (this would naturally result in two different indices; each of which is representative of the cost of living for 1984 based on 1983). The table above shows the 1984 data under the heading Weights (w); using these weights, we obtain the following cost of living index:

Cost of Living Index $= \frac{\Sigma wx}{\Sigma w}$

$$= \frac{180 \times 120 + 30 \times 120 + 75 \times 125 + 29 \times 116 + 36 \times 120 + 55 \times 137.5}{180 + 30 + 75 + 29 + 36 + 55} = \frac{49821.5}{405}$$

$= \underline{123.0}$

Chain based indices

So far, we have only considered indices based upon a fixed and specified year (in the above example, this was 1983); however, if we wish to obtain an index which reflects changes during the *preceeding year only*, the base year would then have to be the preceding year. Naturally, as time goes on, the base year would move from year to year, as opposed to a *fixed* base year.

Consider the data below concerning the cost of books during the years 1982, 1983 and 1984.

	Average Cost (£)			
Type of Book	1982	1983	1984	Sales (100000s)
Paperback Fiction	1.20	1.30	1.45	12.2
Hardback Fiction	2.30	2.60	2.75	3.1
Paperback Non-Fiction	3.70	4.20	4.55	7.3
Hardback Non-fiction	8.40	8.80	9.50	2.1

Suppose we wish to calculate chain-based indices for 1983 and 1984; we simply use the preceding year as base year and follow the same method as before. We must first calculate the percentage relatives for 1983 based on 1982, and for 1984 based on 1983, as below.

	Percentage Relatives		Sales (100000s)
Type of Book	1983	1984	
Paperback Fiction	108.3	111.5	12.2
Hardback Fiction	113.0	105.8	3.1
Paperback Non-Fiction	113.5	108.3	7.3
Hardback Non-Fiction	104.8	108.0	2.1

Using the sales figures as weights (the more books sold, the greater relative importance of that type of book), we obtain the two chain based indices below.

Chain-based index for 1983

$$= \frac{12.2 \times 108.3 + 3.1 \times 113.0 + 7.3 \times 113.5 + 2.1 \times 104.8}{12.2 + 3.1 + 7.3 + 2.1} = \frac{2720.19}{24.7}$$

$= 110.13$

Chain-based index for 1984

$$= \frac{12.2 \times 111.5 + 3.1 \times 105.8 + 7.3 \times 108.3 + 2.1 \times 108.0}{12.2 + 3.1 + 7.3 + 2.1} = \frac{2705.67}{24.7}$$

$= 109.54$

Thus, we can see that the cost of books went up by 10.13% between 1982 and 1983, and by 9.54% between 1983 and 1984.

13.5 Death rates

This is another application of weighted averages.

There are two types of death rate, viz *crude death rate* and *standardised death rate.*

The crude death rate for a population is simply the number of people who die per 1,000 in one year; thus, if 250 people die out of a population of 50,000 in one year, the crude death rate for that year is:

$$\frac{250}{50} = 5 \text{ deaths per } 1{,}000$$

However, this does not take into account the differing death rates for different age groups, ie infants and elderly people will make up most of the above 250 deaths. In order to reflect truly the *overall* death rate, we must take into account the ages of those who die; we do this when calculating the so-called *standardised death rate.* Consider the data below, giving the number of deaths in a town in a year within each of 5 age groups, together with the number of people in each age group.

Age Group	No. of Deaths	Population
0 - 5	100	15,000
5 - 15	70	17,000
15 - 40	180	28,000
40 - 60	220	24,000
60 and over	200	16,000
Total	770	100,000

We must first calculate the *crude death rate* for each age group.

Age Group	No. of Deaths	Population	Crude Death Rates
0 - 5	100	15,000	100/15 = 6.67
5 - 15	70	17,000	70/17 = 4.12
15 - 40	180	28,000	180/28 = 6.43
40 - 60	220	24,000	220/24 = 9.17
60 and over	200	16,000	200/16 = 12.50

As can be seen, the crude death rates for the '60 and over' age group is highest, and yet this group comprises relatively little of the population; thus, it would be unrepresentative to use this single crude death rate alone to describe the *overall* death rate in the town. Indeed, even averaging the 5 crude death rates would still bias the outcome towards this last age group. Thus, we must weigh each crude death rate, using the population in each group as weight, to obtain the *weighted average* of the crude death rates, which is called the *standardised death rate.*

$$\text{Standard Death Rate} = \frac{\Sigma wx}{\Sigma w}$$

$$= \frac{15000 \times 6.67 + 17000 \times 4.12 + 28000 \times 6.43 + 24000 \times 9.17 + 16000 \times 12.50}{15000 + 17000 + 28000 + 24000 + 16000}$$

= 7.70 deaths per thousand

The above standardised death rate was obtained using entirely data for the *town*, however, this may differ from the national death rate owing to the ages of the inhabitants, eg there are relatively more elderly people in Bournemouth. If we were given the breakdown of the *national* population into age groups though, we could use the data from the row to calculate a standardised death rate for the *whole country.* Suppose the population of the *country* can be broken down as follows:

Age Group	Population of Country (millions)
0 - 5	5
5 - 15	8
15 - 40	18
40 - 60	14
60 and over	10

Using these new figures as weights, we would obtain the standardised death rate for the *country*, based on the town's data, as below:

$$\text{Standardised death rate} = \frac{\Sigma wx}{\Sigma w}$$

$$= \frac{5 \times 6.67 + 8 \times 4.12 + 18 \times 6.43 + 14 \times 9.17 + 10 \times 12.50}{5 + 8 + 18 + 14 + 10}$$

= <u>7.92 deaths per thousand</u>

Exercise 13

1. State briefly the purpose of moving averages.

   The figures given below are observations of a variable taken at equal consecutive intervals of time.

   3.4, 4.0, 1.6, 5.2, 5.9, 4.5, 4.9, 5.2, 3.7, 7.5, 8.0, 6.8, 6.8,

   7.0, 5.9, 9.7, 9.9, 8.5, 8.6, 8.8, 7.4, 10.8, 11.7, 10.5.

   Illustrate these figures on a graph.

   Using the most appropriate number of observations, calculate moving averages and illustrate these on the same graph. Estimate the next value of the observed variable.

2. a) Explain what is meant in a time series by
      i) seasonal variation,
      ii) general trend.

   b) The table below gives the monthly values of the index of industrial production from January 1961 to June 1962 for all industries. (Average 1958 = 100). Represent this information graphically.

      Smooth the series by calculating the three-monthly moving averages and plot these moving averages on the same graph. State any deductions that can be made.

Month	Index	Month	Index
Jan	112	Oct	118
Feb	117	Nov	119
March	118	Dec	112
April	115	Jan	111
May	114	Feb	117
June	118	March	120
July	110	April	113
August	96	May	119
Sept	116	June	116

3. Determine the value of the second average in a three point moving average for the values

   13 21 17 10 15 20 26 18 14

4. An industrial concern maintains a record of the number of day's work lost through illness. The records for three years are summarised below, each figure representing the number of day's work lost during a two-month period.

Year	Number of days work lost					
1975	217	242	161	118	102	159
1976	226	241	180	120	109	160
1977	232	245	214	126	105	177

Draw a graph to illustrate these figures and on it superimpose a graph of the moving average using the most appropriate number of observations.

Draw the straight line which indicates the general trend and use it to estimate the next two values of the moving average. Hence estimate the number of day's work lost in the second period of 1978.

5. Explain what is meant by seasonal variation in a time series.

6. The following table gives the takings (in £1000s) of a shopkeeper in each quarter of four successive years. Draw a graph to illustrate the data and on it superimpose a graph of the four-quarterly moving average. State the conclusions to be drawn from the graphs.

		Quarter		
Year	1st	2nd	3rd	4th
1974	13	22	58	23
1975	16	28	61	25
1976	17	29	61	26
1977	18	30	65	29

Estimate the takings for the four quarters of 1985.

7. Find the value of the second average in a five-point moving average for the following series:

21 23 24 27 28 30 35 34

8. Distinguish between secular and seasonal variation.

9. The information in the following table is obtained from a report of a Medical Officer of Health of a local authority.

Year	Number of deaths from Lung cancer	Tuberculosis
1967	22	71
1968	54	45
1969	52	54
1970	53	38
1971	46	36
1972	71	28
1973	90	24
1974	105	24
1975	82	18
1976	105	18
1977	112	12

Calculate the five-year moving averages for each disease and draw a trend curve or line for each of the sets of averages. From your curves estimate the number of deaths from each disease in the year 1978.

10. The table below shows the figures for the number of articles sold by a manufacturing firm over the three-year period 1976-1977:

Year	Quarter 1	Quarter 2	Quarter 3	Quarter 4
1975	670	630	680	700
1976	740	700	760	780
1977	820	800	840	888

a) Calculate the four-quarterly moving averages. Using a scale of 2 cm to represent 25 articles on the vertical axis, and allowing for values from 600, represent the above data graphically and mark clearly on your graph the four-quarterly moving averages.

b) Draw the trend line on the graph, and use it to calculate the output for the first quarter of 1978.

Estimate the number of articles sold in the four quarters of 1985.

11. Observations of a variable are made once a month for 5 years. How many values of a twelve-point moving average would these give?

12. The figures below are observations of a variable taken at equal consecutive intervals of time. Using the most appropriate number of observations, calculate the value of the fifth moving average.

3.7 4.1 4.5 4.2 3.9 3.6 2.8 3.2 3.6 3.3 3.0 2.7

13. The average weekly sales in £s of two firms P and Q during the period 1971 to 1973 were as follows:

Firm	1971			1972			1973		
	Jan Apr	May Aug	Sept Dec	Jan Apr	May Aug	Sept Dec	Jan Apr	May Aug	Sept Dec
P	275	250	300	500	475	525	725	700	750
Q	500	365	410	575	440	485	650	515	560

a) Using 2 cm to represent £100 and to represent 4 months, plot both sets of data on the same graph.

b) For each firm, calculate an appropriate moving average to eliminate the seasonal variation in the sales.

c) Plot the two moving averages on your graph.

d) Use your graph to determine the period in which the moving average for firm P first equalled the moving average for firm Q.

e) State, with reasons, which firm experienced the greater seasonal variation.

f) Estimate the average weekly sales of firm P during the first four months of 1974.

14. Calculate the weighted averages of the following sets of values of x.

a)

x	1	2	3	4	5
w	60	28	19	52	41

b)

x	0.1	0.2	0.3	0.4
w	15.2	39.1	14.8	19.3

c)

x	94	28	69	31	22	59
w	1.2	3.5	9.1	2.5	6.1	2.1

d)

x	19	16	15	28	13	100
w	492	384	295	784	129	59

15. Explain briefly what is meant by a weighted mean.

At the beginning of a school year the children of a particular class investigated the pattern of spending in their area. Each of six items was given a certain weight and an initial price relative of 100. During the school year various increases in price occured. The weights and the percentage increases in price are shown below.

Item	Weight	Percentage increase in price
Food	350	25
Household goods	110	21
Clothing	95	12
Housing	90	8
Fuel and light	65	18
Miscellaneous	100	13

What is the value of the Cost-of-Living Index at the beginning, and at the end, of the school year?

16. The table below shows the number of deaths occuring in a particular year in each of two towns, X and Y, together with the population of each town and of the country. All are classified by age group.

Age group (years)	Number of deaths		Population		Country
	Town X	Town Y	Town X	Town Y	(millions)
0-	21	48	3000	8000	15
20-	7	15	3500	7500	14
40-	32	77	2500	3500	13
60 and over	88	80	1000	1000	8

Calculate crude and standardised death-rates for each town.

17. The table below gives the price relative for a commodity for the four years 1965-1968. The price relative is calculated on the chain-base method, that is the value of the item is calculated as a percentage of its value in the previous year.

Year	1965	1966	1967	1968
Price relative (chain base)	97.5	103	98	102

If the commodity cost £12.50 in 1964 calculate its value in each of the years 1965 to 1968.

18. The 1956 and 1961 index numbers for volume of imports were 100 and 135 (1954 = 100).

a) What are the 1954 and 1961 index numbers if 1956 = 100?

b) What are the 1954 index numbers of 1961 = 100?

19. Explain why a standardised death-rate is usually preferred to the crude death-rate.
In the first three columns of the table below are shown the age groups, the corresponding death-rates, and the number of deaths for a particular town in a given year. The final column gives the age distribution of the population of the county in which the town is situated.

Age group (years)	Death-rate (per 1000)	Number of deaths	Population of county (100,000s)
0 - 4	4.3	16	3.4
5 - 14	0.6	5	7.7
15 - 24	0.8	8	6.2
25 - 34	1.2	13	7.2
35 - 44	2.8	35	5.8
45 - 54	7.4	90	6.7
55 - 64	20.7	248	6.2
65 and over	122.3	1284	4.8

Calculate:

a) the standardised death-rate for the town,

b) the population, to the nearest 100, of the town by age groups,

c) the crude death-rate for the town.

20. The index number of a commodity in 1972 was 125, taking 1970 as base year. If the 1972 price was £67.50 calculate the price in 1970.

21. The prices of a commodity in 1971 and 1974 were 35p and 45p respectively. What is the price relative for the commodity in 1974 if 1971 = 100?

22.

Commodity	A	B	C	D
Index	101	103	105	102
Weight	2	3	4	2

Calculate a suitable index for the combined commodities.

23. The index of retail prices of an article in 1966 was 110, and in 1968 was 120. If the price of the article in 1968 was £24, what was the price in 1966?

A. £19.20

B. £20

C. £21.60

D. £22.

24.

Year	1968	1969	1970	
Index number	104	106	108	(1962 = 100)

Taking 1968 = 100, what is the index number for 1969?

A. 98.1

B. 101.9

C. 102.0

D. 106.0

25. Calculate an index number, using the prices and weights given in the table below, for a change in the cost-of-living between 1967 and 1977.

Commodity	Price (p) 1967	1977	Weight
Milk (per pint)	6	12	32.5
Bread (per loaf)	7	20	42
Meat (per pound)	35	60	49
Vegetables (per pound)	6	16	35

26. a) Contrast crude and standardised death-rates.

b) The following table gives the relevant vital statistics for a certain town in 1965.

Age group	Number of deaths	Population (thousands)	Population in United Kingdom (millions)
0 and under 15	216	8	12
15 and under 25	200	26	7
25 and under 45	288	24	14
45 and under 65	156	6	13
65 and over	105	1	6

Calculate the crude death-rate for each age group and the standardised death-rate for the town.

27. The index numbers, calculated on the chain-base method, for a particular commodity are shown below. Calculate the 1970 index, to the nearest whole number, using 1968 as base.

1968	1969	1970
100	107	108

28. The price of a certain commodity in each of the years 1970 to 1973 is shown below.

Year	1970	1971	1972	1973
Price	£2.00	£2.40	£3.00	£3.30

Starting with 100 for 1970 calculate index numbes for 1971, 1972 and 1973 using the chain-base method.

29.

Commodity	A	B	C	D
Index	103	-	112	115
Weight	3	3	2	1

Given that the weighted index for the four commodities shown above is 105, find the index for commodity B.

30. The sales figures of a company marketing five different products are given in the table below for the years 1980 to 1982. The percentage of the total output for each product is also given.

Calculate an index of output for:

a) 1981 based on 1980.

b) 1982 based on 1980.

Calculate a chain based index for 1982; compare the usefulness of this with the index in b).

Product	% of Total output	Sales (£1,000)		
		1980	1981	1982
A	28	70	82	85
B	16	51	50	57
C	42	48	63	52
D	10	22	14	12
E	4	17	25	29

ANSWERS

Chapter 1

1. Male smokers = 90; Female smokers = 110;
   Male non-smokers = 135; Female non-smokers = 165.
2. Relevant variates are age and sex; hair colour is unimportant.

	20-40	40-50	50-60
Male	135	90	45
Female	165	110	55

3.

	A	B	C
B12 Deficient	12	8	4
Not Deficient	18	12	6

Assuming that B12 deficiency is related to virus susceptibility.

4. i) Choose numbers between 0 and 10 first, then subtract 5 from each.
5. To calculate ratios of stratified categories, count up number of data in each category; ratio is 6 : 3 : 2 : 1.
6. Repeat process for no. 5, first part, five times, renumbering each time.

Chapter 2

1. a) cont., quant.; b) disc., quant.; c) qual.; d) qual.; e) qual.;
   f) qual.; g) cont., quant.; h) disc., quant.; i) disc., quant.; j) disc., quant.;
   k) cont., quant.; l) disc., quant.
2. Negatively skewed; a) 69, b) 31, c) 34, d) 35.
3. i) a) disc., d) symmetrical;
   ii) a) cont., b) 5, c) 2.5, d) reverse J-shaped;
   iii) a) disc., b) 3, c) 4, d) negatively skewed;
   iv) a) disc., b) 0.3; c) 0.2, d) symmetrical;
   v) a) disc., b) unknown, c) 127, d) positively skewed;
   vi) a) cont., b) 0.58, c) 7.31, d) symmetrical;
   vii) a) disc., d) positively skewed.
5. a) 112; b) 196; c) 108.
6. c) the percentage bar charts can only be used to compare changes in a *relative* manner, whereas ordinary sectional bar charts also show the actual productions for each year.
7. a) i) bar chart, ii) pie chart, ii) histogram
   b) three comparative sectional bar charts.
8. a) i) data comprising a whole made up of a number of components,
   ii) similar data that varies over a period of time,
   iii) grouped frequency distribution.
10. 56.1%
11. a) 21.9%, b) 6.7%
12. 44.8%
14. 4.82%
15. 67.5%
17. Symmetrical.

Chapter 3

1. a) $\bar{x} = 4.18$, $M = 4$, male = 1 ;
   b) 3.82, 3, 2;
   c) 2.44, 2.15, all values are modes;
   d) 116.25, 93.5, all values are modes.
2. i) 2.19, 2, 2;
   ii) 7.17, 5.06, modal class is 0-5;

iii) 14.29, 14.71, 12-16;

iv) 1.01, 0.95, 0.7-0.9;

v) impossible to estimate mean because of no MCV for last class, 136.14, 135-139;

vi) 7.94, 7.86, 7.50-7.92;

vii) 13.73, 13, 13.

3. a) 95.0; b) 705; c) 59.5; d) £2,690;

   e) 47.0; f) 42

4. b) a) 34.1, b) 33.2

5. 7.0 (diag.), 7.0 (calculation)

6. 6

7. 15 yrs 142 days;

8. 1.36;

9. 8;

10. 1;

11. 15 yrs;

12. a) 2, b) 2, c) 2, $2\frac{1}{3}$

13. $11\frac{2}{3}$ hrs per day;

14. 2, grade of 1;

15. i) mean; ii) median; iii) median; iv) mean;

    v) median; vi) mean; vii) median.

Chapter 4

1. a) R = 8; SIR = 2; mean deviation from mean = 2.24,
   from median = 2.21, from mode = 2.5. $S^2 = 6.74$, S = 2.60.

   b) 0.4, 0.095, 0.09, 0.09, 0.10, 0.014, 0.12

   c) 13.3, 2.25, 2.58, 2.23, 2.22, 15.36, 3.92

   d) 52, 6.5, 11.08, 10.00, 10.00, 219.16, 14.80.

2. i) 6, 2, 0.93, 0.88, 0.88, 1.39, 1.18;

   ii) $20 < R < 30$, 1.13, 4.64, 4.66, impossible, 38.29, 6.19;

   iii) $12 < R < 16$, 2.37, 2.35, 2.5, impossible, 11.38, 3.37;

   iv) $1.5 < R < 1.9$, 0.31, 0.37, 0.36, impossible, 0.19, 0.44;

   v) $26 < R < \infty$, 4.04, impossible, impossible, impossible, impossible, impossible

   vi) $0.92 < R < 1.88$, 0.36, 0.40, 0.39, impossible, 0.21, 0.46;

   vii) 6, 1, 1.30, 1.12, 1.12, 2.28, 1.51.

3. a) SIR = 8.0;
   b) SIR = 143;
   c) SIR = 2.9;
   d) SIR = £2,905;
   e) SIR = 56.0;
   f) SIR = 5.1;
   g) SIR = 34.
4. 2.33;
5. 1.3;
6. 1;
7. 48, 18.44, overall mean = 50.7, overall standard deviation = 15.92;
8. 1;
9.

x	16	17	18	19	20	21	22
f	3	7	8	11	12	6	1

a) 18.90; b) 1.23; c) 1.50;

10. a)

x	0	1	2	3	4	5
f	14	26	12	6	0	2

positively skewed;

b) 1.3, 1.13, MD = 0.87; c) 5 ;

11. 8;
12. 3.6;
13. 5;
14. 6.27;
15. 16.42", 2.97"
16. 0, 3.045, 2, 6.68, 2.59;
17. 12.08 yrs; 1.83 yrs;

18. i) 17.1 kg, 4.1 kg; ii) 17.85 kg, 5.57 kg;

    Median and SIR because of positive skew.

19. 6.55, 1.74;

20. i) variance; ii) SIR; iii) SIR; iv) variance;

    v) SIR; vi) variance vii) SIR.

Chapter 5

1. a) 1, 2, 3, 4, 5, 6; b) H, T; c) (H,H), (H,T), (T,H), (T,T);

   d) (1,1), (1,2), ... (1,6), (2,1), (2,2), ... (2,6), ..., (6,1), (6,2) ... (6,6).

   e) 0, 1, ... , 9;

f) R, Y;

g) (W,B), (B,B).

2.

(a)

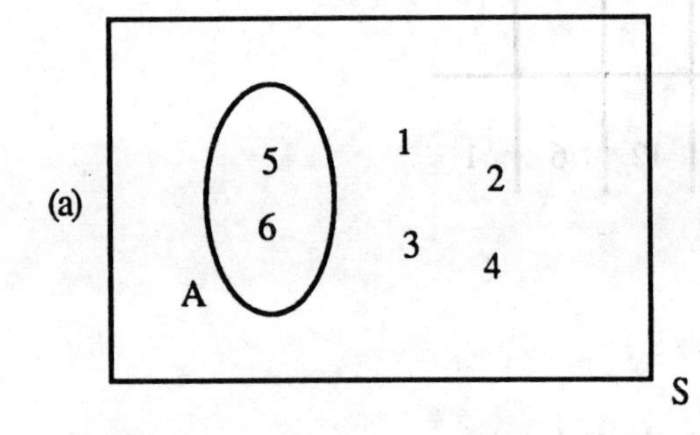

(b)

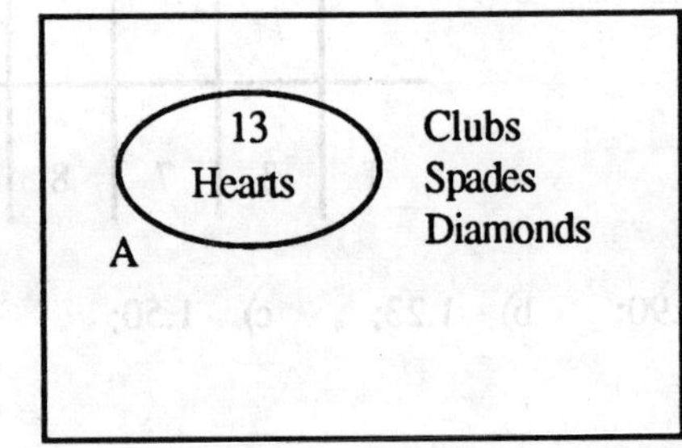

(c)

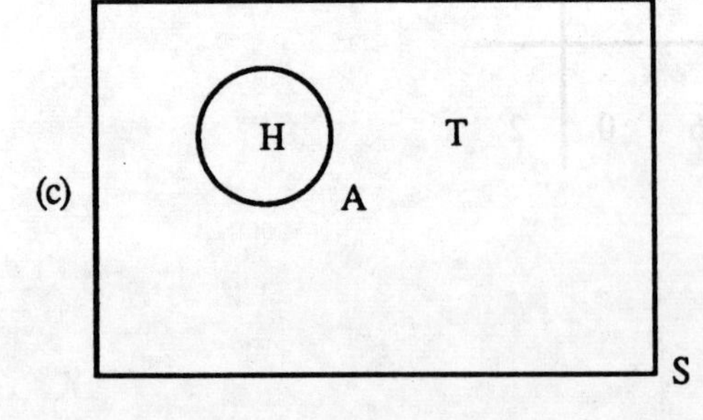

(d)

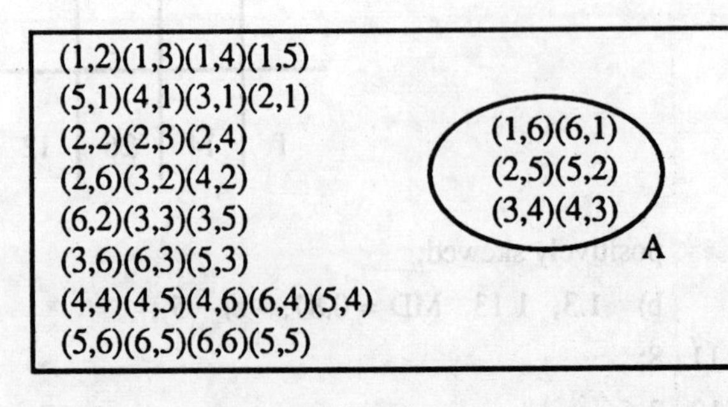

(e)

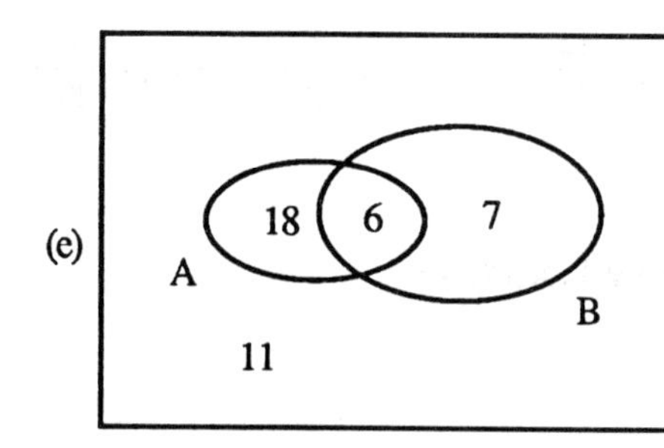

(f)

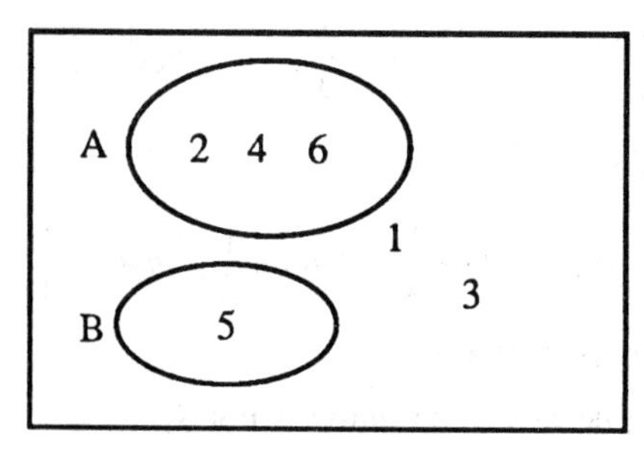

(g)

(4,1)(1,4) A (1,3)(3,5)(3,6)
(2,5)(5,3)(6,3)
(3,2)(2,3) (5,6)(4,2)(2,4)
(6,5)(5,1)(1,5)
(6,1,(1,6) (5,2)(1,2)(3,1)(3,3)

B (6,6)(4,6)
(1,3)(4,5)(1,2)(1,1)(2,1)(3,1)
(5,4)(2,1)(2,2)(6,4)(2,6)(6,2)

S

3. a) $\frac{1}{3}$, b) $\frac{1}{4}$, c) $\frac{1}{2}$, e) $P(A) = \frac{6}{13}$, $P(B) = \frac{1}{4}$,

f) $P(A) = \frac{1}{6}$, $P(B) = \frac{1}{2}$ 0, $\frac{5}{6}$, $\frac{1}{2}$, $\frac{5}{6}$, $\frac{1}{2}$, $\frac{1}{6}$

g) $P(A) = \frac{1}{9}$, $P(B) = \frac{1}{9}$; $\frac{1}{18}$, $\frac{1}{6}$, $\frac{8}{9}$, $\frac{5}{6}$, $\frac{1}{18}$, $\frac{1}{18}$

4.

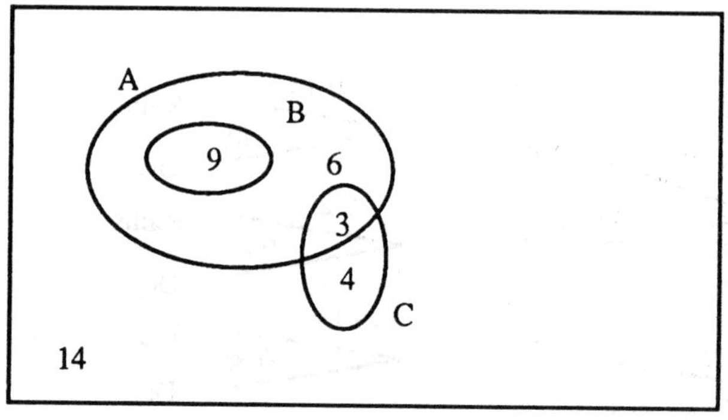

i) $\frac{1}{4}$; ii) $\frac{1}{2}$; iii) f(1,12); iv) $\frac{11}{18}$, v) 0; vi) $\frac{4}{9}$;

vii) $\frac{11}{18}$; viii) $\frac{1}{2}$, ix) $\frac{3}{4}$, x) 0, xi) $\frac{4}{9}$; xii) 0,

xiii) $\frac{11}{12}$; xiv) $\frac{7}{18}$; xv) 1; xvi) $\frac{1}{2}$; xvii) $\frac{3}{7}$, xviii) 0;

xix) 0 ; xx) $\frac{1}{3}$; xxi) $\frac{2}{3}$; xxii) $\frac{14}{29}$; xxiii) $\frac{15}{29}$; xxiv) $\frac{9}{29}$;

xxv) $\frac{7}{27}$.

5. i) No: $P(A \cap B) = \frac{1}{4}$, $P(A) = \frac{1}{2}$, $P(B) = \frac{1}{4}$ ... $P(A \cap B) \neq P(A)P(B)$;
   ii) No: $A \cap B \neq \emptyset$;
   iii) No, because their union is not S;
   iv) No: $P(B \cap C) = 0$, $P(B) = \frac{1}{4}$, $P(C) = \frac{7}{36}$ $\therefore$ $P(B \cap C) \neq P(B).P(C)$
   v) No: $P(A \cap C) = \frac{1}{12}$, $P(A) = \frac{1}{2}$, $P(C) = \frac{7}{36}$ $\therefore$ $P(A \cap C) \neq P(A).P(C)$
   vi) Yes: $A \cap B \neq \emptyset$;
   vii) Yes, because $A' \cup B' \cup C' = S$.
6. a) $\frac{1}{3}$, b) $\frac{2}{3}$; Yes, because P(B *or* W *and* W *or* Y) = $\frac{2}{9}$ = P(B *or* W). P(W *or* Y)
7. 3326400 ;
8. 120, 720, 288, $\frac{1}{2}$;
9. a) 0.032 , b) $\frac{1}{32}$, c) 0.031 , d) 0.231 , e) 0.019,
   f) 0.421 , g) $\frac{1}{36}$.
10. $\frac{1}{2}$, $\frac{1}{8}$, a) 0.00035 , b) 0.0013 , c) 0.079 .
11.

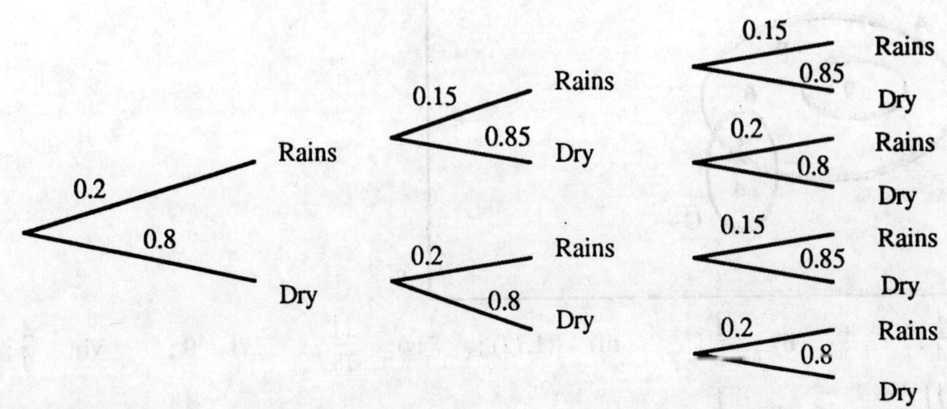

a) 0.0045 b) 0.4 ; 0.176

12. 1260, 36, $\frac{2}{7}$, $\frac{2}{3}$;

13. $\frac{1}{343}$, $\frac{120}{343}$;

14. $\frac{1}{12}$, $\frac{1}{12}$

15. a) $\frac{9}{24}$, b) 296, 0.9610 ;

16. i) $\frac{25}{49}$ ii) $\frac{13}{25}$

17. $\frac{3}{10}$, $\frac{9}{10}$, $\frac{7}{10}$, $\frac{3}{5}$; i) No, because $P(A \cap B) = \frac{3}{10}$ and

$P(A) = \frac{1}{2}$ and $P(B) = \frac{7}{10}$ $\therefore$ $P(A).P(B) = \frac{1}{2} \times \frac{7}{10} = \frac{7}{20} \neq \frac{3}{10} = P(A \cap B)$;

ii) No, because $P(A \cap B) = \frac{3}{10} \neq 0$.

18. a) c) d) all fair, b) unfair.

19. a) 0.467, b) 0.68, ii) 0.88, iii) $\frac{20}{68}$.

20. i) $\frac{4}{27}$, ii) $\frac{16}{75}$, iii) $\frac{20}{59}$;

21. a) 1320, 56 b) i) ${}^{52}C_4$, ii) $\frac{11}{4165}$, iii) $13/{}^{52}C_4$, iv) 24

c) i) $\frac{12}{51}$, ii) $\frac{4}{51}$

22. a) i) 720, ii) 120, iii) 24, 15;

b) i) 0.0225 , ii) 0.2112, iii) 0.1536, iv) 0.8649

23. a) $\frac{8}{27}$, b) $\frac{19}{27}$, c) $\frac{20}{27}$;

24. a) $\frac{1}{3}$, b) $\frac{4}{9}$

25. $\frac{169}{330}$

26. a) $\frac{1}{36}$, $\frac{2}{36}$, $\frac{3}{36}$, $\frac{4}{36}$, $\frac{5}{36}$, $\frac{6}{37}$, $\frac{5}{36}$, $\frac{4}{36}$, $\frac{3}{36}$, $\frac{2}{36}$, $\frac{1}{36}$;

b) 7 ; c) $\frac{1}{81}$; d) $\frac{125}{216}$

27. a) i) 20 , ii) $\frac{9}{100}$, iii) $\frac{3}{25}$, iv) $\frac{27}{100}$

b) i) 20 , ii) $\frac{1}{15}$, iii) $\frac{2}{15}$ iv) $\frac{4}{15}$

28. a) 0.088 , b) 0.115 , c) 0.115 , d) 0.829

29. a) $\frac{26}{32}$, b) 0.219 ;

30. i) $\frac{1}{216}$, ii) $\frac{1}{36}$, iii) 0.579 , iv) $\frac{1}{36}$

31. a) i) $\frac{1}{27}$, ii) $\frac{2}{9}$, iii) $\frac{20}{27}$; b) 0.339.

32. a) i) $\frac{4}{17}$, ii) $\frac{1}{17}$, iii) $\frac{5}{17}$;

b) i) $\frac{1}{4}$, ii) $\frac{1}{13}$, iii) $\frac{4}{13}$; c) $\frac{17}{108}$.

33. a) $P(j) = \frac{j}{21}$, $j = 1, 2, 3, 4, 5$ or 6; 0.166;

b) i) 0.246, ii) 0.246, iii) 0.410

34. a) $P(9) = \frac{25}{216} = 0.1157$, $P(10) = \frac{27}{216} = 0.1250$;

b) P(at least one six) = 0.5177, P(at least one double six) = 0.4914 ;

c) P(at least one six) = 0.6651, P(at least two sixes) = 0.6186.

35. a) 0.82 , b) $\frac{47}{66}$

36. a) $\frac{2}{3}$, b) $\frac{33}{128}$

37. a) i) P(no officers meet) = $\frac{2}{9}$, ii) P(all officers meet) = $\frac{1}{9}$;

b) P(6 different residents involved) = 0.64, P(no return to originator) = 0.88.

38. a) $\frac{3}{11}$, b) $\frac{201}{448}$

Chapter 6

1.

r	1	2	3	4	5
P(r)	0.2	0.5	0.2	0.08	0.02

2.

r	0	1	2	3	4	Total
f	4	8	4	2	2	20

3. a) $P(r) = {}^5C_r \left(\frac{1}{3}\right)^r \left(\frac{2}{3}\right)^{5-r}$, r = 0, ..., 5 ;

   b) $P(r) = \frac{1}{4} \left(\frac{3}{4}\right)^{r-1}$, r = 1, 2, ... ;

   c) $P(r) = \frac{1}{9}$, r = 1, ..., 9.

4. a) 0.330 ; b) 0.017 ; c) 0.285 ; d) 0.735.
5. a) 5.9 x 10-6 ; b) 0.998 ; c) 0.028.
6. a) 0.3 ; b) 0.103
7. a) 0.041 ; b) 0.015 ; c) 3.1×10^{-7}
8. a) 0.3 ; b) 0.00015 ;
9. i) 4.2, 4, 4, 1.56, 1.25 ;
   ii) 127.5, 100, 100, 3369, 58 ;
   iii) 1.875, 1, 1, 1.11, 1.05 ;
   iv) -0.45, -1, -1, 2.85, 1.69
10. Unfair, average loss = 4p per game.
11. a) $\frac{2}{3}$, b) 3p.
12. No; he loses 1.6p per game on average.
13. 6
14. 4
15. a) 6 , b) 9
16. i) a) 9.4, 6.24; b) -4.2, 1.56; c) -23, 39 d) 57.6, 94.5
   ii) a) 256, 13475; b) -127.5, 3369; c) -637.5, 84219; d) 58875, 4.6×10^8 ;
   iii) a) 3.75, 4.436; b) -1.875, 1.109; c) -9.375, 27.725; d) 4.625, 227.1;
   iv) a) -0.9, 11.39 b) 0.45, 2.85; c) 2.25, 71.19; d) 9.16, 27.45.
17. $15\frac{1}{2}$p
18. i) 7, 5.8; ii) 23, 52.2; iii) -38, 21267; iv) 1190; v) 109.6.
19. i) 0.5905; ii) 0.3281; iii) 0.4095.
20. 2.8p, -0.04p, -2.8p
21. i) 0.328; ii) 0.410; iii) 0.262.

22.

x	0	1	2	3	4
P(x)	0.05	0.30	0.45	0.18	0.02

, 1.82, 0.73 .

23. Yes; 0.341, 0.98; Increase at 14%.

24. 7.2, 9;

25. E [X] = 2.5, V [X] = 2, E [C] = 10, V [C] = 8.

26. E [X] = 1.5, V [X] = 1, E $[X_1 - X_2]$ - 0, $P(X_1 = X_2)$ = 0.430

27. a) £2.25, b) £0.857

Chapter 7

1. a)

Y \ X	0	1	2	3	4	5
3	1/30	1/30	1/30	1/30	1/30	1/30
4	1/30	1/30	1/30	1/30	1/30	1/30
5	1/30	1/30	1/30	1/30	1/30	1/30
6	1/30	1/30	1/30	1/30	1/30	1/30
7	1/30	1/30	1/30	1/30	1/30	1/30

b)

Y \ X	1	2	3	4	5
2	1/18	1/18	1/18	1/9	1/18
4	1/18	1/18	1/18	1/9	1/18
6	1/18	1/18	1/18	1/9	1/18

c)

T \ S	3	4	5	6	7	8	9	10	11
2	1/18	0	0	0	0	0	0	0	0
4	0	1/18	1/18	0	0	0	0	0	0
6	0	0	1/18	0	1/18	0	0	0	0
8	0	0	0	1/6	0	0	0	0	0
10	0	0	0	0	1/18	0	0	0	0
12	0	0	0	0	1/18	1/18	0	0	0
16	0	0	0	0	0	1/9	0	0	0
18	0	0	0	0	0	0	1/18	0	0
20	0	0	0	0	0	0	1/18	0	0
24	0	0	0	0	0	0	0	1/9	0
30	0	0	0	0	0	0	0	0	1/18

d)

Y \ X	0	1	2	3
0	0.002	0.007	0.004	0.0005
1	0.038	0.114	0.069	0.008
2	0.086	0.257	0.154	0.017
3	0.038	0.114	0.069	0.008
4	0.002	0.007	0.004	0.0005

e)

Y \ X	0	1	2	3	4
0	0	0	0.0065	0.017	0.0065
1	0	0.017	0.104	0.104	0.017
2	0.0065	0.104	0.234	0.104	0.0065
3	0.017	0.104	0.104	0.017	0
4	0.0065	0.017	0.0065	0	0

f)

Y \ X	1	2	3	4	5
1	0	0.0036	0.044	0.183	0.488
2	0	0	0.0036	0.044	0.183
3	0	0	0	0.0036	0.044
4	0	0	0	0	0.0036
5	0	0	0	0	0

2. a) Yes; b) Yes; c) No; d) Yes; e) No; f) No.
   If P(X and Y) = P(X)P(Y) for all X and Y then X and Y are independent; in examples c), e), and f) this is *not* true for *all* X and Y pairs.
3. a) i) 2.5, 5, 2.92, 2; ii) 7.5, 4.92; iii) 12.25;
   iv) 17.5, 42.5; v) -2.5, 4.92; vi) 167.17.
   b) i) 3.17, 4, 1.81, 2.67; ii) 7.17, 4.48; iii) 12.68;
   iv) 17.51, 44.53; v) -0.83, 4.48; vi) 135.6.
   c) i) 7.17, 12.67, 4.47, 60.44; ii) 19.84 not allowed to use rule;
   iii) not allowed; iv) 45.85, not allowed;
   v) -5.5 not allowed; vi) not allowed.
3. d) i) 1.21, 2, 0.57, 0.57; ii) 3.21, 1.14; iii) 2.42;
   iv) 7.63, 18.89; v) -0.79, 1.14; vi) 30.
   e) i) 2, 2, 0.73, 0.73; ii) 4, not allowed; iii) not allowed;
   iv) 10, not allowed; v) 0, not allowed; vi) not allowed.
   f) i) 4.66, 4.66, 0.34, 0.34; ii) 9.32, not allowed; iii) not allowed;
   iv) 23.3, not allowed; v) 0, not allowed; vi) not allowed.
4. No, because, for example, P(X = 2 and Y = 3) = 0.1 yet P(X = 2).P(Y = 3) = 0.3 x 0.4 = 0.12
5. No; eg if X were 4, then Y could not possibly be 5, say, and so the value of one affects the value of the other.

Chapter 8

2. a) $\frac{1}{2}$; b) $\frac{3}{7}$; c) 1 ; d) $\frac{1}{(e - 1)}$; e) 2

3. Q.1 a) 1 ; b) 0 ; c) 0.44 ; d) 2.5 ; e) 1 ;

   Q.2 a) $\frac{2}{3}$; b) $\frac{75}{28}$; c) $\frac{\pi}{2} - 1$; d) $\frac{1}{e^{-1}}$; e) $\frac{5}{3}$

4. Q.2 a) 0.88 ; b) 0 ; c) 0.46 ; c) 1 .

   Q.2 a) 0.59 ; c) 0.52 ; d) 0.62 ; c) $\sqrt{3}$.

5. Q.1 a) 0 ; b) -1, +1 ; c) 0 ; d) 2.5 ; e) 1

   Q.2 a) 0 ; b) 3 ; c) 0 ; d) 1 ; e) 2

6. Q.1 a) $\frac{1}{2}$; b) $\frac{3}{10}$; d) 3.8 ; e) $\frac{1}{6}$

   Q.2 a) $\frac{2}{9}$; b) 0.054 ; c) 0.14 ; d) 0.38 ; e) 0.39

7. i) $\frac{1}{3}$; ii) $\frac{1}{2}$; iii) 0.59 ; iv) 0 ; v) (α) $\frac{1}{8}$; (β) $\frac{7}{8}$

   vi) $\frac{3}{4}$; vii) 0.77 ; viii) 0.37 ; (ix) 0.62 ; x) (α) $\frac{1}{6}$; (β) 0.21.

8. mean < median < mode, $1\frac{1}{5}$ < median < $1\frac{1}{3}$

9. $\mu = 55$, $\sigma^2 = 675$; i) $\frac{1}{3}$; ii) 0.224 ;

   a) 110π, $2700\pi^2$; b) 93220 ; c) 0.54 ; d) 0.516

10. School A: $\mu = 5.5$ yrs, $\sigma^2 = \frac{25}{12}$ years^2

    School B: $\mu = 8$ yrs, $\sigma^2 = \frac{64}{12}$ years^2

    a) 20.5, $\frac{91}{4}$; b) -3, $\frac{64}{12}$; c) 35, $\frac{676}{12}$;

    d) -2.5, $\frac{89}{12}$; E [T] = 32.5 years, V [T] = $\frac{481}{12}$ years^2 .

11. $f(t) = \frac{1}{500} e^{-t/500}$, $t \geq 0$; V [T] = 25000 hrs^2 ;

    a) 0.247 ; b) 0% ; c) 1498 hrs ; positively skewed;

    mode = 0, mean = 500, median = 346.6; 0.63.

12. $a = \frac{1}{5000}$; $\mu = 5000$ km, $\sigma^2 = 25000000$ km^2 ;

    a) 0.165 ; b) 0.134 ; c) 8047 km.

14. $\alpha = \frac{3}{32}$; $\mu = 2$; $\sigma 2 = \frac{14}{5}$; $\frac{5}{32}$.

15. $\mu = \frac{1}{a}$, $\sigma^2 = \frac{1}{a^2}$; 0.223 ;

16. $c = \frac{1}{\sigma}$;

18. $A = \frac{1}{108}$; $\mu = 2.4$; $\sigma^2 = 1.44$; $\sigma = 1.2$; mode = 2 ;

19. b) 0.262 ; c) 0.4 ; d) £44.82.

Chapter 9

1. a) 0.8413; b) 0.0228; c) 0.9495; d) 0.01; e) 0.242;
   f) 0.99; g) 0.5446; h) 0.3413; i) 0.2.
2. a) 1.64; b) -1.28; c) -0.255; d) -0.125;
   e) 1.64; f) 1.375.
3. a) -1.5; b) -0.33; c) 0.032; d) -1; e) 2.44;
   f) -0.18.
4. a) 115; b) -0.86; c) 54.4; d) 13.53; e) 3a/2.
5. a) i) 0.75; ii) 0.8413; iii) 0.28;
   b) i) 0.15; ii) 0.89; iii) 0
   d) i) 0.63; ii) 0.26; iii) 0.07
6. a) i) 124.6; ii) 96.3;
   b) i) 16.6; ii) 23.5;
   c) i) 2.64; ii) -2.28.
7. i) $\sigma = 1.56$; ii) $\mu = 6.92$; iii) $\mu = 0.094$ $\sigma = 0.011$.
8. i) 0.41; ii) 0.18; iii) 0.43; 0.68; 0.07; 2.73m; 0.5.
9. a) $\mu = 6.003$cm, $\sigma^2 = 0.994$;
   b) i) 0.42; ii) 0.31; iii) 0.66
   c) 4.72cm, 7.28cm.
10. a) 401; b) 81; c) 242; 39.5; 100.
11. 0.641 secs; 94.2%; $\mu = 0.491$ secs; $s^2 = 0.00125$ sec^2; changes from 59.9% to approx 100%.
12. 63 plants; 7.16 kg; 218
13. i) $\mu = 6$, $\sigma^2 = 81$; ii) $\mu = 4$, $\sigma^2 = 101$; iii) $\mu = 4$; $\sigma^2 = 21$;
    iv) $\mu = 0$, $\sigma^2 = 128$; v) $\mu = 2$, $\sigma^2 = 26$; vi) $\mu = -1$, $\sigma^2 = 6.5$.
    Assume variables are independent.
14. a) 24 cm, 0.23; b) 99.8%, 56.5 cm; c) 0.55.

15. a) $\mu = 0.5$mm, $\sigma = 12.5$mm; b) 0.29; c) 0.75.

16. a) $\mu = 0.05$mm, $\sigma = 0.009$; b) 0.73; c) 0.43

17. i) a) N(3, 1.8), b) N(15, 45);

ii) a) N(0, 3.33), b) N(0, 3000);

iii) a) $\mu = 12, \sigma^2 = 1.4$, not necessarily Normal; b) $\mu = 60, \sigma^2 = 60$, not necessarily Normal;

iv) a) N(-2, 0.006), b) N(-100, 15), by CLT;

v) a) N(100, 10), b) N(1000, 1000);

vi) a) N(9, a^2/100) b) N(100a, 100a^2), by CLT

18. i) 0.11, ii) virtually 1

19. a) 0.42, b) 0.235, c) 0.0054, 1.43, 8.57.

20. $\frac{1}{9}$, 0.08.

21. 0.08, 4 choices, 0.65

22. i) 0.32, ii) 0.075, 0.17, 392.

23. a) 0.34; b) 98.71g; c) 15.9%

24. a) 1.4%; b) 0.00079; c) 326.3 hrs; d) 0.145; e) 39.

25. $\mu = 100$, $\sigma^2 = 75$; i) 0.409, iii) 0.68.

26. i) 0.045; ii) 0.604; 93.1g, 106.9g; i) 10 kg; ii) 9.95kg;

Yes, 95% of the time; 59.25kg; underlying distribution is Normal.

27. 25g; 0.13; 173.392g, 176.08g.

28. i) 0.024; ii) 0.948; iii) 0.9977.

No assumptions necessary, since sample size is large.

29. i) 0.405; ii) 369; see 27 for assumptions.

30. $\mu = 10.24$, $\sigma^2 = 10.89$.

The CLT has been used unjustifiably since the samples are not large.

31. i) $\mu = 4$, $\sigma^2 = 3$; ii) $\mu = 4$, $\sigma^2 = \frac{3}{50}$; 0.18; 0.04.

32. i) 0.382; ii) 3.2×10^{-7}, 865

33. a) $11.37 \leq \mu \leq 12.63$; b) $0.2986 \leq \mu \leq 0.3014$

c) $-0.583 \leq \mu \leq 0.583$; d) $-2.2 \leq \mu \leq -1.8$

34. a) $0.143 \leq \beta \leq 0.206$; b) $0.22 \leq \beta \leq 0.28$; c) $0.286 \leq \beta \leq 0.377$

35. a) $48.1 \leq \mu \leq 55.9$; b) $0.156 \leq \beta \leq 204$.

36. a) $23.25 < \mu \leq 25.17$; b) $129.4 < \mu < 147.2$.

37. $0.510 < \beta < 0.574$; Yes.

38. a) $50.06 < \mu < 54.58$; b) $109.4 < \mu < 122.8$

39. a) see text; b) $49.00 < \mu < 50.08$

40. mean = np, variance = np(1-p); Bopper: $0.245 < p_B < 0.455$;
Shooter: $0.352 < p_S < 0.548$. Confidence intervals overlap, in which case it is possible that $p_B = 0.42$, say, and $p_S = 0.40$, say, which implies $p_S < p_B$; there is insufficient evidence that the sales of Shooter are higher than sales of Bopper.

41. $6.84 < \mu < 7.53$

42. $10.48 < \mu < 11.70$

43. 0.1

44. 0.161; a) 0.9999; b) 0.24

45. a) Not likely to come from a Normal distribution
b) Normal; $\mu = 12.5$, $\sigma = 1.75$
c) Normal; $\mu = 13.4$, $\sigma = 2.1$
d) Normal; $\mu = 24$, $\sigma = 1.4$
e) Not likely to have come from a Normal distribution.
f) Normal; $\mu = 23.3$, $\sigma = 1.7$
g) Normal; $\mu = 50$, $\sigma = 2.6$
h) Normal; $\mu = 3.1$, $\sigma = 1.3$
i) Not likely to have come from a Normal distribution

47. a) i) 34.3, ii) 3.34, 26.05 years old;
b) i) $E[x+y] = E[x] + E[y]$, $V[x+y] = V[x] + V[y]$
ii) $E[x-y] = E[x] - E[y]$, $V[x-y] = V[x] + V[y]$

48. 0.082; 0.0019.

49. $\mu = m_1\mu_1 + m_2\mu_2$, $\sigma^2 = m_1^2\sigma_1^2 + m_2^2\sigma_2^2$; 0.4% for 16 year olds.
31.85% for 15 year olds; $\mu = 6$; $\sigma = 5$; 0.115

50. $\mu = 0$, $\sigma^2 = \frac{n}{12}$; 2634.

51. 0.01; 0.97; 6.186m.

52. 90, 165; 0.73.

53. $\mu_1 + \mu_2$, $\sigma_1^2 + \sigma_2^2$; 0.5, $\frac{1}{12}$, 32.1 years, 9.693 years^2.

54. $a_1\mu_1 + a_2\mu_2 + a_3\mu_3$, $a_1^2\sigma_1^2 + a_2^2\sigma_2^2 + a_3^2\sigma_3^2$; 0.0526; 0.1496

55. b) 76.2, 7.29.

56. P(support 1300 units) = 0.1056; Breaking load to be quoted = 1242.78

57. i) $P(40 \leq S \leq 60) = 0.6247$;
ii) 93.32%;
iii) No: $\bar{y}$ is measured on a restricted scale 0 - 10.

58. i) 0.3292, ii) 0.5391; 0.8042, assuming a Normal approximation.

59. i) 0.0294, ii) 0.3325, iii) 0.9955, iv) 0.7128.

60. i) 0.9938, ii) 4730, iii) 0.1056, iv) 0.9998.

61. 137, 149.5.

62. $\frac{16}{37}$; 0.323.

63. a) 6, 6.5, 7, 7.5, 8, 8.5; b) variance = 0.73;

c) $\mu = 7.25$, $\sigma^2 = 2.19$; d) $E\bar{X} = \mu$, $V\bar{X} = \sigma^2$

64. $E[\bar{X}] = 50.5$, $V[\bar{X}] = 160$; 0.23.

65. $\mu = 6$, $\sigma = 4$; (0, 2), (0, 4), ..., (6, 10), (8, 10);

means are: 1, 2, 3, 3, 4, 4, 5, 5, 5, 6, 6, 6, 7, 7, 7, 8, 8, 9, 9, 10, 11;

$\bar{x}$	1	2	3	4	5	6	7	8	9	10	11
$P(\bar{x}$	$\frac{1}{21}$	$\frac{1}{21}$	$\frac{2}{21}$	$\frac{2}{21}$	$\frac{3}{21}$	$\frac{3}{21}$	$\frac{3}{21}$	$\frac{2}{21}$	$\frac{2}{21}$	$\frac{1}{21}$	$\frac{1}{21}$

Standard deviation of this probability distribution is 2.58; using the formula on p175, we see that $\sigma_{\bar{x}} = 2.58$ also.

66. $E[\bar{X}] = \mu$, $V[\bar{X}] = \frac{\sigma^2(N-n)}{n(N-1)}$; $\frac{m(n+1)}{2}$, $\frac{m(n^2-1)}{12}\left[\frac{n-m}{n-1}\right]$

Chapter 10

1. a) $\chi^2 = 5$, $\nu = 9$, $\chi^2_{crit} = 16.92$; Yes;

   b) $\chi^2 = 1.9$, $\nu = 3$, $\chi^2_{crit} = 7.81$; Yes.

   c) $\chi^2 = 33.75$, $\nu = 2$, $\chi^2_{crit} = 5.99$; No.

2. $\chi^2 = 19.36$, $\nu = 1$, $\chi^2_{crit} = 3.84$ at 5%; Yes.

3. $\chi^2 = 5.2$, $\nu = 10$, $\chi^2_{crit} = 18.31$ at 5%; Fair.

4. $\chi^2 = 15.47$, $\nu = 11$, $\chi^2_{crit} = 19.68$ at 5%; Yes.

5. $\chi^2 = 1.09$, $\nu = 1$, $\chi^2_{crit} = 3.84$; Yes.

6. $\chi^2 = 5.36$, $\nu = 1$, $\chi^2_{crit} = 3.84$; No.

7. $\chi^2 = 54.4$, $\nu = 6$, $\chi^2_{crit} = 12.59$; strong association; $\chi^2 = 5.8$, $\nu = 2$, $\chi^2_{crit} = 5.99$; no association; $\chi^2 = 1.95$, $\nu = 1$, $\chi^2_{crit} = 3.84$; no association. (Yates' Correction must be used here.)

8. $\chi^2 = 11.95$, $\nu = 6$, $\chi^2_{crit} = 10.64$; not fair.

9. $\mu = 3.75$; expected frequencies are: 0.28, 2.8, 11.59, 25.75, 32.19, 21.48, 5.96. $\chi^2 = 5.87$, $\nu = 5$, $\chi^2_{crit} = 11.07$; Binomially distributed.

10. $\chi^2 = 8.39$, $\nu = 6$, $\chi^2_{crit} = 12.59$ for 5% level; raio is 3 : 1.

11. $\chi^2 = 1.20$, $\nu = 4$, $\chi^2_{crit} = 9.49$; good fit.

12. $\chi^2 = 0.105$, $\nu = 2$ (two parameters estimated), $\chi^2_{crit} = 4.61$ at 10% level; very good fit because χ^2 is very small.

13. $\chi^2 = 4.37$, $\nu = 4$, $\chi^2_{crit} = 9.49$ at 5%; No evidence.

14. $\chi^2 = 52.64$, $\nu = 2$ (estimating two parameters and combining last two classes), $\chi^2_{crit} = 5.99$ at 5%; not Normally distributed.

15. $\chi^2 = 5.91$, $\nu = 4$ (combine last 5 classes together), $\chi^2_{crit} = 9.49$ at 5%; accept hypothesis.

16. 645, $\nu = 4$, VHS.

17. 12, 24, 36, 48, 60, 72, 60, 48, 36, 24, 12; $\chi^2 = 14.5$, $\nu = 10$, not significant; $\chi^2 = 2.12$, too good to be true; no evidence of bias.

18. a) $\chi^2 = 4.40$, $\chi^2_{crit} = 11.07$ with $\nu = 5$; results consistent with manager's claim;

    b) $\chi^2 = 0.47$, $\nu = 1$, $\chi^2_{crit} = 3.84$ at 5%.

    No evidence of association (*any* upper tail points could be used with correct degrees of freedom, as number % levels were specified).

19. mean = np, variance = np(1 - p)

x	0	1	2	3	4	5	6
f	16.5	42.4	45.4	25.9	8.3	1.4	0.1

20. b) 76.2, 7.29.

21. i) $\chi^2 = 3.43$ (given), $\nu = 2$, $\chi^2_{crit} = 5.99$ at 5%. There is no association between the factors;

    ii) for males: $\chi^2 = 2.97$, $\nu = 2$, $\chi^2_{crit} = 5.99$ at 5% - no association; for females: $\chi^2 = 12.58$, $\nu = 2$, $\chi^2 = 5.99$ at 5% - associated;

iii) A true difference was masked originally by not taking account of an important factor, viz sex. The attitudes of females do vary on this issue from party to party.

$\bar{x}$ = 50.3 mm, S = 2.65, $\chi^2 = 4$, $\nu = 1$ (after combining classes)

22. $\bar{x}^2_{crit} = 6.63$ at 1%; the Normal model for these data is supported.

Chapter 11

1.

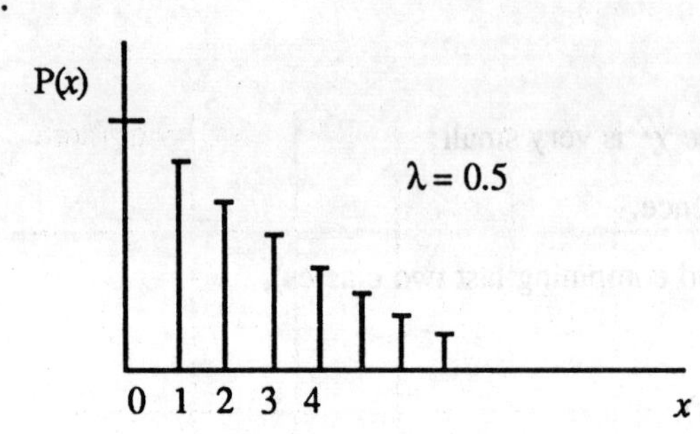

P(x)

λ = 1.0

0 1 2 3 4 x

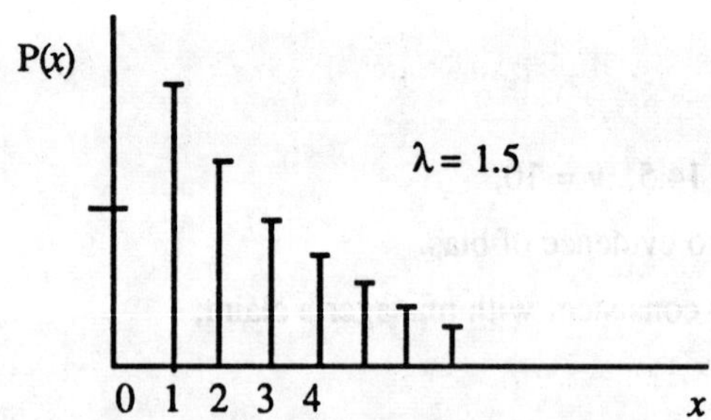

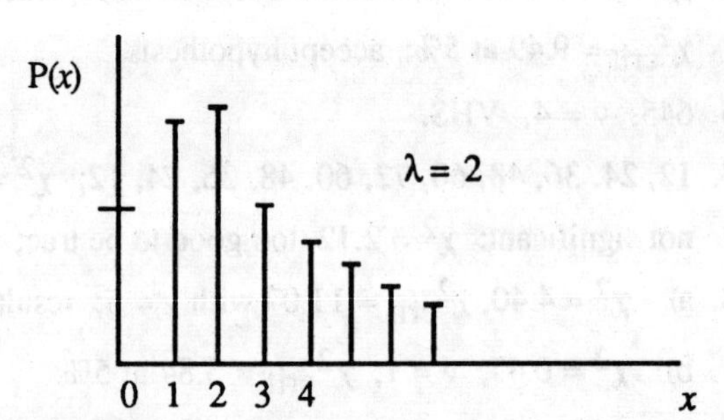

a) 0.135, b) 0.017, c) 0.762.

2. a) 0.00034, b) 0.029.
3. 0.221.
4. a) 0.22, b) 0.934, c) 0.034.
5. a) 0.00075, b) 0.844, c) 0.294.
6. 308.
7. a) 0.567, b) 0.763, c) 0.0082.
8. a) 0.368, b) 0.0006, c) 0.067. 3.00m^2.
9. a) 0.92, b) 0.66, 0.92.
10. a) 0.0007, b) negligible.
11. $\chi^2 = 23.3$, $n = 3\chi^2_{crit} = 7.815$ reject Poisson distribution (expected frequencies below 5must be combined)

12. Poisson distribution, $\lambda \approx 1.59$ aphids per plant, 0.0004.

13. $n \to \infty$ with $n\beta$ constants; $\mu = \lambda$, $\sigma^2 = \lambda$; i) 0.195, ii) 0.762, iii) 0.147, iv) 0.073.

14. $\chi^2 = 3.96$, $v = 4$, $\chi^2_{crit} = 9.49$ at 5% level; good model.

15. i) 0.165, ii) 5, iii) 0.027, iv) 5.

16. See No. 1; $\chi^2 = 0.74$, $v = 4$, $\chi^2_{crit} = 9.49$ at 95% level; good Model.

17. a) 13.5, b) 14.3, c) 40.6.

18. See No. 1; $\chi^2 = 0.31$, $v = 2$, $\chi^2_{crit} = 5.99$, at 95% level, good argument.

19. See bookwork page 234, and No. 1; $\mu = 4.8$; at 95% level.

No. of Accidents	0	1	2	3	4	5	6	7	8	9 or more
Observed Frequency	2	7	12	13	10	15	14	10	10	7
Expected Frequency	0.82	3.95	9.48	15.17	18.2	17.4	13.98	9.59	5.75	5.66

20. 0.109, 0.185.

21. i) $P(x = 0)$, expected frequency 99.40; $x = 1$, 119.26,

 $x = 2$, 71.58, $x = 3$, 28.61; $x = 4$, 8.58; $x = 5$, 2.05, $x \geq 6$, 0.53;

 ii) No evidence to suggest that this is not a reasonable model.

22. 0.632, 0.1548.

23. Mean ≈ 3;

 i) P(more than 2 unplanned stoppages) = 0.5768,

 ii) P(even number) = e^{-3}.Cosh(3) = 0.501;

 iii) P(odd number) = 1 - P(even) = 0.499.

24. i)

No. of hits x	0	1	2	3	4	5	6 or more
Expected Frequencies	182.96	164.66	74.10	22.23	5.00	0.90	0.15

 ii) $\chi^2 = 1.67$, $v = 3$, $\chi^2_{crit} = 6.25$ at 90% level; Poisson distribution adequate.

25. $1 - e^{-\theta}$; $\dfrac{e^{-\theta}.\theta^x}{(1-e^{-\theta})x!}$ for $x = 1, 2, ...$

26. Not likely to be Poisson distributed because of first value.

27. Approximately Poisson; $\lambda \approx 0.7$; 0.00016.

Chapter 12

1. a) Strong negative correlation; b) Strong positive correlation;
   c) Weak negative correlation; d) Fairly strong positive correlation.
2. a) -0.82; b) 0.89; c) -0.44; d) 0.78.
3. a) -0.61; b) 0.43; c) 0.76; d) 0.06
   e) -0.65; f) 0.51; g) -0.12; h) -0.57.
4. a) 0.93; b) -0.93; c) -0.696.
5. a) -0.87; b) 0.86; c) -0.28; d) 0.84.
6.

Q1 a) i) $y = -0.05x + 2.10$; ii) $y = -0.07x + 2.44$;
b) i) $y = 0.62x - 9.56$; ii) $y = 0.78x + -27.67$;
c) i) $y = -34.79x + 35.33$ ii) $y = -175.72x + 84.91$;
d) i) $y = 0.06x + 0.26$; ii) $y = 0.10x + 0.02$

Q3 a) i) $y = -0.38x + 8.16$; ii) $y = -1.03x + 10.83$;
b) i) $y = 2.74x + 22.17$; ii) $y = 15.05x + 16.35$;
c) i) $y = 0.81x + 6.65$; ii) $y = 1.39x + 7.88$;
d) i) $y = 0.05x + 106.02$; ii) $y = 13.83x + 84.18$;
e) i) $y = -0.21x + 4.65$; ii) $y = -0.49x + 7.07$'
f) i) $y = 0.05x - 2.96$; ii) $y = 0.18x + 19.50$
g) i) $y = -0.10x + 3.33$; ii) $y = -6.53x + 39.23$;
h) i) $y = -0.001x + 0.58$; ii) $y = -0.003x + 1.25$.

Q4 a) i) $y = 2.09x + 15.44$; ii) $y = 2.49x + 13.63$;
b) i) $y = -1.1x - 6.9$; ii) $y = -1.27x - 6.73$;
c) i) $y = -73.42x + 15.00$ ii) $y = -126.38x + 17.10$.

7. a) 0.51; b) 0.60; c) i) $y = 0.06x - 4.27$;
   ii) $y = 0.22x - 22.07$; iii) 118.5 days; iv) 3.23 offspring.
   Coefficients of regression are 0.22 and 0.06 respectively.
   Estimates are likely to be poor in view of small correlation.
8. $r_{s_{AB}} = 0.62$, $r_{s_{BC}} = 0.31$, $r_{s_{AC}} = 0.082$; only judges A and B mark similarly.
9. 0.375.

10. a) -0.38; b) -0.45; c) i) $y = -29.50x + 408.64$;
    ii) $y = -209.58x + 1644.48$; iii) £5,367; iv) 196000.
11. 0.29
12. a) i) $y = 0.81x + 14.05$; ii) $y = 1.28x - 17.08$; c) 0.80.
13. $y = 1.58x + 62.77$; b) 0.93.
14. i) $y = 0.62x + 19.21$, ii) $y = 1.28x - 17.52$; r = 0.70.
15. i) $y = 0.78x + 20.83$, ii) $y = 0.90x + 10.64$; r = 0.94; 86.35.
16. $y - 52 = 0.53(x - 146.8)$; $x - 146.8 = 1.008(y - 52)$; 0.73, 53.7m.
17. ii) $y = 2.75x + 48.35$; iii) 0.79.
18. i) r = -0.374, ii) $r_s = -0.256$,
    iii) Little linear relationship; however looking at the data reveals a possibly quadratic relationship.
19. i) $y = 0.486x - 2.397$, ii) r = 0.997, iii) No: y is measured on a restricted scale, 0 - 10.
20. 0.58.
21. 0.88.
22. 0.78.
23. 0.71.
24. 0.551.
25. i) $y = -0.51x + 105.15$, 105.15, ii) -0.976.

Chapter 13

1. To make the overall trend more easily discernible; the graph repeats itself every 6 units of time, therefore use 6-point moving averages which are (correct to one decimal place): 4.1, 4.4, 4.6, 4.9, 5.3, 5.6, 6.0, 6.3, 6.6, 7.0, 7.4, 7.7, 8.0, 8.3, 8.6, 8,8, 9.0, 9.3 and 9.6; 10.8.
3. 16
4. Graph repeats itself every 6 units of time, therefore use 6-point moving averages which are (correct to the nearest whole number); 167, 168, 168, 171, 171, 173, 174, 174, 180, 181, 180 and 183; 183.5 and 185.0; 254.
6. The 4-point moving averages (correct to one decimal place) are: 29.0, 29.8, 31.3, 32.0, 32.5, 32.8, 33.0, 33.0, 33.3, 33.5, 33.8, 34.8 and 35.5. The takings show marked seasonal variation, being high in summer and low in winter; the overall trend is that takings are gradually increasing.
7. 26.4
9. 5-year moving averages are: 45.4, 55.2, 62.4, 73.0, 78.8, 90.6 and 98.8 (lung cancer); 48.8, 40.2, 36.0, 30.0, 26.0, 22.4 and 19.2 (tuberculosis); 136 and 8.
10. a) Four-quarterly moving averages are: 670, 687.5, 705, 725, 745, 765, 790, 810, and 837
    b) 942.
11. 49
12. 3.4

13. b) 3-point moving averages are: 275, 350, 425, 500, 575, 650 and 725 (P); 425, 450, 475, 500, 525, 550 and 575 (Q) d) May/August 1972

    e) Firm Q: much greater deviation from the moving average

    f) £950

14. a) 2.930

    b) 0.243

    c) 47.931

    d) 23.073

15. 100; 119.0

16. For X: crude death-rate = 14.8, standardised death-rate = 20.1;

    for Y: crude death-rate = 11.0, standardised death-rate = 20.9

17. £12.19, £12.56, £12.31, and £12.56

18. a) 90.9 and 122.7 b) 74.1 and 81.5

19. a) 17.0 b) 3,700, 83,000, 10,000, 10,800, 12,500, 12,200, 12,000 and 10,500

    c) 21.2

20. £54

21. 128.6

22. 103.2

23. D

24. C

25. 229

26. b) 27, 7.7, 12, 26 and 105; 29.5

27. 116

28. 120, 125 and 110

29. 99

30. a) 115.9;

    b) 109.7; 95.1;

    c) gives the change in sales since 1980, whereas the chain-based index reflects the change in sales during the last year, and so is of more immediate importance, especially in view of the overall *decline* in the past year, even though sales are more than in 1980.

APPENDIX

The use of log-log paper

A question which almost always comes up in the 'A' level requires the verification of a specified law relating two variables. This is really a Pure Mathematics topic and is covered in the Pure Mathematics textbook. However, when the question appears on the *Statistics paper*, a different method must be used, involving the use of log-log paper. Consider the following example.

Example: In a scientific experiment, the energy output, E, of a machine was recorded as the amount of fuel input, I, was varied; the table below summarises the data obtained. It is thought that E and I may be related by the formula:

$$E = a\,I^b$$

where a and b are numerical constants. Verify this to be the case using log-log paper, and use your graph to estimate the values of a and b.

Amount of Fuel I	2	3	4	6	8
Energy Output, E	24.6	62.6	121.3	308.1	597.1

Now, since log-log paper is being used, this implies that the relationship to be verified must be logarithmic in nature; manipulating the above formula, by taking logarithms, base 10, of both sides, we see that:

$$\log_{10}(E) = \log_{10}[a\,I^b]$$

ie $$\log_{10}(E) = \log_{10}(a) + \log_{10}(I^b)$$

ie $$\log_{10}(E) = b.\log_{10}(I) + \log_{10}(a) \qquad (1)$$

Thus, if we plotted $\log_{10}(E)$ against $\log_{10}(I)$ on *ordinary graph paper*, we would obtain a straight line (see Pure Mathematics textbook). This would require taking logarithms, base 10, of each value of E and I. By using log-log graph paper we can avoid this, and plot the values of E against I *directly*, by choosing suitable scales on the vertical and horizontal axes.

Let us first consider the I values, to be plotted on the horizontal axis. Looking at the horizontal axis on the log-log graph paper, we see that the major divisions are not of equal size - they gradually diminish. Furthermore, this pattern of decreasing intervals is repeated twice; each pattern contains 10 intervals, and is referred to as a *cycle*. Now, each cycle spans a power of 10, which we can choose according to the range of values to be plotted along the axis. In our case, the values lie in the range 1 to 10; thus, we will only require the use of one cycle (the first) labelled from 1 to 10 (see diagram overleaf).

The E values, however, lie in the range 10 to 1,000, and therefore require *two* cycles - 10 to 100 and 100 to 1,000 - span them; we therefore label these two cycles accordingly (see diagram).

We can now proceed to plot the E values against the I values, obtaining an approximate straight line; and draw the line of best fit through these (see diagram). Thus, we have verified that E and I are related by the given formula.

To estimate the value of a, simply read off the intercept of the straight line with the vertical axis; *this is approximately 5*.

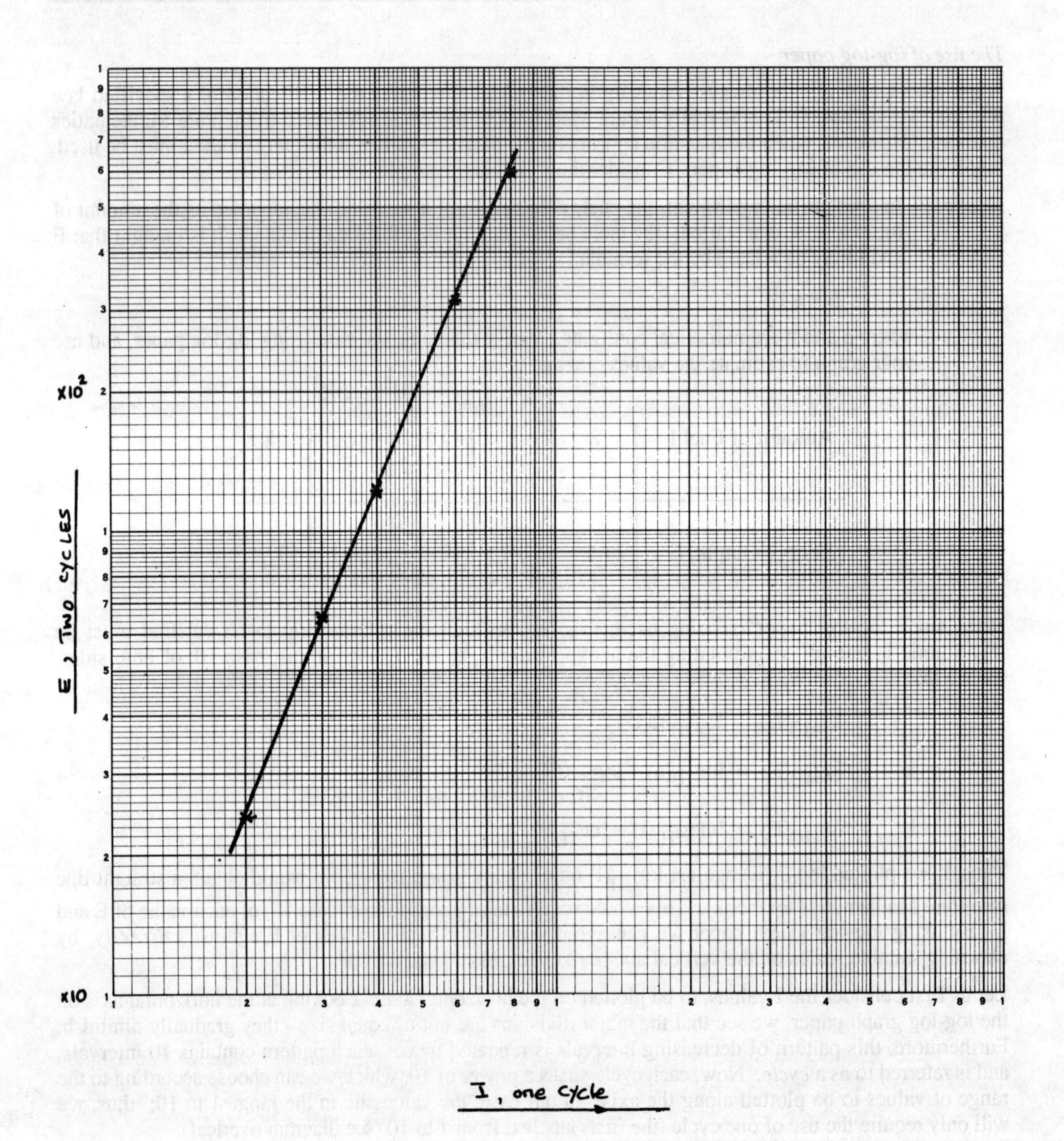

To estimate the value of b is slightly more complicated. Looking at equation (1), we see that, had we plotted $\log_{10}(E)$ against $\log_{10}(I)$ as described in the Pure Mathematics textbook, b would be the gradient of the straight line. In our case, b is still the gradient of the straight line; *however*, when calculating the gradient, we must use the *logarithms* (base 10) of the values of E and I. Thus, using the right-angled triangle ΔABC in the diagram, we see that:

$$b \approx \frac{AB}{BC} = \frac{\log_{10}(308.1) - \log_{10}(62.6)}{\log_{10}(6) - \log_{10}(3)} = 2.30$$

The use of log-linear paper

Suppose the variables y and x are related by the equation

$$y = ab^x$$

then

$$\begin{aligned} \log y &= \log(ab^x) \\ &= \log a + \log b^x \\ &= \log a + x \log b \\ \log y &= (\log b)\, x + \log a \end{aligned}$$

Compare this with '$y = mx + c$'

Plotting the values of log y against x on ordinary graph paper would produce a straight line of gradient log b, y intercept log a.

Using log-linear paper, the values of y are plotted on the log scale the values of x on the linear scale.

The gradient of the line is log b (note that when calculating the gradient log of the y-values are used) and the y-intercept is a (not log a, because log-linear paper is being used).

eg

x	0	1	2	3	4	5
y	3	6	12	24	48	96

$$\log b = \frac{\log 48 - \log 6}{4 - 1} = 0.30103$$ (only two points on the line could be used)

$$b = 10^{0.30103} = 2$$

$a = 3$ (the value of y when $x = 0$)

so $y = 3(2^x)$

If the value of a does not fit on the graph then find b first and use one point on the line to find a, eg $b = 2$ and when $x = 4$, $y = 48$.

so $48 = a(2^4)$ $\quad a = \frac{48}{16} = 3$

Log-linear $y = ab^x$

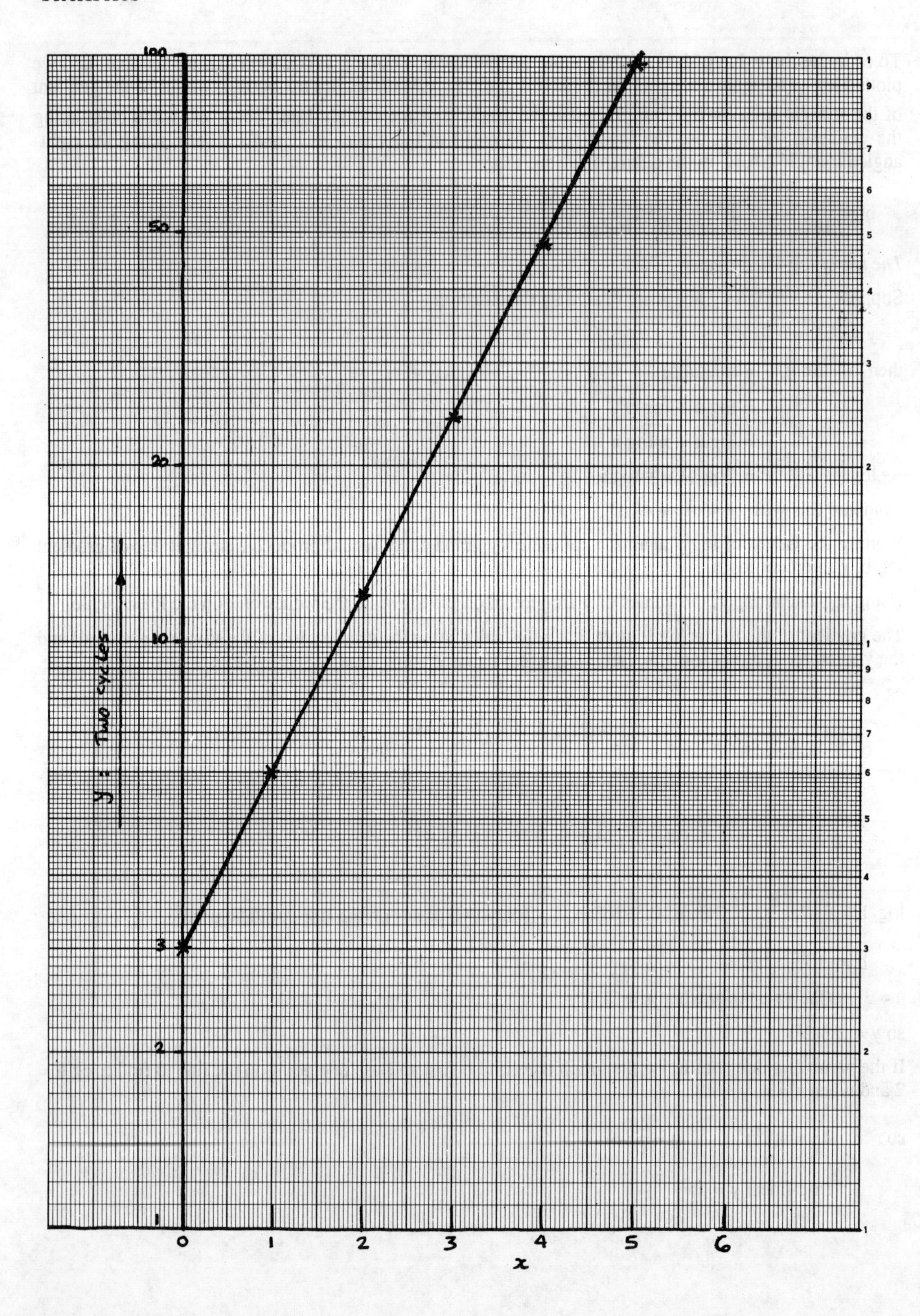
100
50
20
10
3
2
1
y : Two cycles
0
1
2
3
4
5
6
x

The use of log-probability paper

If a random variable X has a normal distribution then plotting it on arithmetic probability paper will produce a straight line from which the mean and variance can be deduced (see page 194 of textbook for details).

If instead log x has a normal distribution, then x has a log-normal distribution. If the same procedure as page 194 is followed but log-probability paper used instead, then a random variable with a log normal distribution will produce a straight line. If the mean and variance are obtained from the graph in the same way then these will be the mean and variance of log x, not of x.

INDEX

ON FROM 'A' LEVEL

One of the great misconceptions surrounding 'A' Level studies is that all subjects must be chosen with a specific career in mind. This fallacy probably arises because most people now know that certain science subjects are stipulated for such careers as medicine, dentistry and pharmacy. Even so, for all science and technological careers, however, 'A' Level subjects do need to be chosen with some care since the subject content of degree and diploma courses is often an extension of 'A' Level work. But, as with most rules, there are always some exceptions, for some universities and polytechnics are now offering a preliminary year to enable students to receive a foundation course in science subjects before starting on a degree course, even though they may have taken non-scientific subjects at 'A' Level.

Advanced Level subjects, therefore, should be seen as a universal currency. The student who takes science 'A' Levels can still apply for Business Studies or Accountancy courses, whilst those who have not taken Law at 'A' Level can still apply for law courses.

Statistics being the collection, analysis and interpretation of numerical data may be studied as an option on many Mathematics degree courses. Similarly, it may be part of a joint honours degree with Applied Mathematics, as at Aberystwyth, Swansea or Warwick, and can be combined with Actuarial Mathematics at Heriot-Watt University (being an ideal preparation for that very highly paid career as an Actuary!).

Career opportunities for statisticians, however, are enormously varied. The skills derived from a degree course in statistics provide considerable flexibility and enable students to respond quickly to the ever-changing economic and technological climate.

Specific details of the destinations of Statistics graduates are not available, however. The following lists of the destinations of Mathematics graduates is taken from *After a Degree* published by the York University Careers Service (£1.00). (This excellent booklet lists the career paths of all York graduates in the various subject areas.) The list will provide the reader with an insight into the range of careers open to numerate graduates.

Details are given of all mathematics graduates who obtained employment or further training within six months of graduation: they number 375 and represent 87 per cent of the total known destinations. The remaining 13 per cent is made up of 1 per cent who were not available for employment, 2 per cent who were in temporary employment, and 10 per cent who were unemployed.

Total number of graduates entering employment 284

FINANCIAL 87

Chartered accountancy	49	Arthur Andersen, Arthur Young (2), Bentley Jefferson (Stafford), Blick Rothenberg & Noble, Brebner Allen & Trapp, Clark Whitehill, Cooper Basden and Adamson, Coopers & Lybrand (2), Deloitte Haskins & Sells (5), Dixon Wilson, Ernst & Whinney (3), Hill Vellacott, Hook (Isle of Wight), Joseph Crossley (Manchester), Kidsons (2), Peat Marwick Mitchell (8), Price Waterhouse, Pulleyn Heselton, Robson Rhodes (2), Spofforth West Hewes (Littlehampton), Spicer & Pegler (2), Thomson McLintock, Thornton Baker (4), Touche Ross (4), Wheawill & Sudworth (Huddersfield)

Actuarial trainee	23	Co-operative Insurance Society, Cubie Wood, Duncan C Fraser, Equitable Life Assurance, Equity & Law Life Assurance (3), Friends Provident Life Office, General Accident, Legal and General (4), National Employers Life Assurance, National Mutual Life, National Provident Institution, Norwich Union (2), Phoenix Assurance (2), Sun Life Assurance (Canada), R Watson (2)
Financial management	5	Bowater Corporation, Chrysler, Lazard Brothers, Sumito Trust (London), Thermal Scientific
Banking traineee	5	Barclays, Lloyds (3), Yorkshire
One each of the following	5	Corporate investment (London), Investment analyst (London Life Association), Public financial management (Avon CC), Stockbroking trainee (Philips & Drew), Tax assistant (Fraser Keen)

MANAGEMENT SERVICES 68

Computer programmer	50	Boots, British Aerospace, Burroughs Machines, Central Computer Agency, Coopers & Lybrand, EPG Computer Services, Fars Computer Services, Ferranti (4), GEC, Geophysical Services International, Great Universal Stores, Honeywell (3), IBM, ITT (2), ICL (5), Leeds Permanent Building Society (3), Leicester Polytechnic, Logica (4), Mars, Metropolitan Pensions Association (Chichester), Microproducts Programming, Midland Bank (3), Ministry of Defence (2), Neve Electronic Laboratories, Royal Insurance, Scicon (2), Software Sciences (2), Standard Telephones & Cables, Systems Designers, Vauxhall Motors
Systems analyst	5	British Aerospace, ICI, Pilkington Group, Plessey, Rowntree Mackintosh
Statistician	4	Department of the Environment, Leeds Permanent Building Society, National Farmers Union Mutual Insurance, World Bureau of Metal Statistics
Computer operator	2	British Aerospace, Burroughs Machines
Data processing	2	Millward Brown (Market Research), Rolls Royce
Operational research	2	East Midlands Gas, Milk Marketing Board
One each of the following	3	Management services trainee (Proctor & Gamble), Market research (ICI), Planning trainee (CV Home Furnishings, Manchester)

SCIENTIFIC/ENGINEERING/TECHNICAL 38

Software engineer	10	GEC (3), Ferranti, ITT, Marconi (4), Smiths Industries
Scientific research	9	British Aerospace, Software House (Cambridge), Civil Service (2), Ministry of Defence (4), UK Atomic Energy Authority

Graduate engineer	5	British Aerospace, GEC (2), Marconi, Plessey
Design and development	2	British Shipbuilders, Inmos (Bristol)
Research assistant	2	British Ship Research Association, York University
Scientific officer	2	GCHQ, Meteorological Office
One each of the following	8	Development engineer (David Brown), Electronics engineer (Marconi), Production engineer (ITT), Research and development (Marconi), Software technician (York University), Systems engineer (British Aerospace), Systems programmer (Hatfield Polytechnic), Test engineer (Northern Engineering Industries)

MANAGEMENT/ADMINISTRATION: INDUSTRY AND COMMERCE 16

Administration	5	Arthur Young, Kensington Employment Agency, Marks & Spencer (3)
One each of the following	11	Advertisement sales (Humberside Publicity, Pocklington), Assistant distribution manager (Henry Oakland, York), Distribution analyst (Watney Mann & Truman), Graduate trainee (Nationwide Building Society), Hotel management (Bath), Management trainee (Greater London Council), Marketing support (ICL), Retail management (Mothercare), Sales representative (Procter & Gamble), Systems engineer (IBM), Traffic superintendent (British Telecom)

EDUCATION 9

Teacher	8	Bolton MDC, Comprehensive school (London), Lambeth LBC, Manchester MDC, North Yorkshire CC, Scarborough College, Southgate School (London), VSO (Nigeria)
Teaching English as a foreign language	1	LSL Tutoring (Hove)

SOCIAL/WELFARE 9

Voluntary work	5	Campus Crusade for Christ (Newcastle), Christian organisation (Reading), Community Service Volunteers, Leicester, India
One each of the following	4	Church work (Mid-Glamorgan Methodist Mission), Personal aide to handicapped person (Hampshire), Project worker (Stonham Housing Association, Chelsea), Residential social worker (Reliance Services)

ADMINISTRATION/MANAGEMENT: 8

Administration	2	London University, Post Office
One each of the following	6	Commissioned officer (Royal Navy)), Executive officer (Civil Service), Graduate trainee (North Thames Gas), Pilot (RAF), Police officer (Hertfordshire), Recreation management (Castlepoint DC)

CREATIVE/ENTERTAINMENT/MEDIA 2

One each of the following	2	Freelance radio presenter/disc jockey, Member of pop group (London)

MISCELLANEOUS 4

One each of the following	4	Information work (Bell Northern Research), Library assistant (Birmingham MDC), Organiser (National Union of Students), President (York University Students' Union)

OTHERS 7

Retail assistant	3	International Marketers (London), Music Centre (Watford), J Sainsbury
Clerical work	2	Surrey AHA, Trent RHA
One each of the following	2	Working in a warehouse (Uxbridge), Working on a family farm
Teacher training	75	*Universities*: Bath, Belfast, Birmingham (2), Bristol, Cambridge (2), Cardiff, Durham, Exeter (2), Hull, Keele (2), Leeds, Leicester (2), Liverpool (2), London, Loughborough (2), Manchester, Newcastle, Oxford (3), Sheffield (2), Sussex, York (28). *Polytechnics*: Middlesex, Oxford. *Colleges*: Avery Hill (London), Christ's and Notre Dame (Liverpool), Homerton (Cambridge) (2), Hull, Matlock, Moray House (Edinburgh), North Cheshire (Warrington), Rolle (Exmouth), Ripon & York St John (2), St Martin's (Lancaster), St Mary's (Newcastle), St Mary's (Twickenham), West Sussex
Higher degrees (PhD/MSc) (research/academic)	33	*Universities*: Arizona (USA), Birmingham (2), Bradford, Brunel, California (USA), Cambridge, Durham, Leeds (2), Lehigh (USA) (2), Liverpool, London (2), Loughborough, Manchester (2), Oxford (4), Reading (4), St Andrews, York (6)

Higher degrees (MSc)	11	Econometrics (London School of Economics), Operational research (2) (*Universities*: Sussex, Warwick), Statistics (8) (*Universities*: Edinburgh, Kent, Leicester, London, Manchester, Sheffield (2), UMIST)
Secretarial course	3	*Colleges*: London, NE Surrey CT, St Helens CT
One each of the following	5	Business studies diploma (Old Dominion University, USA), Graduate certificate in mathematics (London University), Mathematical statistics diploma (Cambridge University), Ordination course (Durham University), Theatre studies diploma (Cardiff University)

The number of applications for degree courses in Statistics is relatively low compared to other subject areas and consequently offers for university places are reasonable, ranging from BCC to CDD and even perhaps lower after the publication of the 'A' Level results when vacancies occur. The most popular universities in the subject are Warwick, Bristol, Bath (ABC offers), (London School of Economics and University College London (BBC offers). Polytechnic offers, however, range from CD down to DD and even EE in some cases (see *The Complete Degree Course Offers*, £11.95, Trotman, 12 Hill Rise, Richmond, Surrey). Statistics is also offered on several joint courses.

In addition to degree courses there are several Higher National Diploma courses offered by the polytechnics for which one 'A' Level pass would be acceptable.

Because of the relatively low number of applications, universities and polytechnics can afford to be very flexible in the offers they make, and anyone who can demonstrate an interest in Statistics on the UCCA/PCAS form or at interview should stand a reasonable chance of a place.

Good luck!

Brian Heap

HLT PUBLICATIONS

All HLT Publications have two important qualities. First, they are written by specialists, all of whom have direct practical experience of teaching the syllabus. Second, all Textbooks are reviewed and updated each year to reflect new developments and changing trends. They are used widely by students at polytechnics and colleges throughout the United Kingdom and overseas.

A comprehensive range of titles is covered by the following classifications.

- **TEXTBOOKS**
- **CASEBOOKS**
- **SUGGESTED SOLUTIONS**
- **REVISION WORKBOOKS**

The books listed overleaf should be available from your local bookshop. In case of difficulty, however, they can be obtained direct from the publisher using this order form. Telephone, Fax or Telex orders will also be accepted. Quote your Access, Visa or American Express card numbers for priority orders. To order direct from publisher please enter cost of titles you require, fill in despatch details and send it with your remittance to The HLT Group Ltd. **Please complete the order form overleaf.**

DETAILS FOR DESPATCH OF PUBLICATIONS

Please insert your full name below

Please insert below the style in which you would like the correspondence from the Publisher addressed to you

TITLE Mr, Miss etc. INITIALS SURNAME/FAMILY NAME

Address to which study material is to be sent (please ensure someone will be present to accept delivery of your Publications).

POSTAGE & PACKING

You are welcome to purchase study material from the Publisher at 200 Greyhound Road, London W14 9RY, during normal working hours.

If you wish to order by post this may be done direct from the Publisher. Postal charges are as follows:

UK - Orders over £30: no charge. Orders below £30: £2.50. Single paper (last exam only): 50p

OVERSEAS - See table below

The Publisher cannot accept responsibility in respect of postal delays or losses in the postal systems.

DESPATCH All cheques must be cleared before material is despatched.

SUMMARY OF ORDER

Date of order:

Add postage and packing:

				£
			Cost of publications ordered:	
			UNITED KINGDOM:	
OVERSEAS:	TEXTS		Suggested Solutions (Last exam only)	
	One	Each Extra		
Eire	£4.00	£0.60	£1.00	
European Community	£9.00	£1.00	£1.00	
East Europe & North America	£10.50	£1.00	£1.00	
South East Asia	£12.00	£2.00	£1.50	
Australia/New Zealand	£13.50	£4.00	£1.50	
Other Countries (Africa, India etc)	£13.00	£3.00	£1.50	
			Total cost of order:	£

Please ensure that you enclose a cheque or draft payable to THE HLT GROUP LTD for the above amount, or charge to ❑ Access ❑ Visa ❑ American Express

Card Number

Expiry Date .... Signature ....

ORDER FORM – 'A' LEVEL AND GCSE PUBLICATIONS

Please indicate on the form below the titles and quantity of each which you require. Then complete the despatch details overleaf where postal charges are set out.

'A' LEVEL TEXTBOOKS	Cost £	Quantity	Total £
Accounting	10.95		
Business Studies	10.95		
Economics	10.95		
Applied Mathematics	9.95		
Pure Mathematics	9.95		
Statistics	9.95		
General Principles of English Law	10.95		
Constitutional Law	9.95		
Sociology	9.95		
'A' LEVEL CASEBOOK			
General Principles of English Law	10.95		
GCSE			
Law Textbook	7.95		
Law Casebook	6.95		
		Total Cost £	

HLT Publications